IET ELECTROMAGNETIC WAVES SERIES 32

Series Editors: Professor P.J.B. Clarricoats
Professor Y. Rahmat-Sammi
Professor J.R. Wait

Electromagnetic Waveguides

theory and applications

Other volumes in this series:

Volume 1 **Geometrical theory of diffraction for electromagnetic waves, 3rd edition** G.L. James
Volume 10 **Aperture antennas and diffraction theory** E.V. Jull
Volume 11 **Adaptive array principles** J.E. Hudson
Volume 12 **Microstrip antenna theory and design** J.R. James, P.S. Hall and C. Wood
Volume 15 **The handbook of antenna design, volume 1** A.W. Rudge, K. Milne, A.D. Oliver and P. Knight (Editors)
Volume 16 **The handbook of antenna design, volume 2** A.W. Rudge, K. Milne, A.D. Oliver and P. Knight (Editors)
Volume 18 **Corrugated horns for microwave antennas** P.J.B. Clarricoats and A.D. Oliver
Volume 19 **Microwave antenna theory and design** S. Silver (Editor)
Volume 21 **Waveguide handbook** N. Marcuvitz
Volume 23 **Ferrites at microwave frequencies** A.J. Baden Fuller
Volume 24 **Propagation of short radio waves** D.E. Kerr (Editor)
Volume 25 **Principles of microwave circuits** C.G. Montgomery, R.H. Dicke and E.M. Purcell (Editors)
Volume 26 **Spherical near-field antenna measurements** J.E. Hansen (Editor)
Volume 28 **Handbook of microstrip antennas, 2 volumes** J.R. James and P.S. Hall (Editors)
Volume 31 **Ionospheric radio** K. Davies
Volume 32 **Electromagnetic waveguides: theory and applications** S.F. Mahmoud
Volume 33 **Radio direction finding and superresolution, 2nd edition** P.J.D. Gething
Volume 34 **Electrodynamic theory of superconductors** S.A. Zhou
Volume 35 **VHF and UHF antennas** R.A. Burberry
Volume 36 **Propagation, scattering and diffraction of electromagnetic waves** A.S. Ilyinski, G. Ya.Slepyan and A. Ya.Slepyan
Volume 37 **Geometrical theory of diffraction** V.A. Borovikov and B.Ye. Kinber
Volume 38 **Analysis of metallic antenna and scatterers** B.D. Popovic and B.M. Kolundzija
Volume 39 **Microwave horns and feeds** A.D. Olver, P.J.B. Clarricoats, A.A. Kishk and L. Shafai
Volume 41 **Approximate boundary conditions in electromagnetics** T.B.A. Senior and J.L. Volakis
Volume 42 **Spectral theory and excitation of open structures** V.P. Shestopalov and Y. Shestopalov
Volume 43 **Open electromagnetic waveguides** T. Rozzi and M. Mongiardo
Volume 44 **Theory of nonuniform waveguides: the cross-section method** B.Z. Katsenelenbaum, L. Mercader Del Rio, M. Pereyaslavets, M. Sorella Ayza and M.K.A. Thumm
Volume 45 **Parabolic equation methods for electromagnetic wave propagation** M. Levy
Volume 46 **Advanced electromagnetic analysis of passive and active planar structures** T. Rozzi and M. Farinai
Volume 47 **Electromagnetic mixing formulae and applications** A. Sihvola
Volume 48 **Theory and design of microwave filters** I.C. Hunter
Volume 49 **Handbook of ridge waveguides and passive components** J. Helszajn
Volume 50 **Channels, propagation and antennas for mobile communications** R. Vaughan and J. Bach-Anderson
Volume 51 **Asymptotic and hybrid methods in electromagnetics** F. Molinet, I. Andronov and D. Bouche
Volume 52 **Thermal microwave radiation: applications for remote sensing** C. Matzler (Editor)
Volume 502 **Propagation of radiowaves, 2nd edition** L.W. Barclay (Editor)

Electromagnetic Waveguides

theory and applications

S.F. Mahmoud

The Institution of Engineering and Technology

Published by The Institution of Engineering and Technology, London, United Kingdom

First published 1991
Reprinted 2006

The Institution of Engineering and Technology
Michael Faraday House
Six Hills Way, Stevenage
Herts, SG1 2AY, United Kingdom

www.theiet.org

British Library Cataloguing in Publication Data
Mahmoud, S.F.
Electromagnetic Waveguides
1. Waveguides
I. Title II. Series
621.381331

ISBN (10 digit) 0 86341 232 7
ISBN (13 digit) 978-0-86341-232-5

Printed in the UK by Short Run Press Ltd, Exeter, Devon
Reprinted in the UK by Lightning Source UK Ltd, Milton Keynes

Contents

Foreword

It is a pleasure to provide a few comments on this remarkable book by Professor Mahmoud. Under the general heading of electromagnetic waveguides, he has touched on a wide variety of physical problems in telecommunications, geophysics and optics. He has made original contributions in all of these fields. The guided wave concept is the underlying one which he presents in a consistent manner. Self-contained derivations of the key mathematical foundations are presented in a forthright style. Each chapter includes relevant references to pertinent papers from the journal literature. While the book is intended for a graduate lecture course, the content will serve as an excellent reference for researchers in the field, not to mention the valuable critical reviews of current journal papers.

Professor James R. Wait
Tucson,
Arizona.

Preface

This text is intended for use in a first graduate course on electromagnetic waveguides.

An attempt is made to present a self-contained analytical approach for a wide variety of waveguides. Stress is laid on applications in several areas such as long distance communication, satellite communication, fiber optics, millimetric wave circuits and tunnel communication. Thus, we cover low attenuation waveguides, low crosspolar field waveguides, dielectric waveguides and natural waveguides including the earth–ionosphere and mine tunnels.

The text is arranged in two major parts, apart from Chapter 1 which is an introduction to the principles of guided waves, boundary conditions and wave solutions in common co-ordinate systems. Chapters 2—4 constitute the first part which covers closed waveguides, constant wall impedance waveguides and open waveguides. In each of these chapters, the questions of mode characterisation, mode excitation, orthogonality and scattering at a discontinuity are addressed. Whenever appropriate, relevance to applications is mentioned. The second part of the text is devoted mainly to applications. Low crosspolar field waveguides, used in fequency reuse telecommunication systems, are covered in Chapter 5, while Chapter 6 discusses communication in tunnels within the earth. Each of these two chapters reviews the state of the art of the topic, with a listing of the appropriate references.

Problems are included in each chapter in order to stress concepts related to the text and to highlight certain phenomena. Some problems address certain mathematical details not dealt with in the text. References are listed at the end of each chapter including those which contain extensive bibliographies.

The author wishes to acknowledge the help and encouragement of Prof. James R. Wait, who, in the first place, suggested that I write such a book. The author is grateful to many of his students who helped, in one way or another, in the preparation of this book. Of these, I mention Sherif A. Mohammed, Dr. Amany K. Farrag, Heba M. Mourad and Sally G. Ibrahim. As a final personal note, I am grateful to my wife and children for bearing with me during the time I devoted to writing this book.

Chapter 1
Introduction to electromagnetic waves

1.1 Introduction

Propagation of electromagnetic waves can be broadly classified into guided and unguided propagation. In guided propagation the waves are transmitted from one point to another following a prescribed path. In unguided propagation the waves are spread or radiated in an open space. Structures that guide electromagnetic waves are usually referred to as waveguides, while structures that allow radiation are termed leaky structures or antennas. This text is devoted to the first class of propagation mechanism; namely, guided propagation and waveguides. Waveguides have undergone considerable development in the last few decades. Thus, besides the simple two copper wire line, the coaxial cable and the hollow metallic pipes that have been used for many years, we have, in addition, the planar strip and microstrip lines, the various millimetric waveguides and the now popular optical fiber lines. Although design details and manufacturing processes of these various types of waveguides are widely different, the basic objectives of any of these types remain almost the same; namely, a waveguiding structure is generally required to transfer electromagnetic power and/or signals from one end to another with the minimum possible attenuation and the minimum signal delay distortion. These are particularly vital objectives in telecommunications systems. In some applications, the waveguide is required to maintain a fixed field polarisation. This capability is often called for in satellite communication systems with frequency reuse, whereby two baseband signals are simultaneously transmitted on two orthogonal polarisations using the same carrier frequency. This obviously doubles the efficiency of use of the frequency spectrum resource.

In this book, waveguides are classified into two broad classes; closed waveguides and open waveguides. Closed waveguides are characterised by being completely bounded with perfectly, or at least highly, reflecting boundaries. Open waveguides, on the other hand, are not completely bounded. However, waves can be guided on these structures as surface waves, but radiation also occurs owing to any non-uniformity of the guide. An important subclass of closed waveguides with many useful applications contains waveguides with constant wall impedance. Examples include waveguides with corrugated walls or dielectric coated walls. Besides, many natural waveguides, such as tunnels, can be considered to belong to this subclass if their walls are modelled as constant impedance walls.

This book is arranged as follows. In this chapter, the basic laws governing electromagnetic propagation are reviewed. The wave equation is thus solved in popular cylindrical coordinate systems and in spherical coordinates along with few illustrative examples. Closed waveguides are treated in chapters 2 and 3. In chapter 2 attention is restricted to waveguides with perfectly conducting boundaries, while in chapter 3 closed waveguides with finite impedance boundaries are studied. Open waveguides are the subject of chapter 4. The study in chapters 2, 3 and 4 concentrates on the main properties of modes such as attenuation, phase velocity, the mode orthogonality property, modal excitation and mode scattering at a longitudinal discontinuity. Chapters 5 and 6 are devoted to specific applications: specific types of low crosspolar waveguides are studied in sufficient detail in chapter 5 while in chapter 6 free and guided propagation in a tunnel is presented.

1.2 Preliminaries

The assumption of time harmonic fields is adopted all along in this text, so that all fields are assumed to vary sinusoidally with time at a fixed temporal frequency. Fields are then represented by complex quantities, or phasors, such that, for example, a time dependent field $e(t)$ is written as $e(t)=\sqrt{2}$ Real (E $\exp(i\omega t)$) where E is the phasor representation of $e(t)$, ω is the angular temporal frequency, t is the time and $i=\sqrt{-1}$. The factor $\sqrt{2}$ is introduced to render the magnitude of E equal to the time root mean square of $e(t)$. The complex fields will then vary with time as $\exp(i\omega t)$ although this factor is usually omitted in time harmonic analysis. In complex notion of fields, it is obvious that differentiation with respect to time t is equivalent to a multiplication by the factor $i\omega$. The time average of the product of two field quantities, say, $a(t)$ and $b(t)$ is simply equal to Real (AB^*) where A and B are the complex or phasor representations of the two fields and the * denotes the process of complex conjugation.

Electromagnetic waves on waveguiding structures are allowable solutions of Maxwell's equations and the boundary conditions imposed by the structure. Most of the material in this text is devoted to cylindrical or spherical conical structures. Thus solutions of Maxwell's equations in cylindrical and spherical coordinate systems are presented in this chapter. The most commonly used cylindrical coordinate systems include the cartesian, circular and elliptical cylindrical coordinate systems. A detailed study of wave functions on these coordinate systems and in spherical coordinates is given in the next few sections. However, it is first necessary to present a discussion on Maxwell's equations and their implications on boundary conditions at interfaces between different media. In the following we adopt the convention that vector quantities are designated by roman letters and scalar quantities by *italic* letters.

1.3 Maxwell's equations and boundary conditions

With phasor representation of fields and sources, Maxwell's equations take the forms:

$$\text{curl } \mathrm{E} = -i\omega\mu \mathrm{H} - \mathrm{M} \tag{1.1}$$

$$\text{curl } \mathrm{H} = i\omega\varepsilon \mathrm{E} + \mathrm{J} \tag{1.2}$$

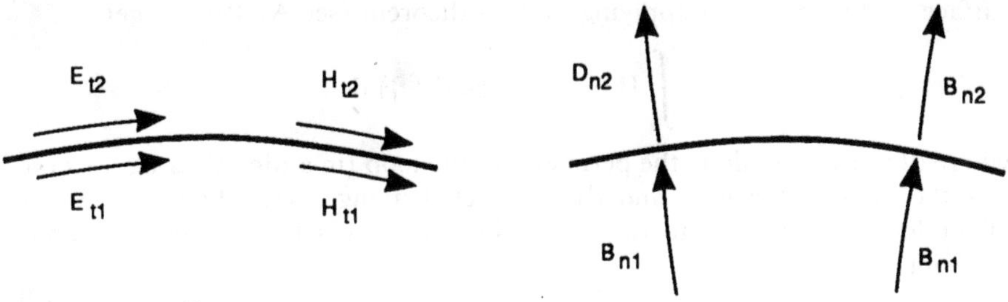

Fig. 1.1
a Boundary conditions for tangential fields at an interface
b Boundary conditions for normal fields at an interface

The terms M and J are magnetic and electric current density vector sources and have the dimensions volt/m^2 and ampere/m^2 respectively. E(volt/m) and H(ampere/m) are the vector electric and magnetic fields respectively. The vectors εE and μH are called the electric displacement and the magnetic induction fields respectively. They take the symbols D ($=\varepsilon$E) and B ($=\mu$H). The quantities μ and ε are the magnetic permeability and the electric permittivity of the medium in which the fields exist. Complex μ or ε accounts for possible magnetic or electric losses in the medium.

Taking the divergence of (1.1) and (1.2) and noting that div.curl of a vector field is identically zero (see A1.15), we get

$$\mathrm{div}(i\omega\mu \mathrm{H} + \mathrm{M}) = 0 \tag{1.3}$$

$$\mathrm{div}(i\omega\varepsilon \mathrm{E} + \mathrm{J}) = 0 \tag{1.4}$$

Equation (1.4) is a statement of the continuity of the total electric current whose driving part is J and the driven part is $i\omega\varepsilon$E. The driving current is related to the free electric charge density ρ (coulomb/m^3) by

$$\mathrm{div}\,\mathrm{J} = -\partial\rho/\partial t = -i\omega\rho \tag{1.5}$$

which is a statement of conservation of charge. Here ρ is the charge injected into the medium by the source. Now combining (1.4) and (1.5), we get Gauss's law of the electric flux

$$\mathrm{div}\,\varepsilon \mathrm{E} = \rho \tag{1.6}$$

Next let us consider equation (1.3). The divergence of each component of the magnetic current, $i\omega\mu$H and M, is separately equal to zero. This is a consequence of the absence of separate magnetic charges in nature such that in any volume, however small, the net magnetic charge is zero. Hence (1.3) can be rephrased as

$$\mathrm{div}\,\mu \mathrm{H} = 0 \tag{1.7}$$

Now consider a boundary surface between two media with (ε_1, μ_1) and (ε_2, μ_2). Let $\hat{n}$ be the unit vector normal to the boundary and directed from medium 1 to 2. With $\hat{t}$ a unit vector directed along the component of H tangential to the boundary, let us form a rectangular loop with two sides parallel to $\hat{n}$ and the other two parallel to $\hat{t}$, as shown in Fig. 1.1*a*. Integrating equation (1.2) on the surface of this loop, and applying Stoke's theorem (see A1.19), we get

$$\int \mathrm{H}.\mathrm{d}l = \int_s (i\omega\varepsilon\mathrm{E} + \mathrm{J}).\mathrm{d}s$$

where dl is a vector along the perimeter of the loop (in anticlockwise direction), s is the area of the loop and ds is directed along $\hat{t} \times \hat{n}$. Allowing the sides of the loop parallel to $\hat{t}$ to diminish, the area tends to zero and the above equation becomes

$$(H_{t1} - \mathrm{H}_{t2}) \times \hat{n} = \mathrm{j} \tag{1.8}$$

where j is a possible surface electric current, or a current sheet, lying on the boundary and has the dimensions of amperes/m; namely it is the surface current per unit width on the surface.

Using an exactly similar procedure, we integrate (1.1) on the loop of Fig. 1.1*a*, and end up with

$$\hat{n} \times (E_{t1} - E_{t2}) = \mathrm{m} \tag{1.9}$$

where m is the surface magnetic current per unit width and has the dimension of volt/m.

Other boundary conditions follow from (1.6) and (1.7) by integrating each of them on an elementary volume as shown in Fig. 1.1*b*. Upon using, the divergence theorem (A1.18) and allowing the side surface, parallel to $\hat{n}$, to diminish, we get

$$\varepsilon_2 E_{n2} - \varepsilon_1 E_{n1} = \rho_s \tag{1.10}$$

$$\mu_2 H_{n2} - \mu_1 H_{n1} = 0 \tag{1.11}$$

where the subscript n signifies the component normal to the boundary surface, and ρ_s is a surface charge density (colomb/m^2).

In words, the boundary conditions (1.10) and (1.11) state that the normal component of the displacement vector εE is discontinuous at the boundary by the amount of surface charge density while the normal component of μH is continuous across the boundary. It is important to note that the boundary conditions (1.10) and (1.11) are not independent of conditions (1.8) and (1.9). This is so since the divergence equations (1.3) and (1.4) follow from Maxwell's equations (1.1) and (1.2). Hence it is sufficient to satisfy the boundary conditions that involve the tangential fields only, that is (1.8) and (1.9), and this guarantees that (1.10) and (1.11) are satisfied.

Finally we note that, in the absence of surface currents or charges, the boundary conditions on the tangential fields reduce to

$$E_{t1} = E_{t2} \tag{1.12}$$

$$H_{t1} = H_{t2} \tag{1.13}$$

which are both necessary and sufficient boundary conditions.

1.4 Solution of Maxwell's equations in cylindrical coordinates

Now we apply Maxwell's equations to cylindrical coordinates (u, v, z) in which the z coordinate lies along the cylindrical axis. A uniform cylindrical structure is assumed and defined as one whose cross section (normal to z) does not change shape or dimensions along the axial direction z. Furthermore, the medium does not change properties along z although it can be inhomogeneous within the cross section. In the following, we confine our discussion to isotropic materials only; that is μ and ε are scalar quantities signifying no directional properties for the medium. A uniform cylindrical structure, as defined above, is depicted in Fig. 1.2.

In general cylindrical coordinates (u, v, z) the metrical coefficient h_3 of z is equal to unity while h_t and h_z are independent of z. The reader is referred to Appendix 1 for a review of generalised coordinate systems. On a uniform cylindrical structure, it is appropriate to seek solutions which behave as travelling waves along z; hence all fields will be assumed to vary as $\exp(-i\beta z)$. A general field variation with z can be decomposed into a spectrum of travelling waves by simple Fourier analysis.

Restricting attention to natural modes of the cylindrical structure, the source terms in (1.1) and (1.2) are set equal to zero. Now, without needing to specify any particular cylindrical coordinate system, let the fields of a given wave be decomposed into longitudinal components E_z and H_z and transverse (to z) components E_t and H_t. Similarly the curl operator is decomposed into a transverse component, denoted by ∇_t and a longitudinal component equal to

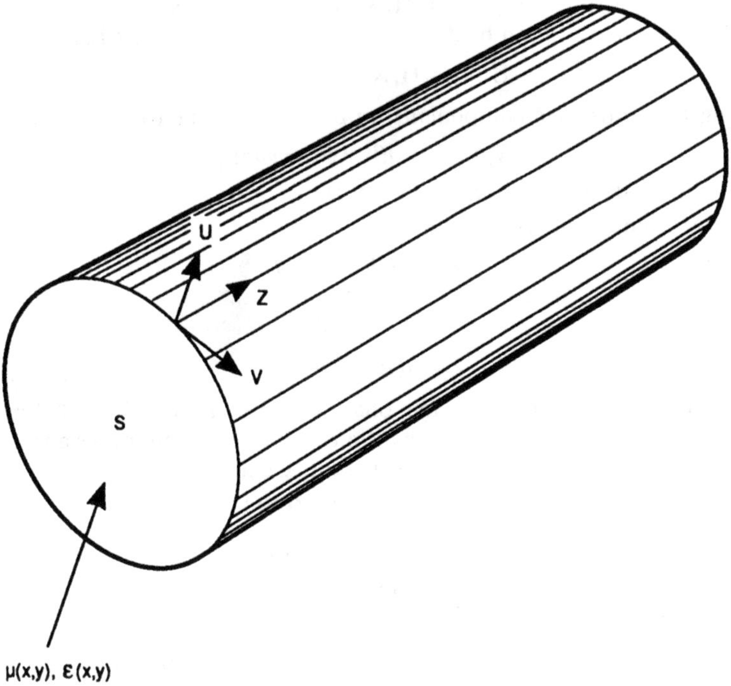

Fig. 1.2 A general uniform cylindrical guiding structure with cross section S and isotropic material

$(\partial/\partial z)\hat{z}\times = -i\beta\hat{z}\times$. Equations (1.1) and (1.2) now become:

$$(\nabla_t - i\beta\hat{z}) \times (E_t + E_z\hat{z}) = +i\omega\mu(H_t + H_z\hat{z}) \quad (1.14)$$

$$(\nabla_t - i\beta z) \times (H_t + H_z\hat{z}) = i\omega\varepsilon(E_t + E_z\hat{z}) \quad (1.15)$$

At this point, it is convenient to classify the field solutions into transverse electric waves (TE) and transverse magnetic waves (TM). In a TE wave the electric field is totally transverse to z, hence $E_z = 0$. Likewise, in a TM wave, $H_z = 0$. An alternative designation to these waves is H waves and E waves respectively. A modal solution can be either TE or TM type. In the most general case the mode is hybrid, i.e. having both E_z and H_z not equal to zero. Hence a hybrid mode can be considered as a superposition of TE and TM parts which are coupled by boundary conditions imposed by the particular structure under consideration. So, in the following we apply (1.14) and (1.15) to TE and TM waves.

TE (or *H*) waves

Setting $E_z = 0$, and separating the transverse and longitudinal components on each side of (1.14), we get

$$\nabla_t \times E_t = -i\omega\mu H_z\hat{z} \quad (1.16)$$

$$-i\beta\hat{z} \times E_t = -i\omega\mu H_t \quad (1.17)$$

Equation (1.17) is an explicit relation between the transverse components E_t and H_t. Clearly E_t is orthogonal to H_t with the vector $E_t \times H_t$ directed along $+z$. A wave impedance Z_{TE} can be defined as the ratio between $\hat{z} \times E_t$ and H_t. Thus

$$E_t = Z_{TE} H_t \times \hat{z};\; Z_{TE} = \omega\mu/\beta \quad (1.18)$$

Equating the transverse components on both sides of (1.15)

$$\nabla_t \times H_z\hat{z} - i\beta\hat{z} \times H_t = i\omega\varepsilon E_t \quad (1.19)$$

From vector calculus $\nabla_t \times H_z\hat{z} = +z \times \nabla_t H_z$, and in view of (1.18), equation (1.19) immediately yields

$$H_t = (-i\beta/\lambda^2)\nabla_t H_z \quad (1.20)$$

where

$$\lambda^2 = k^2 - \beta^2 \quad (1.21)$$

$$k = \omega(\mu\varepsilon)^{1/2} = \omega/v \quad (1.22)$$

k is the wavenumber and v the wave phase velocity in the unbounded medium. The variable λ may be designated by the transverse wavenumber and β by the longitudinal wavenumber or phase constant. Using (1.18) and (1.20)

$$E_t = (-i\omega\mu/\lambda^2)\nabla_t H_z \times \hat{z} \quad (1.23)$$

Equations (1.20) and (1.23) are explicit expressions relating the transverse components of TE waves to the longitudinal component H_z.

TM (or *E*) waves

Following similar steps, we deduce for TM waves ($H_z = 0$) that

$$E_t = (-i\beta/\lambda^2)\nabla_t E_z \quad (1.24)$$

$$H_t = (-i\omega\varepsilon/\lambda^2)\hat{z} \times \nabla_t E_z \quad (1.25)$$

A *TM* wave admittance Y_{TM} is defined as

$$Y_{TM} = \omega\varepsilon/\beta \tag{1.26}$$

which is the ratio between H_t and $\hat{z} \times \mathrm{E}_t$.

Hybrid waves

In the general case of a hybrid wave, the relations between transverse and longitudinal field components are obtained by superposition of *TE* and *TM* waves. Thus, combining (1.20)—(1.26) we get

$$\mathrm{E}_t = \left(\frac{-i\omega\mu}{\lambda^2}\right) [\nabla_t H_z \times \hat{z} + (\beta/\omega\mu)\nabla_t E_z] \tag{1.27}$$

$$\mathrm{H}_t = \left(\frac{-i\omega\varepsilon}{\lambda^2}\right) [\hat{z} \times \nabla_t E_z + (\beta/\omega\varepsilon)\nabla_t H_z] \tag{1.28}$$

It is worth recalling that these formulae are valid for any cylindrical coordinate system (e.g., cartesian, circular, elliptical, bipolar . . .).

1.5 Wave equation

Now we derive the wave equation satisfied by the longitudinal field components H_z and E_z. Both μ and ε are allowed to be functions of the transverse coordinates, but not of z. Consider first the *TE* waves, and recall the divergence equation for H, (1.7), which is repeated here for convenience:

$$\nabla.\mu\mathrm{H} = 0 \tag{1.29}$$

Decomposing both ∇ and H into their respective transverse and longitudinal components, and noting that μ is independent of z, we get:

$$\nabla_t \mu \mathrm{H}_t = i\beta\mu H_z \tag{1.30}$$

From (1.20) we also have

$$\mu\mathrm{H}_t = -i\beta(\mu/\lambda^2)\nabla_t H_z$$

Taking the divergence ∇_t of both sides and noting that both μ and λ^2 may vary with the transverse coordinates, we get

$$\nabla_t.\mu\mathrm{H}_t = -i\beta\nabla_t.[\mu/\lambda^2)\nabla_t H_z] = (-i\beta\mu/\lambda^2)\nabla^2 H_z - i\beta\nabla_t(\mu/\lambda^2).\nabla_t H_z \tag{1.31}$$

where identities (A1.11 and (A1.5) in Appendix A1 have been used. Comparing (1.30) and (1.31):

$$\nabla_t^2 H_z + \lambda^2 H_z = -(\lambda^2/\mu)\nabla_t(\mu/\lambda^2).\nabla_t H_z \tag{1.32}$$

which is the wave equation satisfied by H_z in a source free region. Similarly for *TM* waves, E_z satisfies the equation:

$$\nabla^2 E_z + \lambda^2 E_z = -(\lambda^2/\varepsilon)\nabla_t(\varepsilon/\lambda^2).\nabla_t E_z \tag{1.33}$$

In the important special case of a homogeneous medium, i.e. both μ and ε are independent of position, both (1.32) and (1.33) reduce to the more familiar form:

$$\nabla^2 f + \lambda^2 f = 0 \tag{1.34}$$

Next, (1.34) is solved for the wave function f in the cartesian, circular and elliptical cylindrical coordinate systems.

1.5.1 Cartesian coordinates: plane wave functions

The simplest frame of cylindrical coordinate systems to be considered is the (x, y, z) frame. In this frame the metrical coefficients $h_1 = h_2 = h_3 = 1$ (see Appendix A1). The wave equation (1.34) then takes the form:

$$\partial^2 f/\partial x^2 + \partial^2 f/\partial y^2 + \lambda^2 f = 0 \tag{1.35}$$

Looking for solutions of (1.35) by the method of separation of variables, f is set equal to a function of x only multiplied by a function of y only; i.e. $f = X(x)\ Y(y)$. This results in:

$$(d^2X/dx^2)/X + (d^2Y/dy^2)/Y + \lambda^2 = 0$$

which requires that

$$(d^2X/dx^2)X = \text{constant} = -k_x^2, \text{ say}$$

and

$$(d^2Y/dy^2)/Y = \text{constant} = -k_y^2$$

such that

$$k_x^2 + k_y^2 = \lambda^2 \tag{1.36}$$

The solutions for $X(x)$ are $\sin(k_x x)$ and $\cos(k_x x)$ or any linear combination of these functions. Similar solutions are valid for $Y(y)$. Hence the general solution for the wave function f is

$$f(x, y) = (A \cos k_x x + B \sin k_x x)(C \cos k_y y + D \sin k_y y) \tag{1.37}$$

where A, B, C and D are arbitrary constants and the separation constants k_x and k_y are governed by (1.36).

It is worth noting that either $\sin k_x x$ or $\cos k_x x$ represents a standing wave. Other solutions formed by linear combinations of these functions are $\exp(\pm ik_x x)$ which represent travelling waves. Furthermore if k_x is purely imaginary, then the solution takes any one of the forms $\exp(\pm \alpha x)$ or $\cosh(\alpha x)$ and $\sinh(\alpha x)$, which are evanescent or growing waves. Whether one chooses one set of solutions or another depends on the particular problem at hand with the boundary conditions to be satisfied.

1.5.2 Circular cylindrical coordinates and wave functions

In the circular cylindrical frame of coordinates (ρ, ϕ, z), $h_1 = h_3 = 1$, and $h_z = \rho$ are the metrical coefficients. Hence the wave equation (1.34) takes the form (see equation A1.6):

$$(1/\rho)(\partial/\partial\rho)(\rho\partial f/\partial\rho) + (1/\rho^2)(\partial^2 f/\partial\phi^2) + \lambda^2 f = 0$$

Taking $f = R(\rho)\Phi(\phi)$, separate equations to be satisfied by R and Φ are obtained:

$$\rho(d/d\rho)(\rho dR/d\rho) + (\lambda^2\rho^2 - n^2)R = 0 \tag{1.38}$$

$$d^2\Phi/d\phi^2 = -n^2\Phi \tag{1.39}$$

where n is a separation constant. Recognising that (1.38) is the Bessel differential equation of order n and argument $\lambda\rho$, the general solution for $f(\rho, \phi)$ may be written as:

$$f(\rho, \phi) = [(AJ_n(\lambda\rho) + BN_n(\lambda\rho)](C \cos n\phi + D \sin n\phi) \tag{1.40}$$

where A, B, C and D are, so far, arbitrary constants; $J_n(.)$ and $N_n(.)$ are the Bessel functions of first and second kinds and order n. These functions are described and tabulated in many standard text books, e.g. Watson (1958), Mclachlan (1955) and Abramowitz and Stegun (1965).

For convenience, we give some of the main properties of the Bessel functions in Appendix A2.

Generally, the separation constant n can take any value, whether real or complex. However, in the case of a homogeneous medium over the whole range of ϕ between 0 and 2π, and in order to have $\Phi(\phi)$ a single valued function of ϕ, n should be restricted to real integer values. To throw some light on the nature of the Bessel function solutions, we note that $J(.)$ and $N(.)$ represent standing circular cylindrical waves just as $\cos kx$ and $\sin kx$ represent plane standing waves. This may also be seen from the asymptotic behaviour of the Bessel functions at large arguments (Appendix A2). Other forms of the solution of (1.38) are given by the Hankel functions of first and second kinds defined by

$$H_n^{(1)}(\lambda\rho) = J_n(\lambda\rho) + iN_n(\lambda\rho) \tag{1.41}$$

$$H_n^{(2)}(\lambda\rho) = J_n(\lambda\rho) - iN_n(\lambda\rho) \tag{1.42}$$

which represents inward and outward travelling waves respectively. Still other forms of the solution are suitable when λ is purely imaginary and are given by

$$I_n(\alpha\rho) = i^n J_n(-i\alpha\rho) \tag{1.43}$$

$$K_n(\alpha\rho) = (\pi/2)(-i)^{n+1} H_n^{(2)}(-i\alpha\rho) \tag{1.44}$$

where λ is substituted for by $-i\alpha$. The $I(.)$ and $K(.)$ functions are the modified Bessel functions as given, for example, by Abramowitz and Stegun (1965). They represent growing and evanescent waves respectively and behave as $\exp(\pm\alpha\rho)/(\alpha\rho)^{1/2}$ as $(\alpha\rho)$ tends to ∞ (see Appendix A2).

1.5.3 Example

As an example of using the circular cylindrical functions described above, let us consider a simple situation. A pair of parallel perfectly conducting plates form a waveguide which is excited by an electric line source, as shown in Fig. 1.3*a*. The current is uniform over the line source, which extends from the lower plate at $z=0$ to the upper plate at $z=d$. Taking the z axis to coincide with the line source results in complete symmetry in the azimuthal direction ϕ; hence the excited fields will be ϕ independent. Because of the uniformity of the current along z, the fields will also be z independent. Anticipating that the fields will be TM to z (since H is transverse to the current), we write

$$E_z = CH_0^{(2)}(k\rho) \tag{1.45}$$

where C is a constant that depends on the current intensity and $k = \omega(\varepsilon\mu)^{1/2}$. The Hankel function $H_0^{(2)}(k\rho)$ is chosen to represent a ϕ independent radially travelling wave emanating from the surce. From (1.24) and (1.25), the only

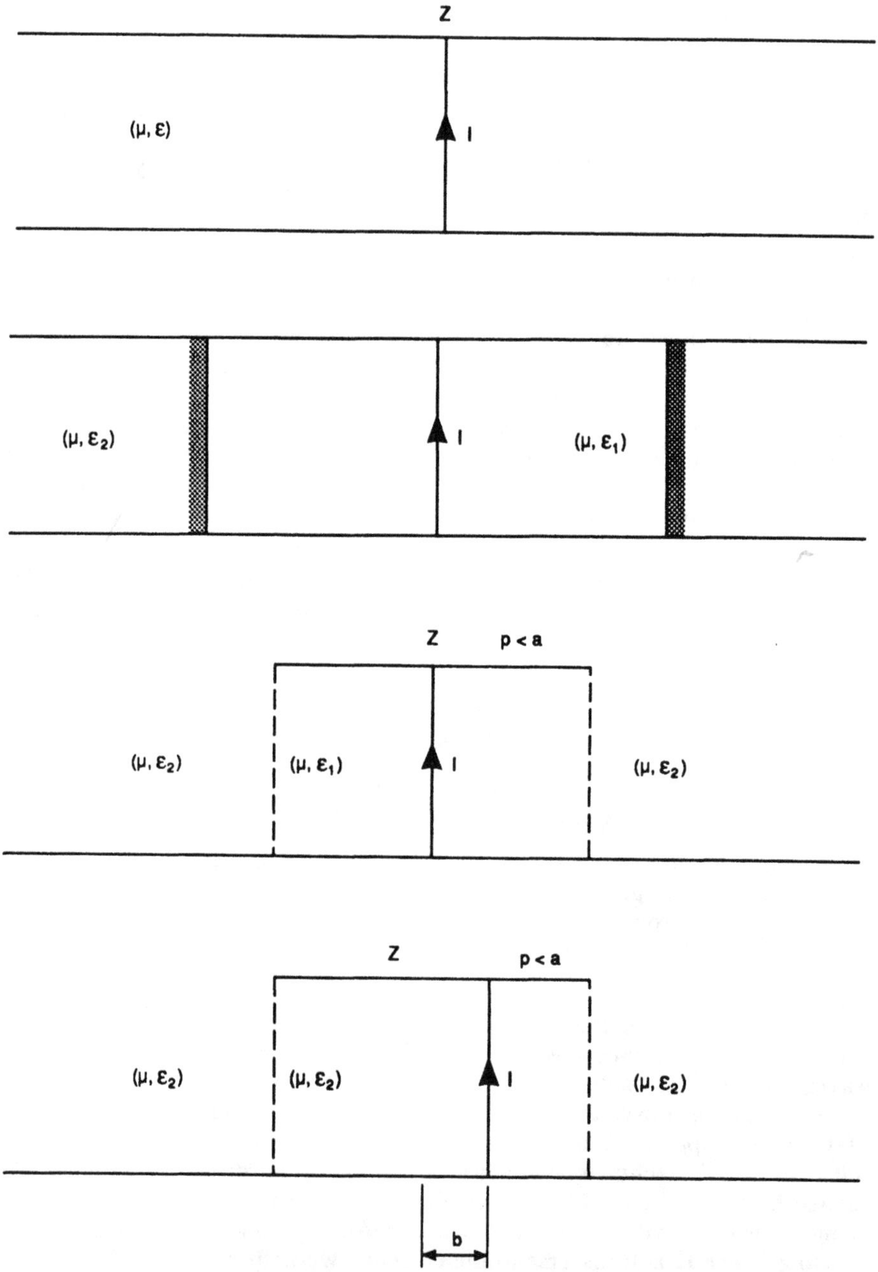

Fig. 1.3
a A two parallel conducting plate waveguide with infinite extent excited by a line source
b The waveguide in *a*, but inhomogeneously filled with two dielectric materials
c As in *b*, but the upper plate is circular with a finite radius equal to *a*
d As in *c*, but with off-axis line source

other non-vanishing field component is H_ϕ and is given by

$$H_\phi=(-i\omega\varepsilon/k^2)\partial E_z/\partial\rho=i(\varepsilon/\mu)^{1/2}CH_1^{(2)}(k\rho) \quad (1.46)$$

where the identity $H_0'^{(2)}(x)=-H_1^{(2)}(x)$ has been used. The constant C is now determined from the condition:

$$\lim_{\rho\to0} 2\pi\rho H_\phi=I \quad (1.47)$$

where I is the value of the current of the line source. Noting that $H_1^{(2)}(k\rho)$ tends to $2i/\pi k\rho$ as ρ tends to zero (see A2.6—A2.13), (1.47) yields C as

$$C=-\omega\mu I/4 \quad (1.48)$$

Next consider a slightly different situation from that given in Fig. 1.3*a*. Let the medium inside the parallel plate guide be inhomogeneous such that the permittivity ε is ε_1 for $\rho<a$ and ε_2 for $\rho>a$, as shown in Fig. 1.3*b*. In this case, the fields in the region $\rho\leqslant a$ have both travelling and standing waves. Hence an appropriate expression for E_{z1} in the region $\rho\leqslant a$ takes the form

$$E_{z1}=CH_0^{(2)}(k_1\rho)+AJ_0(k_1\rho) \quad (1.49)$$

where C is still given by (1.48) to account for the singularity at the source. The added term involving $J_0(k\rho)$ accounts for reflection at the boundary $\rho=a$ and is finite at the origin. The associated magnetic field $H_{\phi1}$ is given by (1.25):

$$H_{\phi1}=i(\varepsilon/\mu)^{1/2}[CH_1^{(2)}(k\rho)+AJ_1(k\rho)] \quad (1.50)$$

In the outer region $\rho\geqslant a$, an outwardly travelling wave solution must be chosen. Hence:

$$E_{z2}=BH_0^{(2)}(k_2\rho) \quad (1.51)$$

$$H_{\phi2}=i(\varepsilon_2/\mu)^{1/2}BH_1^{(2)}(k_2\rho) \quad (1.52)$$

where $k_{1,2}=\omega(\varepsilon_{1,2}\mu)^{1/2}$ are the wavenumbers in the inner and outer dielectric media respectively. Now we apply the boundary conditions (1.12) and (1.13) which require the continuity of E_z and H_ϕ at $\rho=a$. Using (1.49—1.52) we get

$$CH_0^{(2)}(k_1a)+AJ_0(k_1a)=BH_0^{(2)}(k_2a) \quad (1.53)$$

$$i(\varepsilon_1/\mu)^{1/2}[CH_1^{(2)}(k_1a)+AJ_1(k_1a)]=i(\varepsilon_2/\mu)^{1/2}BH_1^{(2)}(k_2a) \quad (1.54)$$

Simultaneous solution of (1.53) and (1.54) yields:

$$A/C=\frac{y_1(a)-y_2(a)}{y_2(a)J_0/H_0+i(\varepsilon_1/\mu)J_1/H_0} \quad (1.55)$$

$$B/C=\frac{y_1(a)J_0+i(\varepsilon_1/\mu)J_1}{[y_2(a)J_0/H_0+i(\varepsilon_1/\mu)J_1/H_0)]H_0^{(2)}(k_2\rho)} \quad (1.56)$$

where $y_1(a)$ and $y_2(a)$ are defined by:

$$y_{1,2}(a)=-i(\varepsilon_{1,2}/\mu)^{1/2}H_1^{(2)}(k_{1,2}\rho)/H_0^{(2)}(k_{1,2}\rho) \quad (1.57)$$

and they physically represent the admittances at $\rho=a$ of radially travelling waves $(-H_\phi/E_z)$ in medium 1 and 2 respectively; namely $y_2(a)$ is the admittance of the outer medium as seen from the boundary surface $\rho=a$. In

(1.55) and (1.56), the following abbreviations have been used:

$$J_{0,1} \equiv J_{0,1}(k_1 a), \quad H_{0,1} \equiv H^{(2)}_{0,1}(k_1 a)$$

As a check on the results (1.55) and (1.56) consider the case of a homogeneous medium throughout the guide; i.e. $\varepsilon_1 = \varepsilon_2$. In this case $k_1 = k_2$ and $y_1(a) = y_2(a)$, hence $A = 0$ and $C = B$ as we should expect.

To continue with this example, let us do another modification on Fig. 1.3*a*, *b* by reducing the dimensions of the upper plate to become a disc of radius *a* as shown in Fig. 1.3*c*. The resulting structure bears a close similarity to the microstrip patch antenna (e.g. Carver and Mink, 1981). The fields on this structure can be analysed if an estimate of the surface admittance $y_2(a)$ is available. The real part of this admittance accounts for the power radiated in the unbounded region $\rho > a$ and the imaginary part for the capacitance due to the fringing fields near $\rho = a$. The estimation of $y_2(a)$ has been the subject of several papers; e.g. Shen (1979), Chew and Kong (1980) and Yano and Ishimaru (1981). Here, let us assume for the moment that this admittance is readily known. Hence the internal fields are completely determined by (1.49) and (1.50) with the constants C and A determined by (1.48) and (1.55). So let us concentrate on obtaining the external fields in the region $\rho \geqslant a$. First we write the E_z field at $\rho = a$ from (1.51):

$$\begin{aligned} E_z(a, z) &= B H^{(2)}_0(k_2 a), \quad |z| < d \\ &= 0 \qquad\qquad\quad\;\, , \quad |z| > d \end{aligned} \tag{1.58}$$

Here we have extended the field to negative z by taking a perfect image below the lower perfectly conducting plate. Our aim now is to extend the known field at $\rho = a$ to all values of $\rho > a$. To this end, expand $E_z(a, z)$ into a spectrum of travelling waves along z by means of Fourier analysis.

Thus

$$E_z(a, z) = \int_{-\infty}^{\infty} F(a, \beta) \exp(-i\beta z)\, d\beta \tag{1.59}$$

where

$$F(a, \beta) = \int_{-\infty}^{\infty} E_z(a, z) \exp(i\beta z)\, dz/2\pi \tag{1.60}$$

On using (1.58) in (1.60)

$$F(a, \beta) = (Bd/\pi) H^{(2)}_0(k_2 a)\, [\sin(\beta d)/\beta d] \tag{1.61}$$

Now we note that, since $E_z(\rho, z)$ is a solution to the wave equation, each wavelet in (1.59) having an $\exp(-i\beta z)$ dependence on z must have the proper dependence on ρ which, for $\rho > a$, is the Hankel function representing a radially travelling wave. Hence, we write for $\rho \leqslant a$

$$E_z(\rho, z) = \int_{-\infty}^{\infty} F(a, \beta)\, [H^{(2)}_0(\lambda\rho)/H^{(2)}_0(\lambda a)] \exp(-i\beta z)\, d\beta \tag{1.62}$$

where $\lambda = (k_2^2 - \beta^2)^{1/2}$.

It is clear that the integrand satisfies the wave equation in circular cylindrical coordinates. Also, it is easy to check that this equation reduces to (1.59) for $\rho=a$. We have thus formally solved the problem for the external fields although, in general, numerical means are needed to obtain the integral in (1.62). However, the integral can be evaluated in the far field by the stationary phase or saddle point of integration (e.g. Clemmow, 1966). Alternatively, using the following formula proved by Harrington (1961) for $\rho\rightarrow\infty$

$$\int_{-\infty}^{\infty} G(\beta)H_n^{(2)}(\rho(k^2-\beta^2)^{1/2}) \exp(-i\beta z)\, d\beta = 2i^{n+1}G(k\cos\theta)\,[\exp(-ikr)/r] \tag{1.63}$$

where (r, θ) are the polar coordinates; $r=(\rho^2+z^2)^{1/2}$, $\theta=\cos^{-1}(z/r)$, we get E_z in the far field

$$E_z(r, \theta)=2i(B/\pi k_2\cos\theta)\sin(k_2 d\cos\theta) \times [H_0^{(2)}(k_2a)/H_0^{(2)}(k_2a\sin\theta)]\,[\exp(-ik_2r)/r] \tag{1.64}$$

Similarly we can derive H_ϕ and E_ρ (see probl. 1.8). It remains to say that, in the far zone, the electric field is dominantly θ oriented. Hence $E_z=-E_\theta\sin\theta$ and $E_\rho=E_\theta\cos\theta$. The radiation conductance of the aperture at $\rho=a$ can be obtained from the power radiated in the far zone (see probl. 1.8). This constitutes the real part of the aperture admittance $y_2(a)$.

So far, we have assumed that the line source lies at $\rho=0$. The case of more practical interest in microstrip patch antennas (Carver and Mink, 1981) is when the source is displaced from the centre of the patch. So, let us consider Fig. 1.3*d* where the line source lies at $\rho=b<a$ and $\phi=0$. The surface current density $j_z(\phi)$ on the cylindrical surface $\rho=b$ can then be written

$$j_z(\phi)=(I/b)\delta(\phi) \tag{1.65}$$

where $\delta(\phi)$ is the impulse function, or delta function, defined by

$$\int_0^{2\pi} \delta(\phi-\phi_a)f(\phi)\, d\phi=f(\phi_a) \tag{1.66}$$

where $f(.)$ is any continuous function of ϕ. In particular, the total surface current is $\int j_z b\, d\phi=I$ (ampere). The next step in deriving the fields of this source is to expand the ϕ dependent surface current into the characteristic circular azimuthal functions $\sin n\phi$ and $\cos n\phi$. This is done by simple Fourier analysis. Thus with the help of (1.66), we get

$$j_z(\phi)=(I/2\pi b)\sum_{n=0}^{\infty} \zeta_n\cos n\phi \tag{1.67}$$

where $\zeta_n=1$ for $n=0$ and 2 for $n>0$. Now define the following transformation for any field function:

$$F(\rho, \phi)=\sum_{n=0}^{\infty} F_n(\rho)\cos n\phi \tag{1.68}$$

where $F_n(\rho)$ is the transformation of $F(\rho, \phi)$. Expressions for the transformed electric and magnetic fields satisfying the wave equation in the different regions of Fig. 1.3*d* are chosen as follows:

$$E_{zn}(\rho) = A_n J_n(k_1 \rho), \; 0<\rho<b$$
$$= C_n H_n^{(2)}(k_1\rho) + B_n J_n(k_1\rho), \; b<\rho<a \qquad (1.69)$$

$$i(\mu/\varepsilon_1)^{1/2} H_{\phi n}(\rho) = A_n J_n'(k_1\rho), \; 0>\rho>b$$
$$= C_n H_n^{(2)\prime}(k_1\rho) + B_n J_n'(k_1\rho), \; b<\rho<a \qquad (1.70)$$

The magnetic field includes also a radial component H_ρ but since this does not enter into the boundary conditions, we can set it aside for now. The boundary conditions at $\rho=b$ can be written in terms of the transformed fields as:

$$E_{zn}(b-) = E_{z,\,n}(b+) \qquad (1.71)$$

$$H_{\phi n}(b+) - H_{\phi n}(b-) = j_{zn} = I\zeta_n/2\pi b \qquad (1.72)$$

Application of these two conditions on (1.69) and (1.70 yields

$$C_n H_n^{(2)}(k_1 b) + (B_n - A_n) J_n(k_1 b) = 0$$

$$C_n H_n^{(2)\prime}(k_1 b) + (B_n - A_n) J_n'(k_1 b) = i(\mu/\varepsilon_1)^{1/2} I\zeta_n/2\pi b$$

Hence

$$C_n = -(I\omega\mu/4) J_n(k_1 b)\zeta_n \qquad (1.73)$$

$$B_n - A_n = (I\omega\mu/4) H_n^{(2)}(k_1 b) \qquad (1.74)$$

where the Wronskian relation $J_n(x)H_n^{(2)\prime}(x) - J_n'(x)H_n^{(2)}(x) = -2i/\pi x$ has been invoked. As a check on (1.73), we note that in the case $b=0$ (the current is in the centre of the patch), all $C_n=0$ except $C_0 = -(I\omega\mu/4)$ in agreement with (1.48). The outer boundary conditions at $\rho=a$ depend on the admittances $y_{2n}(a) \equiv -H_{\phi n}(a+)/E_{zn}(a+)$, which are determined by the behaviour of fields in the external region $\rho>a$. For the structure of Fig. 1.3*b*, these are given by

$$y_{2n}(a) = i(\varepsilon_2/\mu)^{1/2} H_n^{(2)\prime}(k_2 a)/H_n^{(2)}(k_2 a)$$

while for the structure of Fig. 1.3*c* or *d*, approximate expressions are found in several references (e.g. Chew and Kong, 1980). Given $y_{2n}(a)$, application of the boundary condition:

$$y_{2n}(a) = -H_{\phi n}(a)/E_{zn}(a) \qquad (1.75)$$

yields:

$$B_n/C_n = [\hat{y}_{2n}(a) - \hat{y}_{1n}(a)] H_n^{(2)}(k_1 a)/[iJ_n'(k_1 a) - \hat{y}_{2n}(a) J_n(k_1 a)] \qquad (1.76)$$

where $\hat{y}_{1n}(a)$ and $\hat{y}_{2n}(a)$ are the normalised admittances:

$$\hat{y}_{1n}(a) = iH_n^{(2)\prime}(k_1 a)/H_n^{(2)}(k_1 a),$$

$$\hat{y}_{2n}(a) = y_{2n}(a)\,(\mu/\varepsilon_1)^{1/2}$$

Of particular interest is the aperture electric field at $\rho=a$. This is obtained from (1.68), (1.69), (1.73) and (1.76)

$$E_z(a, \phi) = -(I\omega\mu/4) \sum_{n=0}^{\infty} \zeta_n J_n(k_1 b) H_n^{(2)}(k_1 a) \frac{iJ_n'(k_1 a) - \hat{y}_{1n}(a) J_n(k_1 a)}{iJ_n'(k_1 a) - \hat{y}_{2n}(a) J_n(k_1 a)}$$
$$\cdot \cos n\phi \qquad (1.77)$$

From this expression, we get the resonant frequencies of the interior region $\rho < a$, considered as a cavity, by equating the denominator by zero. For pure imaginary admittances $y_{2n}(a)$, the resonant frequencies are real, but for complex admittances, the resonant frequencies are complex, reflecting the presence of radiation losses.

1.5.4 Elliptical coordinates and wave functions

Elliptical cylindrical coordinates are defined by (u, v, z) which are related to the familiar cartesian coordinates (x, y, z) by

$$x = c \cosh u \cos v \tag{1.78}$$

$$y = c \sinh u \sin v \tag{1.79}$$

from which one can easily deduce that

$$x^2/(c^2 \cosh^2 u) + y^2/(c^2 \sinh^2 u) = 1$$

Here u = constant, say u_0, is an ellipse whose focii lie at $(x = \pm c, y = 0)$ and the semimajor and semiminor axes are given by $c \cosh u_0$ and $c \sinh u_0$ respectively. Likewise (1.78) and (1.79) imply that

$$x^2/(c^2 \cos^2 v) - y^2/(c^2 \sin^2 v) = 1$$

Hence v = constant, say v_0, is a hyperbola cofocal with the ellipses $u = u_0$. The two cofocal families of ellipses and hyperbolas are depicted in Fig. 1.4. The

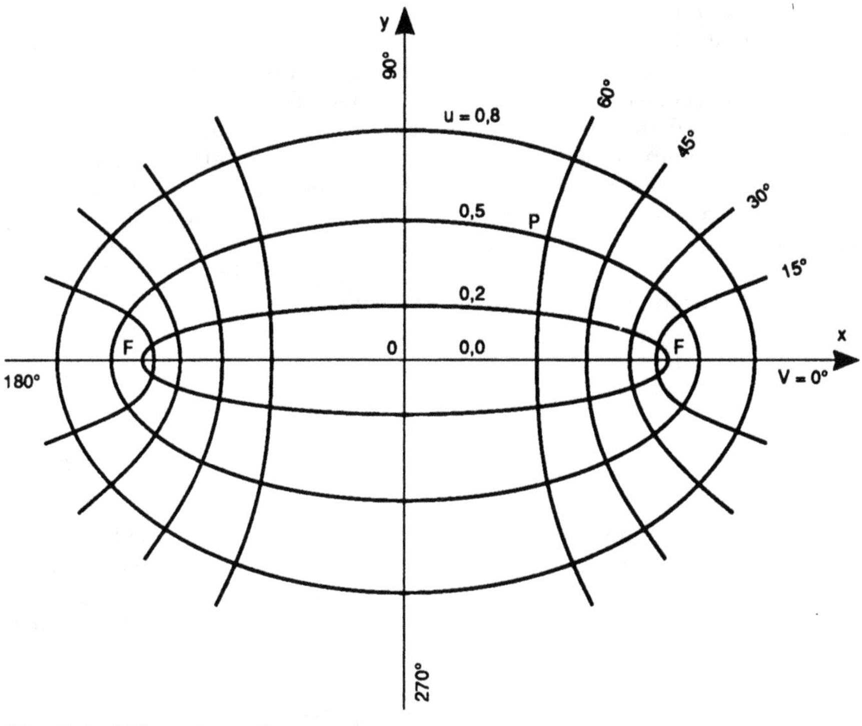

Fig. 1.4 Elliptical coordinate system

coordinate u is a radial coordinate and takes all values between 0 and ∞. The ellipse collapses into a piece of straight line joining the two focii when $u_0 = 0$. The coordinate v is an angular coordinate taking the values between 0 and 2π as shown in Fig. 1.4. The transition to the circular case occurs as the limit c tends to zero and u tends to infinity, whence y/x tends to tan v showing clearly that v is readily the angular coordinate.

The metrical coefficients h_1 and h_2 are obtained from (1.78) and (1.79) by applying equation (A1.1). This results in

$$h_1 = h_2 = (c/\sqrt{2})\,(\cosh 2u - \cos 2v)^{1/2} \equiv h \tag{1.80}$$

The wave equation (1.34) then takes the form

$$\partial^2 f/\partial u^2 + \partial^2 f/\partial v^2 h^2 f = 0$$

or

$$\partial^2 f/\partial u^2 + \partial^2 f/\partial v^2 + 2q(\cosh 2u - \cos 2v) f = 0 \tag{1.81}$$

where $4q = \lambda^2 c^2$. Now, using the method of separation of variables, we set $f = U(u)V(v)$. Upon substitution in (1.81), we get separate equations for the functions V and U:

$$d^2 V/dv^2 + (a - 2q \cos 2v)V = 0 \tag{1.82}$$

$$d^2 U/du^2 + (2q \cosh 2u - a) = 0 \tag{1.83}$$

which are known as the Mathieu and modified Mathieu differential equations. The constant a is a separation constant which, in general, can take any arbitrary value. However, it turns out that, if $V(v)$ is required to be a periodic function of v, then a can assume a denumerable, but discrete, set of characteristic values. This set is divided into two subsets: one subset corresponding to even functions of v and the other to odd functions of v. Let these two subsets of characteristic values be denoted respectively by $a_m^{(e)}$, $m = 0, 2, \ldots$ and $a_m^{(0)}$, $m = 1, 2, \ldots$. Formulae for these values in terms of q are given in Abramowitz and Stegun (1965). Corresponding to each characteristic value a, there exists a characteristic function $V(v)$. Let us denote the characteristic function corresponding to $a_m^{(e)}$ by $C_m(q, v)$ and that corresponding to $a_m^{(0)}$ by $S_m(q, v)$. Using the notation of Stratton (1941), these functions are given by:

$$C_m(q, v) = \sum_n{}' A_n^m(q) \cos nv \tag{1.84}$$

$$S_m(q, v) = \sum_n{}' B_n^m(q, v) \sin nv \tag{1.85}$$

where the primed summation sign signifies that the summation extends over even values of n if m is even and over odd values of n if m is odd. The coefficients $A_n(q)$ and $B_n(q)$ can be obtained recursively by introducing (1.84) and (1.85) into (1.82); see Appendix A2. Normalisation of C_m and S_m can be chosen such that $C_m(q, 0) = 1$ and $dS_m(q, v)/dv = 1$ at $v = 0$. Hence:

$$\sum_n{}' A_n^m = 1 \text{ and } \sum_n{}' nB_n^m = 1$$

Explicit expressions for the coefficients in (1.84) and (1.85) are found in Abramowitz and Stegun (1965).

An important difference between angular elliptic wave functions and angular circular wave functions can now be made clear. While the circular functions $\cos m\phi$ and $\sin m\phi$ are independent of the medium constants of frequency, the elliptic functions $C_m(q, v)$ and $S_m(q, v)$ do depend on these quantities through the parameter q. This difference will be of major concern in solving problems involving boundaries between different media, as will be seen in examples to follow in this Section.

Orthogonality relationships among the characteristic functions can be established by using (1.84)—(1.85). These are summarised as follows (Stratton, 1941):

$$\int_0^{2\pi} C_m(q, v)C_n(q, v)\, dv = 0, \; m \neq n \tag{1.86}$$

$$\int_0^{2\pi} S_m(q, v)S_n(q, v)\, dv = 0, \; m \neq n \tag{1.87}$$

$$\int_0^{2\pi} C_m(q, v)S_n(q, v)\, dv = 0 \tag{1.88}$$

These relations greatly simplify the decomposition of any periodic function of v, with period 2π, into an equivalent set of characteristic functions. Thus, if $G(v)$ is such a function, it is permitted to write

$$G(v) = \sum_m A_m C_m(q, v) + B_m S_m(q, v) \tag{1.89}$$

To obtain the coefficients A_m and B_m, multiply both sides successively by $C_n(q, v)$ and $S_n(q, v)$, then by integrtating over the period $(0, 2\pi)$ and using (1.86)—(1.88), one gets

$$A_n = \int_0^{2\pi} G(v)C_n(q, v)\, dv/N_{e,n} \tag{1.90}$$

$$B_n = \int_0^{2\pi} G(v)S_n(q, v)\, dv/N_{0,n} \tag{1.91}$$

where $N_{e,n}$ and $N_{0,n}$ are normalisation factors for even and odd characteristic functions respectively, and are defined by:

$$N_{e,n} = \int_0^{2\pi} C_n^2(q, v)\, dv \tag{1.92}$$

$$N_{0,n} = \int_0^{2\pi} S_n^2(q, v)\, dv \tag{1.93}$$

and can be easily obtained in closed forms by using (1.84) and (1.85). The special case of an impulsive, or delta, function $\delta(v - v_0)$ defined by (1.66) is of common interest. Using this function in place of $G(v)$ in (1.89)—(1.91), we get the expansion of the delta function

$$\delta(v-v_0)=\sum_m \{C_m(q, v_0)C_m(q, v)/N_{e,m}+S_m(q, v_0)S_m(q, v)/N_{0,m}\} \quad (1.94)$$

where the summation is over all integer values of m.

Next, we turn our attention to the radial function $U(u)$ satisfying (1.83). To each characteristic value of a, there exist two possible radial function solutions. Even radial function solutions, corresponding to $a_m^{(e)}$ are given by (Stratton, 1941; Abramowitz and Stegun, 1965):

$$U(u)\equiv\left\{\begin{matrix} R_{e,m}^{(1)} \\ R_{e,m}^{(2)} \end{matrix}(q, u)\right\}=(\pi/2)^{1/2}\sum_{\nu}{}' i^{m-n}A_n^m(q)\left\{\begin{matrix} J_n \\ N_n \end{matrix}(\lambda\rho_e)\right\} \quad (1.95)$$

and odd radial solutions corresponding to $a_m^{(0)}$ are:

$$U(u)\equiv\left\{\begin{matrix} R_{0,m}^{(1)} \\ R_{0,m}^{(2)} \end{matrix}(q, u)\right\}=(\pi/2)^{1/2}\sum_{\nu}{}' B_n^m(q)\left\{\begin{matrix} J_n \\ N_n \end{matrix}(\lambda\rho_0)\right\} \quad (1.96)$$

where $\rho_{e,0}$ are defined by: $\rho_e=c\cosh u$, $\rho_0=c\sinh u$, and J_n and N_n are the familiar Bessel functions. Just as $J_n(.)$ and $N_n(.)$ represent standing waves in circular cylindrical coordinates, the $R_{e,m}^{(1)}$, $R_{e,m}^{(2)}$, $R_{0,m}^{(1)}$ and $R_{0,m}^{(2)}$ functions represent cylindrical standing waves in elliptical coordinates. A main difference, however, is that, while the J and N functions are the same for even and odd angular solutions, the even R and odd R functions are not the same. Inward and outward travelling waves are formed by combining the two standing wave functions. Thus, we define $R^{(3)}$ and $R^{(4)}$ by:

$$\begin{aligned} R_m^{(3)}&=R_m^{(1)}+iR_m^{(2)} \\ R_m^{(4)}&=R_m^{(1)}+iR_m^{(2)} \end{aligned} \quad (1.97)$$

which apply to both even and odd functions.

The growing and evanescent solutions corresponding to imaginary λ or negative q are given by $I(q, u)$ and $K(q, u)$ defined by:

$$I_m(q, u)=e^{-i\pi m/2}R_m^{(1)}(-q, u) \quad (1.98)$$

$$K_m(q, u)=(\pi/2)e^{-i\pi(m+1)/2}R_m^{(4)}(-q, u) \quad (1.99)$$

for either even or odd functions.

Thus, valid solutions of $f(u, v)$ in (1.81) may take any of the forms:

$$f(u, v)=C_m(q, v)R_{e,m}^{(1\to4)}(q, u) \quad (1.100)$$

$$f(u, v)=S_m(q, v)R_{0,m}^{(1\to4)}(q, u) \quad (1.101)$$

which are even and odd wave functions in v respectively.

1.5.5 Example

To give an example on the use of the elliptic wave functions, let us reconsider the structure of Fig. 1.3*b*, but with the inner medium occupying an elliptic rather than a circular region. The new structure is shown in Fig. 1.5 in which

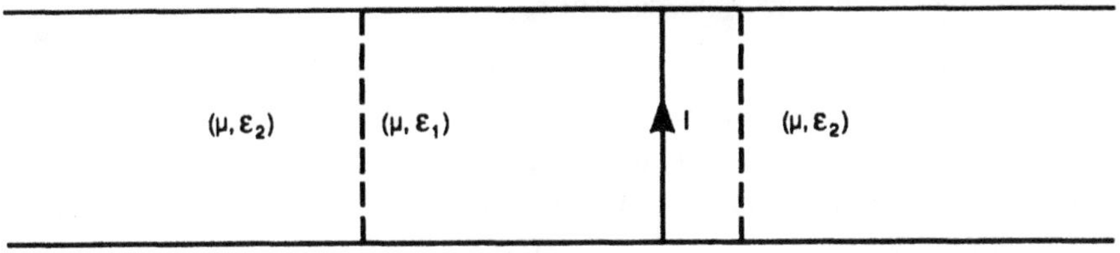

Fig. 1.5 A two parallel conducting plate waveguide with inner elliptic region.

the dielectric medium with ε_1 occupies the elliptical cylinder with semimajor and semiminor axes a and b. A line source with a uniform electric current I is connected across the two parallel plates at the point (u_s, v_s) in an elliptical coordinate system. Our purpose is to find the electromagnetic field everywhere within the parallel plates. First we express the given current as a surface current over the elliptic cylindrical surface $u = u_s$. Thus we write:

$$j_z(v) = [I/h(u_s, v)]\,\delta(v - v_s) \tag{1.102}$$

where $\delta(.)$ is the delta function. Using (1.66), one can readily verify that the total current on the surface $\oint j_z h\,dv$ is equal to I. Next, we expand this current distribution as a sum over the angular Mathieu functions of the inner medium. Thus, by invoking (1.94), one gets

$$h(u_s, v) j_z(v) = I \sum_{r,n} V_{r,n}(q_1, v_s) V_{r,n}(q_1, v)/N_{r,n} \tag{1.103}$$

where the summation is over even $(r = e)$ and odd $(r = o)$ angular modes of all orders $n = 0, 1, 2 \ldots$; namely $V_{e,n} \equiv C_n$, $V_{o,n} \equiv S_n$ and $q_1 = k_1^2(a^2 - b^2)/4$. The fields are expected to be TM to z, and since there is no variation with z, the nonvanishing field components are E_z, H_v and H_u, as deduced from (1.24)—(1.25). Appropriate expressions for the fields E_{1z} and H_{1v} in the inner region $u < u_0 = \cosh^{-1}(a/(a^2 - b^2)^{1/2})$ are:

$$E_{1z}(u, v) = \sum_{r,n} E_{r,n}(u) V_{r,n}(q_1, v) \tag{1.104}$$

$$h(u, v) H_{1v} = (I/i\omega\mu) \sum_{r,n} H_{r,n}(u) V_{r,n}(q_1, v) \tag{1.105}$$

with

$$\begin{aligned} E_{r,n}(u) &= A_{r,n} R^{(1)}_{r,n}(q_1, u), & u < u_s \\ &= C_{r,n} R^{(4)}_{r,n}(q_1, u) + B_{r,n} R^{(1)}(q_1, u), & u_s < u < u_0 \end{aligned} \tag{1.106}$$

$$\begin{aligned} H_{r,n}(u, v) &= A_{r,n} R'^{(1)}_{r,n}(q_1, u) & u < u_s \\ &= C_{r,n} R'^{(4)}_{r,n}(q_1, u) + B_{r,n} R'^{(1)}_{r,n}(q_1, u), & u_s < u < u_0 \end{aligned} \tag{1.107}$$

The boundary conditions at $u=u_s$ read

$$E_{1z}(u_s^-,v)=E_{1z}(u_s^+,v)$$

$$H_{1v}(u_s^+,v)-H_{1v}(u_s^-,v)=j_z(v)$$

Using (1.103)–(1.105) in the above two equations and noting the orthogonality of the $V_{r,n}$ functions over the full range of v, the summation signs can be removed; i.e. the equality is made on a term by term basis. This results in two equations for each pair of coefficients $C_{r,n}$ and $(B_{r,n}-A_{r,n})$, with solutions:

$$C_{r,n}=(i\omega\mu I)V_{r,n}(q_1,v_s)R^{(1)}_{r,n}(q_1,u_s)/WN_{r,n} \tag{1.108}$$

$$B_{r,n}-A_{r,n}=-C_{r,n}R^{(4)}_{r,n}(q_1,u_s)/R^{(1)}_{r,n}(q_1,u_s) \tag{1.109}$$

where W is the Wronksian of the $R^{(1)}_{r,n}(q_1,u)$ and $R^{(4)}_{r,n}(q_1,u)$ functions which are independent of the argument u.

In the outer region $u>u_0$, field expansions which represent outwardly travelling waves are chosen, thus

$$E_{2z}(u,v)=\sum_{r,n}F_{r,n}(u)V_{r,n}(q_2,v) \tag{1.110}$$

$$h(u,v)H_{2,v}(u,v)=(I/i\omega\mu)\sum_{r,n}G_{r,n}(u)V_{r,n}(q_2,v) \tag{1.111}$$

where

$$F_{r,n}(u)=D_{r,n}R^{(4)}_{r,n}(q_2,u) \tag{1.112}$$

$$G_{r,n}(u)=D_{r,n}R'^{(4)}_{r,n}(q_2,u) \tag{1.113}$$

Now it is necessary to satisfy the boundary conditions at the interface $u=u_0$. This requires that the field components E_z and H_v be continuous across this interface. Looking at equations (1.104) and (1.110), we immediately discover a basic problem in satisfying these boundary conditions; the two sets of angular functions $V(q_1,v)$ and $V(q_2,v)$ are different; hence they are not mutually orthogonal. The implication is that the equality of field expansions cannot be made on a term by term basis. In fact this is a fundamental difficulty which is encountered in any cylindrical coordinate system other than the circular (Wait, 1967). To clarify, let us write down the boundary conditions at $u=u_0$:

$$\sum_{r,n}E_{r,n}(u_0)V_{r,n}(q_1,v)=\sum_{r,n}F_{r,n}(u_0)V_{r,n}(q_2,v) \tag{1.114}$$

$$\sum_{r,n}H_{r,n}(u_0)V_{r,n}(q_1,v)=\sum_{r,n}G_{r,n}(u_0)V_{r,n}(q_2,v) \tag{1.115}$$

Since the V functions on both sides of these equations are different, it is obvious that we cannot equate the individual terms of the summations. We can at most remove one of the summations by invoking the orthogonality relationships (1.86–1.88) with $q=q_1$ or q_2. For example, multiplying (1.114) and (1.115) by

$V_{r',m}(q_2, v)$ and integrating over v between 0 and 2π, we get:

$$\sum_{r,n} E_{r,n}(u_0) \int_0^{2\pi} V_{r,n}(q_1, v) V_{r',m}(q_2, v)\, dv = F_{r',m}(u_0) = D_{r',m} R^{(4)}_{r',m}(q_2, v) \quad (1.116)$$

$$\sum_{r,n} H_{r,n}(u_0) \int_0^{2\pi} V_{r,n}(q_1, v) V_{r',m}(q_2, v) dv = G_{r',m}(u_0) = D_{r',m} R'^{(4)}_{r',m}(q_2, v) \quad (1.117)$$

with $m=0, 1, 2 \ldots, r'=e$ or 0 and (1.112) and (1.113) have been used. Dividing (1.116) and (1.117) and invoking (1.106–1.107), we get an infinit set of linear equations in the coefficients $B_{r,n}$ and $C_{r,n}$. Since $C_{r,n}$ are known from (1.108), the set of equations are solved for $B_{r,n}$, possibly by a suitable truncation of the set. The field coefficients $D_{r,n}$ in the outer medium will then follow from (1.116) or (1.117). This basically completes the solution. An alternative approach to the solution of (1.114) and (1.115) is to use the triogeometric expansions of the angular functions $V(q_1, v)$ and $V(q_2, v)$ and then equate the coefficients of cos nv and sin nv on both sides of the equations to obtain an infinite set of linear equations in the unknown coefficients. Again this set is solved, in any practical situation, by a judicious truncation of the equations.

1.6 Spherical wave functions

A spherical coordinate system (r, θ, ϕ) is the natural choice for many structures such as conical horns and conical antennas. The free space itself is often treated as a spherical waveguide. Waves radiated from any source of finite dimensions possess spherical wavefronts at sufficiently far distances from the source. In this section we drive solutions to Maxwell's equations in the spherical coordinate system. Our starting point is the solution of the scalar wave functions in spherical coordinates. This is followed by vector solutions to Maxwell's equations which are *TE* and *TM* to the radial vector r. Finally orthogonality properties of the spherical wave functions are derived at the end of the section.

1.6.1 Scalar wave functions

Let $f(r, \theta, \phi)$ be a solution to the scalar wave equation

$$\nabla^2 f + k^2 f = 0$$

In spherical coordinates this equation reads [see (A1.5)]

$$\frac{1}{r^2}\frac{\partial}{\partial r}\left(r^2\frac{\partial f}{\partial r}\right) + \frac{1}{r^2 \sin\theta}\frac{\partial}{\partial\theta}\left(\sin\theta\frac{\partial f}{\partial\theta}\right) + \frac{1}{r^2\sin^2\theta}\frac{\partial^2 f}{\partial\phi^2} + k^2 f = 0 \quad (1.118)$$

Solving by separation of variables, we set

$$f(r, \theta, \phi) = R(r)\Theta(\theta)\Phi(\phi)$$

and we get the following separate equations for R, Θ and Φ functions:

$$(d/dr)(r^2 dR/dr) + [(kr)^2 - \nu(\nu+1)]R = 0$$

$$(d/d\theta)(\sin\theta\, d\Theta/d\theta)/\sin\theta + [\nu(\nu+1) - m^2/\sin^2\theta]\Theta = 0 \quad (1.119)$$

$$d^2\Phi/d\phi^2 + m^2\Phi = 0$$

where ν and m are arbitrary separation constants. The first of these equations can be transformed to the form of the Bessel equation by using the substitution $R = B/(kr)^{1/2}$. Doing this exercise, we end up with the following definition for the spherical Bessel functions:

$$R(r) \equiv b_\nu(kr) = (\pi/2kr)^{1/2} B_{\nu+1/2}(kr) \tag{1.120}$$

where $B_{\nu+1/2}$ is any valid solution of the Bessel equation of order $\nu + 1/2$; namely $B_{\nu+1/2} = J_{\nu+1/2}$, $N_{\nu+1/2}$, or $H^{(1),(2)}_{\nu+1/2}$. In the special case when ν is an integer n, the Bessel functions $B_{n+1/2}$ have the important property that their asymptotic expansions in powers of $(1/kr)$ have finite number of terms and they are exact [see (A2.17)]. For example, the following are exact expressions for the lower order spherical Bessel functions:

$$\begin{aligned} j_0(kr) &= (\pi/2kr)^{1/2} J_{1/2}(kr) = \sin(kr)/kr \\ j_1(kr) &= (\pi/2kr)^{1/2} J_{3/2}(kr) = -\cos(kr)/kr + \sin(kr)/(kr)^2 \\ h_0^{(1),(2)}(kr) &= (\pi/2kr)^{1/2} H^{(1),(2)}_{1/2}(kr) = \pm e^{\pm ikr}/ikr \\ h_1^{(1),(2)}(kr) &= (\pi/2kr)^{1/2} H^{(1),(2)}_{3/2}(kr) = -(e^{\pm ikr}/kr)(1 \pm i/kr) \end{aligned} \tag{1.121}$$

The higher order functions are obtainable from the recurrency relation

$$b_n(kr) = [kr/(2n+1)][b_{n-1}(kr) + b_{n+1}(kr)] \tag{1.122}$$

and the derivatives may be obtained from

$$b_n'(kr) = [n b_{n-1}(kr) - (n+1) b_{n+1}(kr)]/(2n+1) \tag{1.123}$$

and

$$b_0'(kr) = -b_1(kr) \tag{1.124}$$

Now turn attention to the second equation of (1.119) which is recognised as the Legendre equation with the argument $\cos\theta$. Hence, solutions for $\Theta(\theta)$ are

$$\Theta(\theta) = P_\nu^m(\cos\theta),\ Q_\nu^m(\cos\theta) \tag{1.125}$$

which are the associated Legendre functions of first and second kind respectively as defined in Magnus *et al.* (1966) and Abramowitz and Stegun (1965). The function P_ν^m is finite on the polar axis $(\theta = 0)$ while Q_ν^m has a logarithmic singularity for integer values of ν. In the special case when $\nu = n$ is an integer and $m = 0$ the Legendre function $P_n^0(\cos\theta) \equiv P_n(\cos\theta)$ is a polynomial of degree n in $\cos\theta$. For example,

$$P_0(\cos\theta) = 1 \text{ and } P_1(\cos\theta) = \cos\theta$$

The higher order functions are obtainable from the recurrence relation:

$$(n+1)P_{n+1}(u) = (2n+1)P_n(u) - nP_{n-1}(u) \tag{1.126}$$

For $m > 0$ the associated Legendre function $P_n^m(\cos\theta)$ is related to the Legendre function $P_n(\cos\theta)$ by

$$P_n^m(\cos\theta) = (-1)^m \sin^m\theta [d^m P_n(\cos\theta)/d\cos^m\theta] \tag{1.127}$$

Finally the third of equations (1.119) has the solutions

$$\Phi(\phi) = \cos m\phi,\ \sin m\phi \tag{1.128}$$

The Legendre function $P_n(\cos\theta)$ is an oscillating function over the whole range of θ; $0\leqslant\theta\leqslant\pi$. Namely it has n zeros in this interval. Likewise, each of the wave solutions $P_n^m(\cos\theta)$ cos θ (sin θ) alternate sign over the surface of a sphere, dividing the surface into adjacent zones of positive and negative signs. Hence they are often called the spherical tesseral harmonics. An important property of the tesseral harmonics is that they form a complete set of orthogonal functions over the surface of a sphere. Expressed mathematically

$$\int_0^{2\pi}\int_0^{\pi} P_{n'}^{m'}(\cos\theta)P_n^m(\cos\theta)\cos m\phi\cos m'\phi\sin\theta\, d\theta\, d\phi=0 \quad (1.129)$$

provided $n\neq n'$ and/or $m\neq m'$. When $n=n'$ and $m=m'$, we have

$$\int_0^{2\pi}\int_0^{\pi}[P_n^m(\cos\theta)\cos m\phi]^2\sin\theta\, d\theta\, d\phi=\frac{2\pi\varepsilon_m}{2n+1}\frac{(n+m)!}{(n-m)!} \quad (1.130)$$

where $\varepsilon_m=2$ for $m=0$ and $\varepsilon_m=1$ for $m>0$. Obviously, similar results apply if cos $m\phi$ is replaced by sin $m\phi$. In addition orthogonality exists between tesseral harmonics having cos $m\phi$ and those having sin $m\phi$ dependence. As a consequence of the above relations an arbitrary function $g(\theta,\phi)$ which, together with its first and second derivatives, is continuous over the surface of a sphere can be expanded as a sum of tesseral harmonics

$$g(\theta,\phi)=\sum_{n=0}^{\infty}\sum_{m=0}^{n}P_n^m(\cos\theta)(a_{mn}\cos m\phi+b_{mn}\sin m\phi) \quad (1.131)$$

where

$$\begin{matrix}a_{mn}\\ b_{mn}\end{matrix}=\frac{2n+1}{2\pi\varepsilon_m}\frac{(n-m)!}{(n+m)!}\int_0^{2\pi}\int_0^{\pi}g(\theta,\phi)P_n^m(\cos\theta)\begin{matrix}\cos m\phi\\ \sin m\phi\end{matrix}\sin\theta\, d\theta\, d\phi \quad (1.132)$$

1.6.2 Vector wave functions

Next we derive the spherical vector wave functions. First it is noted that none of the spherical components E_r, E_θ or E_ϕ satisfies the scalar wave equation (prove!). In order to get vector solutions for E and H in a source free region in spherical coordinates, we resort to Maxwell's equations (1.1) and (1.2) and notice that both E and H satisfy the following vector equation:

$$\nabla\times\nabla\times\mathrm{m}=k^2\,\mathrm{m} \quad (1.133)$$

and

$$\nabla\cdot\mathrm{m}=0 \quad (1.134)$$

Following Stratton (1941) we construct two vector field functions m and n as follows:

$$\mathrm{m}=\nabla\times(f\mathrm{r})\quad\text{and}\quad \mathrm{n}=k^{-1}\nabla\times\mathrm{m} \quad (1.135)$$

where f is a scalar function and r is the radial position vector. Since the divergence of a curl is identically zero, then both m and n satisfy (1.134).

Furthermore one can show that m satisfies the vector wave equation (1.133) if f satisfies the scalar wave equation (1.118); see probl. 1.14. It then follows that n also satisfies (1.133). Therefore we can construct electromagnetic vector fields satisfying Maxwell's equations in a source free and homogeneous region as follows:

$$(\mathrm{E}, \mathrm{H})_{TE} = (\mathrm{m}, i\mathrm{n}(\varepsilon/\mu)^{1/2}) \tag{1.136}$$

$$(\mathrm{E}, \mathrm{H})_{TM} = (\mathrm{n}, i\mathrm{m}(\varepsilon/\mu)^{1/2}) \tag{1.137}$$

Since m is, by definition (1.135), transverse to r, the fields given by (1.136) are transverse electric to r. Similarly the fields of (1.137) are transverse magnetic to r. More explicit expressions of the vector fields m and n can be obtained by using (1.135) and substituting for f by the general scalar wave solution:

$$f(r, \theta, \phi) = P_n^m(\cos\theta) \begin{matrix} \cos m\phi \\ \sin m\phi \end{matrix} \, b_n(kr) \tag{1.138}$$

Therefore

$$\mathrm{m}_{\substack{e\\o}mn} = \left\{ (\mp m P_n^m(\cos\theta)/\sin\theta) \begin{matrix} \sin m\phi \\ \cos m\phi \end{matrix} \hat{\theta} - (\partial P_n^m(\cos\theta)/\partial\theta) \begin{matrix} \cos m\phi \\ \sin m\phi \end{matrix} \hat{\phi} \right\} b_n(kr) \tag{1.139}$$

$$\mathrm{n}_{\substack{e\\o}mn} = (n(n+1)/kr) P_n^m(\cos\theta) \begin{matrix} \cos m\phi \\ \sin m\phi \end{matrix} b_n(kr)\hat{\mathrm{r}}$$
$$+ \left\{ (\partial P_n^m(\cos\theta)/\partial\theta) \begin{matrix} \cos m\phi \\ \sin m\phi \end{matrix} \hat{\theta} \mp (m P_n^m(\cos\theta)/\sin\theta) \begin{matrix} \sin m\phi \\ \cos m\phi \end{matrix} \hat{\phi} \right\}$$
$$(\partial \hat{B}_n(kr)/\partial kr)/kr \tag{1.140}$$

where

$$\hat{B}_n(kr) = kr\, b_n(kr) \tag{1.141}$$

and the subscripts e and o denote even and odd wave functions with respect to field dependence on ϕ.

Orthogonality relationships between pairs of (m, n) of different degrees n and/or orders m stem from the general reciprocity relation between any two solutions of the source free Maxwell's equations; namely if (E, H) and (E′, H′) are two different solutions of source free Maxwell's equations, then

$$\nabla \cdot (\mathrm{E} \times \mathrm{H}' - \mathrm{E}' \times \mathrm{H}) = 0$$

as can be easily verified by expanding the LHS of the equation and applying Maxwell's equations. Integrating over a closed spherical volume and using the divergence theorem [see (A1.18)], we get:

$$\int_0^{2\pi} \int_0^{\pi} (\mathrm{E} \times \mathrm{H}' - \mathrm{E}' \times \mathrm{H}) \cdot \hat{\mathrm{r}} \sin\theta \, d\theta \, d\phi = 0 \tag{1.142}$$

which is the general reciprocity relationship. Taking (E, H) and (E′, H′)

according to the general form (1.136) of a *TE* wave (1.142) yields:

$$\int_0^{2\pi}\int_0^{\pi} (\mathrm{m}_{\substack{e\\o}mn} \times \mathrm{n}_{\substack{e\\o}m'n'} - \mathrm{m}_{\substack{e\\o}m'n'} \times \mathrm{n}_{\substack{e\\o}mn}) \cdot \hat{\mathrm{r}} \sin\theta \, \mathrm{d}\theta \, \mathrm{d}\phi = 0 \qquad (1.143)$$

It is useful to note that a new valid electromagnetic field is obtained from (m, $in(\varepsilon/\mu)^{1/2}$), by reversing the transverse components of n, while keeping m and the radial component of n. So, by using this new field in place of the unprimed field (E, H) in (1.143), this equation becomes

$$\int_0^{2\pi}\int_0^{\pi} (\mathrm{m}_{\substack{e\\o}mn} \times \mathrm{n}_{\substack{e\\o}m'n'} + \mathrm{m}_{\substack{e\\o}m'n'} \times \mathrm{n}_{\substack{e\\o}mn}) \cdot \hat{\mathrm{r}} \sin\theta \, \mathrm{d}\theta \, \mathrm{d}\phi = 0 \qquad (1.144)$$

which requires however that $m \neq m'$ and/or $n \neq n'$. By adding (1.143) and (1.144), we get

$$\int_0^{2\pi}\int_0^{\pi} (\mathrm{m}_{\substack{e\\o}mn} \times \mathrm{n}_{\substack{e\\o}m'n'}) \cdot \hat{\mathrm{r}} \sin\theta \, \mathrm{d}\theta \, \mathrm{d}\phi = 0, \; m \neq m', \; n \neq n' \; \ldots \qquad (1.145)$$

When $m = m'$ and $n = n'$, a direct substitution from (1.139)–(1.140) and the use of (1.130) yields

$$\int_0^{2\pi}\int_0^{\pi} (\mathrm{m}_{\substack{e\\o}mn} \times \mathrm{n}_{\substack{e\\o}mn}) \cdot \hat{\mathrm{r}} \sin\theta \, d\theta \, d\phi = \gamma_{mn} \hat{B}_n(kr) \hat{B}'_n(kr) / (kr)^{1/2} \qquad (1.146)$$

where

$$\gamma_{mn} = 2\pi\varepsilon_m \frac{n(n+1)}{2n+1} \frac{(n+m)!}{(n-m)!}, \quad \varepsilon_m = 2 \text{ for } m = 0, \; \varepsilon_m = 1 \text{ otherwise}$$

Other relations are obtained if (E, H) is taken according to (1.136) and (E′, H′) is taken according to (1.137):

$$\int_0^{2\pi}\int_0^{\pi} (\mathrm{m}_{\substack{e\\o}mn} \times \mathrm{m}_{\substack{e\\o}m'n'}) \cdot \hat{\mathrm{r}} \sin\theta \, \mathrm{d}\theta \, \mathrm{d}\phi = 0 \qquad (1.147)$$

A similar relation exists with m replaced by n.

Now, a specified field distribution over a spherical surface can be expanded in terms of the spherical wave functions m and n by using the above orthogonality relationships. To this end, let $\mathrm{E}_a(\theta, \phi)$ be an aperture electric field distribution over the spherical surface $r = r_1$. Owing to the completeness of the set of vector functions m and n over a spherical surface, it is permissible to write

$$\mathrm{E}_a(\theta, \phi) = \sum_{j,\,l} (a_{jl}\mathrm{m}_{ojl} + b_{jl}\mathrm{n}_{ejl}) \qquad (1.148)$$

where the θ component of E_a has been assumed, arbitrarily, to be an even function of ϕ (recall (1.139)–(1.140)). If the outside region $r \geqslant r_1$, is an open space, then the radial dependence of m and n must be chosen according to: $b_l(kr) = h_l^{(2)}(kr)$. To obtain the a and b coefficients in (1.148), cross multiply the two sides of the equation by $\mathrm{n}_{omn} \times$ and $\mathrm{m}_{emn} \times$ successively and integrate over the spherical surface $r = r_1$. Upon using the orthogonality relations

(1.145)—(1.147) one obtains a_{jl} and b_{jl}:

$$\begin{matrix} a_{mn} \\ b_{mn} \end{matrix} = (kr_1)^2 \int_0^{2\pi} \int_0^{\pi} \left\{ \begin{matrix} \mathbf{n}_{omn} \\ \mathbf{m}_{emn} \end{matrix} \times \mathbf{E}_a \right\} \cdot \hat{\mathbf{r}} \sin\theta \, d\theta \, d\phi / c_{mn} \tag{1.149}$$

where

$$c_{mn} = \frac{2\pi\varepsilon_m n(n+1)(n+m)!}{(2n+1)(n-m)!} H_n^{(2)}(kr_1) H_n^{\prime(2)}(kr_1) \tag{1.150}$$

The vector magnetic field associated with the electric field of (1.148) in the region $r \geqslant r_1$ is obtained simply from (1.136)–(1.137) as

$$\mathbf{H} = i(\varepsilon/\mu)^{1/2} \sum_{j,l} a_{jl}\mathbf{n}_{ojl} + b_{jl}\mathbf{m}_{ejl} \tag{1.151}$$

Applications of this field expansion appears in problems of radiation from conical horns as will be discussed in chapter 5.

1.7 Problems

1.1 Show that the continuity of the tangential electric field at an interface between two media implies the continuity of the normal magnetic induction (*Hint*: use the first of Maxwell's equations; (1.1)).

Similarly, show that the continuity of the tangential magnetic field implies that of the normal electric displacement vector.

1.2 Derive equations (1.24) and (1.25) for *TM* to z waves.

1.3 Deduce equation (1.33) which is the wave equation satisfied by E_z in inhomogeneous media in a general cylindrical coordinate system.

1.4 Consider a *TM* wave in a dielectric slab in which $\varepsilon = \varepsilon(x)$ is a function of x only. Take $H = H_y\hat{y}$, the fields are independent of y and are travelling waves along z, i.e. behaving as $\exp(-i\beta z)$. Show that H_y satisfies the wave equation:

$$\frac{\partial^2 H_y}{\partial x^2} - \frac{\partial H_y}{\partial x} \frac{\mathrm{d}\varepsilon(x)/\mathrm{d}x}{\varepsilon(x)} + (k^2(x) - \beta^2)H_y = 0$$

where $k^2(x) = \omega^2\mu_0\varepsilon(x)$.

Show also that the associated electric field component E_z satisfies equation (1.33).

1.5 Repeat problem (1.4) for a *TE* wave with $E = E_y\hat{y}$ and show that E_y satisfies the wave equation:

$$\frac{\partial^2 E_y}{\partial x^2} + (k^2(x) - \beta^2)E_y = 0$$

Show also that the associated H_z satisfies (1.32).

The reader is referred to Wait (1970) for the solution of the wave equations in problems (1.4) and (1.5) for several ε profiles.

1.6 Find the far field of a line source $I \exp(i\omega t)$ lying on the z axis. Show that $E_z/H_\phi = -\eta_0 \equiv -(\mu/\varepsilon)^{1/2}$.

1.7 Find the fields of a line source with a travelling wave current; $I \exp(i\omega t - i\beta z)\hat{z}$ lying on the z axis.

1.8 Derive the radiation pattern of the magnetic field component H_ϕ for the structure of Fig. 1.3*c* by using equation (1.63). Find the radiation conductance of the aperture $\rho=a$, $|z|\leqslant$d.

1.9 Consider a conducting cylinder of radius a and infinite length. The cylindrical surface has an aperture defined by the area:

$$|\phi|<\phi_a/2 \text{ and } |z|<d/2,$$

and is excited by an axially directed electric field given by:

$$E_z=E_0\cos(\pi\phi/\phi_a).$$

Find the fields in the open region $\rho>a$. Apply to the far zone and obtain the radiation pattern.

Hint: Appropriate expressions for the fields in the region $\rho>a$ are:

$$E_z=\sum_m e^{im\phi}\int_{-\infty}^{\infty} A_m(\beta)H_m^{(2)}(\lambda\rho)\mathrm{e}^{-i\beta z}\,d\beta$$

$$H_z=\sum_m e^{im\phi}\int_{-\infty}^{\infty} B_m(\beta)H_m^{(2)}(\lambda\rho)\mathrm{e}^{-i\beta z}\,d\beta$$

Use these to find E_z and E_ϕ on $\rho=a$, and equate by the known E_z and $E_\phi(=0)$ on the aperture to find the unknown coefficients $A_m(\beta)$ and $B_m(\beta)$.

1.10 Consider the situation in problem (1.9) again, but with the aperture being excited by a ϕ directed electric field such that:

$$E_\phi=E_0\cos(\pi z/d) \quad \text{and} \quad E_z=0$$

Find all fields in the open region $\rho>a$.
(Use the hint of problem 1.9, but notice that $E_z=0$ everywhere.)

1.11 Derive equations (1.76) and (1.77).

1.12 Derive equations (1.108) and (1.109).

1.13 Show that the spherical wave functions $\hat{B}_\nu(kr)$ derived in (1.141) satisfy the differential equation:

$$[d^2/dr^2+k^2-\nu(\nu+1)/r^2]\hat{B}_\nu=0$$

1.14 Show that the vector wave function m defined by (1.135) has the spherical components $m_r=0$, $m_\theta=(\partial f/\partial\phi)/\sin\theta$ and $m_\phi=-\partial f/\partial\theta$. Hence, show the m satisfies the vector wave equation $\nabla\times\nabla\times\mathrm{m}=k^2\mathrm{m}$ if f satisfies the scalar wave equation $\partial^2 f+k^2 f=0$. Next, show that n defined by (1.135) satisfies the vector wave equation.

1.15 A radial phase propagation parameter β may be defined for the vector wave functions m and n defined in (1.139)–(1.140) by:

$$\beta=ik\hat{H}_\nu'^{(2)}(kr)/\hat{H}_\nu^{(2)}(kr)$$

Show that when kr is sufficiently larger than ν^2, we may approximate β by:

$$\beta\sim k[1-\nu(\nu+1)/2(kr)^2]$$

1.16 The aperture field on a conical horn excited by a $TE_{m\nu}$ mode is given by:

$$E_\theta = m[P_\nu^m(\cos\theta)/\sin\theta]\cos m\phi$$

$$E_\phi = -[\partial P_\nu^m(\cos\theta)/\partial\theta]\sin m\phi$$

for $\theta \leqslant \theta_1$, and zero for $\theta > \theta_1$, where θ_1 is the semiflare angle of the conical horn. Expand this aperture field in spherical wave functions for the open region $r \geqslant r_1$; r_1 being the radius of the cone aperture; namely equate the aperture fields by the expansion:

$$\sum_{m,\,n} a_{mn}\mathrm{m}_{omn} + b_{mn}\mathrm{n}_{emn}$$

and show that

$$a_{mn} = [(kr_1)^2\pi/c_{mn}]\int_0^{\theta_1}\left\{\frac{(mP_\nu^m(\cos\theta))^2}{\sin^2\theta} + P_\nu'^m P_n'^m\right\}\sin\theta\, d\theta$$

$$b_{mn} = [(kr_1)^2\pi/c_{mn}]\,[mP_\nu^m(\cos\theta)P_n^m(\cos\theta)]_0^{\theta_1}$$

where c_{mn} is given by (1.150) and $P_\nu'^m \equiv \partial P_\nu^m(\cos\theta)/\partial\theta$.
Find the radiation pattern by using the asymptotic formula for $\hat{H}_\nu^{(2)}(kr)$:

$$\hat{H}_\nu^{(2)}(kr) \underset{2kr \gg \nu^2}{\rightarrow} \exp(-i(kr - \nu\pi/2 - \pi/2))\left[1 - i\frac{\nu(\nu+1)}{2kr}\cdots\right]$$

1.8 References

ABRAMOWITZ, M., and STEGUN, I.A. (1965): 'Handbook of mathematical functions' (Dover Publications, New York)

CARVER, K.R., and MINK, J.W. (1981): 'Microstrip antenna technology (invited)' *IEEE Trans.* **AP–29**, pp. 2–24

CHEW, W.C., and KONG, J.A. (1980): 'Effects of fringing fields on the capacitance of circular microstrip disk'. *IEEE Trans.* **MIT-28,** pp. 98–104

CLEMMOW, P.C. (1966): 'The plane wave spectrum representation of electromagnetic fields'. (Pergamon Press)

HARRINGTON, R.F. (1961): 'Time harmonic electromagnetic fields' (McGraw-Hill, New York, Toronto, London)

MAGNUS, W., and OBERHETTINGER, F. (1949): 'Formulas and theorems for the special functions of mathematical physics' (Chelsea Publishing Co., New York, NY) Chap. 4

McLACHLAN, N.W. (1955): 'Bessel functions for engineers' (Oxford University Press)

STRATTON, J.A. (1941): 'Electromagnetic theory' (McGraw-Hill)

WAIT, J.R. (1967): 'A fundamental difficulty in the analysis of cylindrical waveguides with impedance walls', *Electron. Lett.* **3,** pp. 87–88

WAIT, J.R. (1970): 'Electromagnetic waves in stratified media' (Pergamon Press)

WATSON, G.N. (1958): 'A treatise on the theory of Bessel functions, 2nd edition (Cambridge Univ. Press)

YANO, S., and ISHIMARU, A. (1981): 'A theoretical study of the input impedance of a circular microstrip disk antenna', *IEEE Trans.* **AP–29,** pp. 77–83

Additional references

COLLIN, R.E. (1960): 'Field theory of guided waves' (McGraw-Hill), New York)
COLLIN, R.E. (1966): 'Foundations for microwave engineering' (McGraw-Hill, New York)
JACKSON, J.D. (1962) 'Classical electrodynamics' (John Wiley, New York)
KONG, A. (1975): 'Theory of electromagnetic waves' (John Wiley, New York)
MARCUVITZ, N. (Ed.) (1965): 'Waveguide handbook' (Dover Publications Inc., New York)
MORSE, P.M., and FESHBACH, H. (1953): 'Methods of theoretical physics' (McGraw-Hill, New York)
WAIT, J.R. (1985): 'Electromagnetic wave theory' (Harper & Row Publishers, New York, London)
WAIT, J.R. (1986); 'Introduction to Antennas & propagation' (Peter Peregrinus Ltd., London)

Chapter 2

Closed waveguides

2.1 Introduction

In the previous chapter, we have found general field solutions to Maxwell's equations on cylindrical structures. Given a particular guiding structure, certain boundary conditions will be imposed on the solution. Such solutions which satisfy the source free Maxwell's equations and the boundary conditions constitute the modes of the waveguiding structure. Each waveguide mode is characterised by an axial propagation constant which is a characteristic eigenvalue of the structure. The corresponding modal field distribution is an eigenvector function.

In this chapter we consider a few examples of waveguiding structures which are characterised by perfectly reflecting walls or boundaries. Such waveguides form a class of closed waveguides since all fields are confined inside the boundaries which are taken as either perfectly electric or perfectly magnetic walls. Waveguides characterised by specified wall impedances form another class of closed waveguides. These will be the subject of chapter 3.

We start by finding the modes in cylindrical waveguides with rectangular, circular and elliptical cross sections. This is followed by a study of propagation in radial waveguides and in curved waveguides. All thse types of waveguides find extensive use in microwave telecommunication systems. Next, the general properties of modes in closed waveguides is considered. Thus, the important property of orthogonality among the fields of different modes is proved. This is followed by a treatment of the problem of modal excitation by impressed current source inside the waveguide or by a prescribed tangential field distribution over a cross section. We conclude the chapter by solving the general problem of a sharp longitudinal discontinuity along the guide.

2.2 Rectangular waveguide

A waveguide with a rectangular cross section of dimensions a and b is shown in Fig. 2.1*a*. The waveguide is assumed to be filled homogeneously with an isotropic medium. The waveguide walls are perfect electric conductors. Thus the boundary conditions are simply specified by: $E_x=0$ at $y=0$ and $y=b$ and $E_y=0$ at $x=0$ and $x=a$. Now the problem at hand is to choose special solutions out of the general solution obtained in section 1.5.1 which satisfy these boundary conditions. Trying first TE to z modes and in view of (1.23),

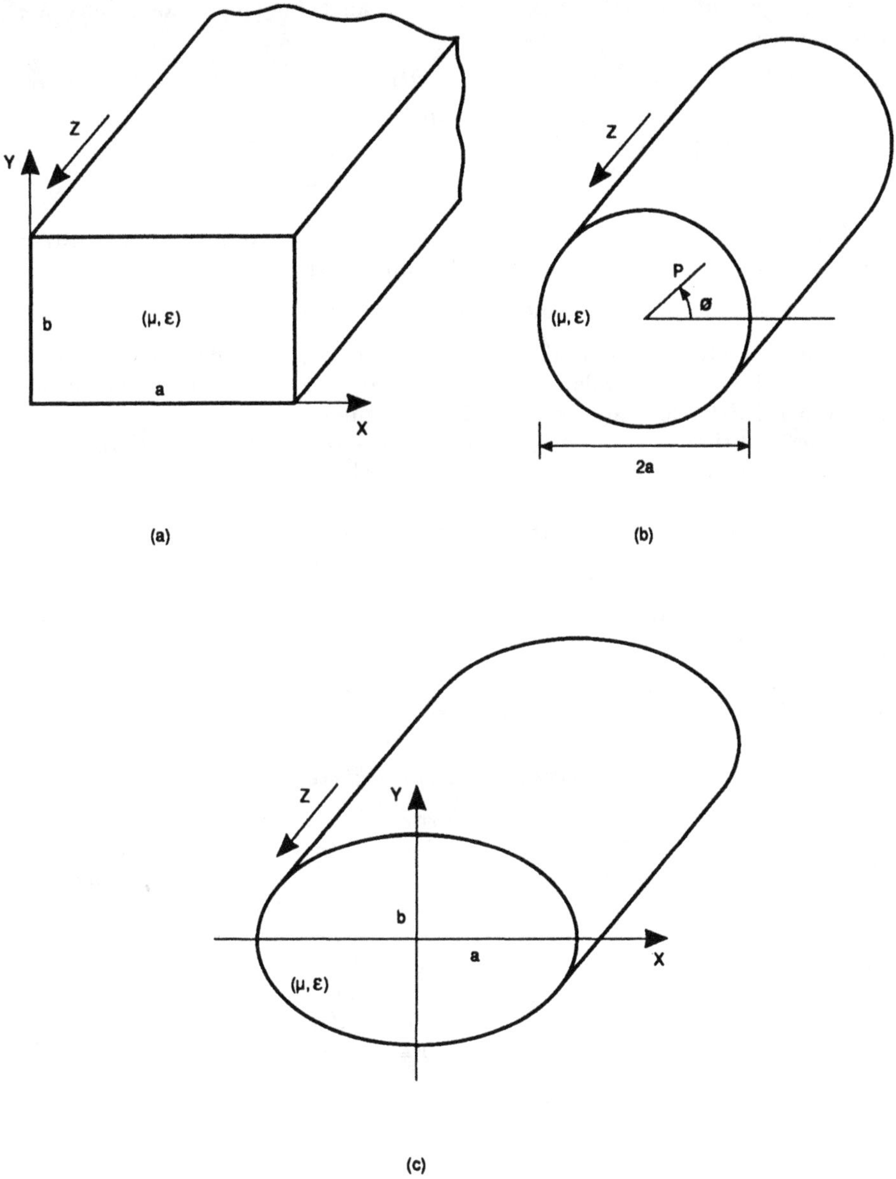

Fig. 2.1 Cyclindrical waveguides with

a Rectangular cross section
b Circular cross section
c Elliptical cross section

$E_x \propto \partial H_z/\partial y$ and $E_y \propto \partial H_z/\partial x$; hence the stated boundary conditions are equivalent to:

$$\partial H_z/\partial y = 0 \text{ at } y=0 \text{ and } y=b \text{ and } \partial H_z/\partial x = 0 \text{ at } x=0 \text{ and } x=a$$

Solutions satisfying the conditions at $x=0$ and $y=0$ walls and conforming with the general solution (1.37) are:

$$H_z = H_0 \cos k_x x \cos k_y y \, e^{-i\beta z} \tag{2.1}$$

The other two boundary conditions at the walls $x=a$ and $y=b$ require that $\sin k_y b = 0$ and $\sin k_x x = 0$; hence k_x and k_y must take the discrete values:

$$k_x = m\pi/a \quad \text{and} \quad k_y = n\pi/b \tag{2.2}$$

where m and n are integers taking the values 0, 1, 2, . . . , excluding the case $m=n=0$ which corresponds to identically zero fields everywhere. This means that the TEM mode cannot exist in the waveguide.

Modes are thus labelled by pairs of integers (m, n) and are designated by TE_{mn} or H_{mn} modes. The transverse fields are obtained from H_z by (1.20) and (1.23).

For TM modes E_z must vanish on the four walls; hence it is given by

$$E_z = E_0 \sin k_x x \sin k_y y \, e^{-i\beta z} \tag{2.3}$$

where k_x and k_y are given by (2.2) such that both m and n are greater than zero (otherwise all fields of the mode become identically zero). TM modes are thus labelled by pairs of integers (m, n) and are designated by TM_{mn}, or E_{mn} modes starting from the TM_{11} mode.

For both TE_{mn} and TM_{mn} modes (1.21), (1.36) along with (2.2) yield

$$\beta^2 + (m\pi/a)^2 + (n\pi/b)^2 = k^2 \equiv \omega^2\mu\varepsilon \tag{2.4}$$

Equation (2.4) implies two basic properties of modes in hollow waveguides. Firstly, the longitudinal phase constant β is a nonlinear function of frequency. This means that multifrequency signals will suffer a delay distortion as they propagate down the guide. This point will be elaborated on later in this chapter. Secondly, each mode has a cutoff frequency below which the mode ceases to propagate. The cutoff frequency is defined as that at which $\beta = 0$. Below that frequency (2.4) tells us that β becomes purely imaginary (for a lossless medium); hence the mode attenuates rather than propagates. The cutoff wavenumber k_c and the cutoff frequency $f_c \equiv \omega_c/2\pi$ are then given by:

$$k_c \equiv \omega_c/v = [(m\pi/a)^2 + (n\pi/b)^2]^{1/2}$$

$$f_c = v((m/2a)^2 + (n/2b)^2)^{1/2}$$

where $v = (\mu\varepsilon)^{-1/2}$ is the phase velocity of a plane wave in the unbounded medium. Equation (2.4) now takes the more general form:

$$\beta^2 = \omega^2\mu\varepsilon - k_c^2 \tag{2.5}$$

which is valid in all waveguides with perfectly conducting walls and isotropic filling, irrespective of the cross sectional shape.

The modes are conventionally ordered according to the values of their cutoff frequencies. The lowest order mode is the one having the least f_c and is obviously the dominant mode since it can propagate, in a certain range of frequencies, when all other modes are cutoff. The dominant mode in a rectangular waveguide having $a>b$ is the TE_{10} mode whose $f_c=v/2a$. The next higher order mode is either the TE_{01} or TE_{20} depending on whether $2b>a$ or $2b<a$ respectively. For $m>0$ and $n>0$, the TE_{mn} and TM_{mn} modes have the same cutoff frequency and therefore the same β at a given frequency. Such modes are then called degenerate modes. Before closing this section, it is worth mentioning that modes whose fields are transverse to either x or y can be equally well defined in a rectangular waveguide. Thus, for example, we can define *TE* to x modes and *TM* to x modes and it can be easily shown that any of these modes is merely a linear superposition of *TE* to z and *TM* to z modes. The cutoff frequency of a *TE* to x mode is the same as its constituent modes. However, the mode impedance for a *TE* to x mode must be obviously different from that of the corresponding *TE* and *TM* to z modes. Namely the mode impedance for a *TE* to x mode is equal to (see prob. 2.5):

$$Z_{TEx}=-E_y/H_x=\omega\mu\beta/(k^2-k_x^2)$$

and the mode admittance for *TM* to x modes is given by

$$Y_{TMx}=H_y/E_x=\omega\varepsilon\beta/(k^2-k_s^2)$$

2.3 Circular cylindrical waveguide

Now consider a cylindrical waveguide with a circular cross section as depicted in Fig. 2.1*b*. The wall at $\rho=a$ is a perfect electric conductor. Hence the boundary conditions are simply $E_z=0$ and $E_\phi=0$ at $\rho=a$. For *TE*(*TM*) modes, the $H_z(E_z)$ is a special case of (1.40), and since the fields must be finite at the origin $\rho=0$, $H_z(E_z)$ must have the form:

$$AJ_m(\lambda\rho)\cos m\phi\, e^{-i\beta z} \tag{2.6}$$

From (1.23), E_ϕ is proportional to $\partial H_z/\partial\rho$ for *TE* modes, hence the boundary condition $E_\phi=0$ at $\rho=a$ for these modes implies that

$$J'_m(\lambda a)=0 \tag{2.7}$$

where the prime stands for differentiation with respect to the argument of the Bessel function. For the *TM* modes the boundary condition $E_z=0$ implies that:

$$J_m(\lambda a)=0 \tag{2.8}$$

Denote the roots of $J'_m(\chi)$ by χ'_{mn} and the roots of $J_m(\chi)$ by χ_{mn} and use (1.21) to get:

$$\beta=(\omega^2\mu\varepsilon-s^2/a^2)^{1/2} \tag{2.9}$$

where s stands for χ'_{mn} or χ_{mn} for *TE* and *TM* modes respectively. The cutoff frequencies f_c at which $\beta=0$ are

$$f_{c,\,mn}\equiv\omega_c/2\pi=vs/2\pi a \tag{2.10}$$

Numerical values of χ'_{mn} and χ_{mn} for the lowest order *TE* and *TM* modes are given in Table 2.1. It is clear that the dominant mode, having the lowest cutoff frequency, is the TE_{11} mode. This is followed by the TM_{01}, then the TM_{11} and the TE_{01} which are degenerate modes.

Table 2.1 Zeros of $J_m(\chi)$ and $J'_m(\chi')$

(m, n)	(0, 1)	(0, 2)	(0, 3)	(1, 1)	(1, 2)	(2, 1)	(2, 2)
χ	2.4048	5.5201	8.6537	3.8317	7.0156	5.1356	8.4172
χ'	3.8317	7.0156	10.174	1.8412	5.3314	3.0542	6.7061

2.4 Elliptical cylindrical waveguides

Cylindrical waveguides with elliptical cross sections are often used in microwave systems as flexible feeders of microwave antennas. They have the advantage of polarisation maintaining capability and ease of bending.

We consider an elliptical cylindrical waveguide with perfect electric conducting walls as depicted in Fig. 2.1*c*. The waveguide boundary coincides with the elliptical surface $u = u_0$ in the elliptical frame of coordinates (u, v, z). Trying *TE* to z modes, appropriate solutions for the longitudinal field H_z for a travelling wave along z are (see (1.100–1.101)):

$$H_z = C_m(q, v) R^{(1)}_{e,m}(q, u) e^{-i\beta z} \tag{2.11}$$

which is an even function of the angular coordinate v. Odd solutions are also valid and give a distinct set of modes. Explicitly, odd solutions are given by:

$$H_z = S_m(q, v) R^{(1)}_{o,m}(q, u)\ e^{-i\beta z} \tag{2.12}$$

The transverse field components are obtainable from (1.20) and (1.23); hence for even modes:

$$H_u = (-i\beta/\lambda^2) C_m(q, v) R'^{(1)}_{e,m}(q, u) \tag{2.13}$$

$$H_v = (-i\beta/\lambda^2) C'_m(q, v) R^{(1)}_{e,m}(q, u) \tag{2.14}$$

$$E_u = (i\omega\mu/\beta) H_v \tag{2.15}$$

$$E_v = (-i\omega\mu/\beta) H_u \tag{2.16}$$

Similar expressions can be written for the odd modes. The boundary condition at $u = u_0$ is that $E_v(u_0) = 0$ *for all values of* v. Hence:

$$R'^{(1)}_{e,m}(q, u_0) = 0 \quad \text{for even modes} \tag{2.17}$$

$$R'^{(1)}_{o,m}(q, u_0) = 0 \quad \text{for odd modes} \tag{2.18}$$

which are the characteristic modal equations for even and odd *TE* modes respectively. Solutions of either equation determines the set of discrete values of $q = \lambda^2 c^2/4$ characterising the respective modes. The cutoff frequency of a mode is given, in terms of its q, by:

$$k_c \equiv \omega_c(\mu\varepsilon)^{1/2} = \lambda = 2q^{1/2}/c$$

It is worth noting here that, unlike circular waveguides, the azimuthally even and odd modes in an elliptical waveguide have different radial functions, cutoff frequencies and phase constants as is clear from the above equations.

TM_z modes are treated in a similar manner with the following results. The characteristic values of even TM_z modes are the roots of:

$$R^{(1)}_{e,m}(q, u_0) = 0 \tag{2.19}$$

and those of odd TM_z modes are the roots of:

$$R^{(1)}_{o,m}(q, u_0) = 0 \tag{2.20}$$

Following the notations of Kretzschmar (1970), even and odd TE modes are labelled respectively as $TE_{c,mn}$ and $TE_{s,mn}$. Here m is the integer giving the order of the angular modal function and n is the order of the root of equation (2.17) or (2.18). The corresponding eigenvalues of q are denoted respectively by $q_{c,mn}$ and $q_{s,mn}$. Similar mode labelling is adopted for TM modes.

Solution of equations (2.17) to (2.20) for the eigenvalues q is rather involved. However, approximate formulae have been given by Kinzer and Wilson (1947) and Kretzschamar (1970). For example the $q_{c,11}$ for the dominant mode $TE_{c,11}$ is given by Kretzschamar as:

$$\begin{aligned} q_{c,11} &= 0{\cdot}847\, e^2 - 0{\cdot}0013\, e^3 + 0{\cdot}0379\, e^4, && 0 \leqslant e \leqslant 0{\cdot}4 \\ &= -0{\cdot}00064\, e + 0{\cdot}8838\, e^2 - 0{\cdot}696\, e^3 + 0{\cdot}820\, e^4, && 0{\cdot}4 \leqslant e \leqslant 1{\cdot}0 \end{aligned} \tag{2.21a}$$

and for the odd mode $TE_{s,11}$:

$$\begin{aligned} q_{s,11} &= +0{\cdot}0018e + 0{\cdot}897e^2 - 0{\cdot}3679e^3 + 1{\cdot}612e^4, && 0{\cdot}05 \leqslant e \leqslant 0{\cdot}5 \\ &= -0{\cdot}1483 - 1{\cdot}0821e + 1{\cdot}0829e^2 + 0{\cdot}3493/(1-e), && 0{\cdot}5 \leqslant e \leqslant 0{\cdot}95 \end{aligned} \tag{2.21b}$$

where e is the eccentricity of the ellipse $u = u_0$, namely, $e = (\cosh u_0)^{-1}$.

The cutoff wavelengths λ_c are plotted, after Kretzschmar [1970], as λ_c/P versus e for the eight lowest order modes in Fig. 2.2. Here P is the perimeter of the ellipse $u = u_0$ .The eccentricity e varies between 0, corresponding to a circular cross section, to unity, corresponding to a zero area cross section. It is seen that the $TE_{c,11}$ is the lowest order mode. The next higher order mode depends on the value of eccentricity e, being the $TE_{s,11}$ for the lower values of e and changing to $TM_{c,01}$ and then to $TE_{c,21}$ as e increases.

The bandwidth of single mode operation is defined as the range of frequencies or wavelengths in which all modes are cutoff except the dominant mode. This is given by the difference between the cutoff frequencies or wavelengths of the first two lowest order modes. The maximum bandwidth of single mode operation in an elliptical waveguide may be deduced from Fig. 2.2 and it occurs around $e = 0{\cdot}866$ which corresponds to major and minor semiaxes related by $a = 2b$. Incidentally, this is the same condition for maximum bandwidth in a rectangular waveguide of dimensions a and b.

An interesting comparison of single mode operation in rectangular, circular and elliptical waveguides is given, after Kretzschamar (1970), in Table 2.2. The axes of the elliptical waveguide $2a$ and a are taken equal to the width and height of the rectangular waveguide. The diameter of the circular waveguide is equal to $2a$. It is clear from the table that the bandwidth of the rectangular waveguide is the highest and is followed by that of the elliptical waveguide. The circular waveguide has nearly half the bandwidth of the elliptical guide.

2.5 Attenuation in closed waveguides

Attenuation in closed waveguides arises primarily from power loss in the walls. While the guide walls so far have been considered perfect conductors, they necessarily have a finite, rather than an infinite, conductivity. Currents flowing

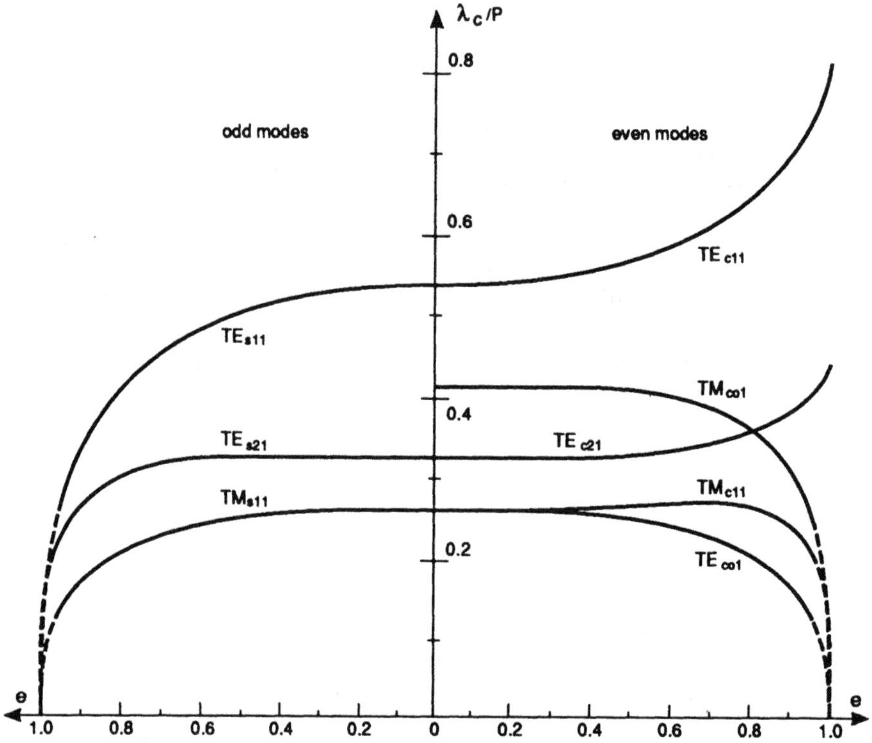

Fig. 2.2 Normalised cutoff wavelengths of modes in an elliptical waveguide versus eccentricity e (after Kretzschmar, 1970)

in the walls will therefore be associated with power loss due to the finite resistance of the walls. We shall calculate the power loss by using the perturbation technique in this section. This technique is based on the assumption that the electric currents in the walls are the same as those occurring in the presence of perfectly conducting walls. In the next chapter we shall present a more rigorous approach for deriving the propagation constant of guides with finite wall impedance.

Table 2.2 Comparison of bandwidth of cylindrical guides (after Kretzschmar, 1970)

Type	Dominant mode	λ_c	1st higher mode	λ_c	Bandwidth
Rectangular	TE_{10}	4·00a	TE_{01} TE_{20}	2·00a	2·00a
Circular	TE_{11}	3·41a	TM_{01}	2·61a	0·80a
Elliptical	TE_{c11}	3·35a	TE_{c21}	1·84a	1·51a

Let P_l watt/m be the power loss per unit length along the axial direction z. Let P_f watt/m^2 be the power flow at a given cross section. The attenuation factor α neper/m is given by:

$$\alpha = (1/2)P_l/P_f \tag{2.22}$$

P_f is simply given by the integration of the axial Poynting vector $\mathbf{E}_t \times \mathbf{H}_t^*$ over the cross section S, namely:

$$P_f = \text{Real} \int_S (\mathbf{E}_t \times \mathbf{H}_t^*) \,.\, \hat{\mathbf{z}}\, dS \tag{2.23}$$

The wall currents per meter are numerically equal, but orthogonal to the magnetic field tangential to the walls. The latter is composed of the longitudinal component H_z plus a transverse component which is denoted here by H_τ where τ is a unit vector tangential to the walls and transverse to z; e.g. $\tau = \phi$ in circular cylindrical coordinates and $\tau = v$ in elliptical coordinates. Hence P_l takes the form:

$$P_l = (1/\sigma\delta) \int_C (|H_z|^2 + |H_\tau|^2) h_\tau \, d\tau \tag{2.24}$$

where C is the contour enclosing the cross section, h_τ is the metric coefficient in the τ direction and δ is the wave penetration depth in the wall. The term $(1/\sigma\delta)$ is simply the resistance in ohm per unit width, normal to the current, and unit length along the current flow. It is explicitly given by $(\omega\mu/2\sigma)^{1/2}$, with σ the conductivity of the wall material.

Now, let us apply (2.23) and (2.24) to calculate the attenuation factor of TE to z modes. In view of (1.20) and (1.23), the power flow is given, in terms of H_z by:

$$P_f = (\omega\mu/\beta) \int_S |H_t^2| \, dS = (\omega\mu\beta/k_c^4) \int_S |\nabla_t H_z|^2 \, dS \tag{2.25}$$

where k_c is the cutoff wavenumber. The power loss may be decomposed into two components, one associated with H_z, or the circumferential current, and the other component with H_r, or the longitudinal current, i.e.:

$$P_l = P_{l_1} + P_{l_2}$$

$$P_{l_1} = (\omega\mu/2\sigma)^{1/2} \int_C |H_z|^2 h_\tau \, d\tau \tag{2.26}$$

$$P_{l_2} = (\omega\mu/2\sigma)^{1/2} (\beta^2/k_c^4) \int_C |\partial H_z/\partial\tau|^2 h_\tau^{-1} \, d\tau \tag{2.27}$$

where (1.20) has been used to obtain the last equation.

Combining (2.25)–(2.27) and denoting the integral terms in these three equations by I_1, I_2, and I_3 respectively, the attenuation factor α takes the form:

$$\alpha = (k_c^4/2\beta) \,.\, (2\omega\mu\sigma)^{-1/2} [I_2 + (\beta^2/k_c^4) I_3]/I_1 \tag{2.28}$$

A close look at (2.28) reveals that the attenuation associated with H_z, or circumferential current, is proportional to $\omega^{-1/2}/\beta$ which in turn is proportional to $\omega^{-3/2}$ when the operating frequency is sufficiently higher than the cutoff frequency. On the other hand, the attenuation associated with H_τ, or the longitudinal current, is proportional to $\omega^{-1/2}\beta$, or $\omega^{+1/2}$. This means that modes having no transverse H will have an attenuation factor which is a decreasing function of frequency when $\omega \gg \omega_c$. Such property is extremely important in telecommunication.

It is easy to check that TE_{0n} modes in circular waveguides do readily have $H_\phi = 0$; hence they enjoy the property of decreasing α with frequency. Conversely, it is noted that modes with zero transverse H do not exist in cylindrical waveguides with any other cross section (noncircular). This gives a definite advantage to the circular cross section over other shapes. However the lowest order TE_{0n} mode, which is the TE_{01} mode is not the dominant mode in a circular waveguide as is clear from section 2.3; namely, the TE_{11}, TM_{01} and TE_{21} modes have lower cutoff frequencies while the TM_{11} mode has the same cutoff frequency as the TE_{01} mode. Furthermore, in order to make use of the reduced attenuation of the TE_{01} mode, the operating frequency should be, perhaps, several times the cutoff frequency of this mode. This assures low TE_{01} attenuation, but certainly allows many other modes to propagate. The waveguide is then an overmoded, or an oversized guide as is usually the case with millimetric waveguides operating at frequencies above 40 GHz. Several problems will then occur. Firstly, the wave launcher will inevitably couple power to modes other than the desired TE_{01} mode. Secondly, any discontinuity in the waveguide will cause mode conversion from the TE_{01} to other modes which amounts to increased attenuation. In particular, waveguide bends cause appreciable conversion to the TM_{11} mode which is degenerate with the TE_{01} mode. Mode reconversion back to the TE_{01} mode will also occur and contribute to signal delay distortion. Furthemore, any distortion in the circular shape of the cross section will increase the attenuation of the TE_{01} mode itself. Analysis of mode conversion in oversized waveguides has been treated by several authors including Nagelberg *et al.* (1965), Falciasecca *et al.* (1977), Boyde *et al.* (1977), Abele *et al.* (1975) and Carlin *et al.* (1977). A good coverage of this work is given by Bhartia and Bahl (1984).

To overcome many of the problems associated with millimetric wave transmission in oversized circular waveguides, a few modified versions of circular guides have been suggested. An example is the helical waveguide in which the guide wall is composed of a helical layer of insulated copper wire backed by a jacket of highly lossy material. The result is an anisotropic wall offering a very low impedance to electrical currents flowing along the helix (almost circumferentially) and a very high impedance to longitudinal currents. Thus all modes other than TE_{0n} modes will suffer very high attenuation while the TE_{0n} modes will have only slightly increased attenuation over that in the regular guide. Analysis of helical waveguides is given by Morgan *et al.* (1956), Unger (1961) and Marcuse (1958) among others. Another example of modified circular waveguide is the dielectric lined waveguide in which the metallic wall of the guide is lined with a thin layer of dielectric material (Carlin *et al.*, 1973). This results in an anisotropic wall impedance as seen at the air-dielectric interface (see chapter 3), hence breaking the degeneracy between the TE_{01} and TM_{11} modes. This reduces coupling between these two modes at waveguide bends, hence reducing bending losses.

2.6 Phase, group and signal velocities

In dealing with a single frequency propagation in a waveguide a quantity of interest is the phase velocity. This is the velocity at which a point of constant phase is moving axially down the guide. From the previous sections the phase constant β of any mode is given in the form

$$\beta = (\omega^2 - \omega_c^2)^{1/2}/c$$

where ω_c is the cutoff frequency of the mode and c is the free space wave velocity, assuming the guide is empty. The wavelength along the guide $\lambda_g = 2\pi/\beta$ and the phase velocity is therefore equal to

$$v_p = \omega\lambda_g//2\pi = \omega/\beta = c/(1 - \omega_c^2/\omega^2)^{1/2} \tag{2.29}$$

The phase velocity for a propagating mode is therefore greater than c; a result which seems to violate the theory of relativity which states that no velocity can be greater than the velocity of light. However, we should remember that this theory applies to physical velocities which, in our context, are the velocities of information and energy flow. On the other hand, the phase velocity is associated with a monochromatic, or a single frequency, wave which carries no information and it is also different from the velocity of energy flow. In addition, the assumption of a monochromatic wave requires that the source be switched on at $t = -\infty$ and that it will remain so for all future times. This is obviously a physically nonrealisable situation; hence it is not surprising that it will lead to nonphysical quantities such as the phase velocity. Let us now consider a more physical situation. An amplitude modulated carrier frequency is applied to the input of the waveguide at $z = 0$. The input waveform is given by the real part of $f_i(t) = s(t)\exp(i\omega_c t)$ where $s(t)$ is the modulating signal and ω_c is the angular carrier frequency. Let $S(\omega)$ be the frequency spectrum of $s(t)$; hence

$$s(t) = \int_{-\infty}^{\infty} S(\omega)\exp(i\omega t)\,d\omega/2\pi \tag{2.30}$$

The frequency spectrum $F_i(\omega)$ of $f_i(t)$ is the same as $S(\omega)$, but centered about ω_c, that is, $F_i(\omega) = S(\omega - \omega_c)$. Assuming a monomode guide, then after travelling a distance z down the guide the frequency spectrum is multiplied by the propagation factor $\exp(-i\beta(\omega)z)$; hence the output waveform is the real part of

$$f_0(t) = \int_{-\infty}^{\infty} S(\omega - \omega_c)\exp(-i\beta(\omega)z)\exp(i\omega t)\,d\omega/2\pi \tag{2.31}$$

Let $\beta(\omega)$ be expanded in a power series about ω_c:

$$\beta(\omega) = \beta(\omega_c) + (\omega - \omega_c)\beta'(\omega_c) + (\omega - \omega_c)^2\beta''(\omega_c)/2 + \cdots \tag{2.32}$$

where the prime denotes differentiation with respect to ω. Let us first assume that the significant spectrum of $F_i(\omega)$ is narrow around ω_c so that we can retain only the first two terms in (2.32); hence (2.31) becomes

$$f_0(t) = \int_{-\infty}^{\infty} S(\omega-\omega_c)\ \exp(-i\beta(\omega_c)z + i\omega_c\beta' z)\ \exp[i\omega(t-\beta' z)]\ d\omega/2\pi$$

$$= \exp[i\omega_c t - i\beta(\omega_c)z]s(t-\beta' z)$$

where $\beta' \equiv \beta'(\omega_c)$. The above result shows that the carrier is phase delayed at a velocity $\omega_c/\beta(\omega_c)$ which is the phase velocity v_p. On the other hand, the modulating signal or the carrier envelope which contains the information is delayed at a velocity equal to $1/\beta'(\omega_c)$. The latter is termed by the group velocity v_g which refers to the group of waves occupying a narrow band around the carrier frequency:

$$v_g = 1/\beta'(\omega_c) = (d\omega/d\beta)_{\omega_c} \tag{2.34}$$

Using (2.29), it follows that

$$v_g = c(1-\omega_c^2/\omega^2)^{1/2}$$

which is readily less than c.

So far as the above approximation is valid, the signal is merely delayed, without undergoing any kind of distortion. This is a consequence of neglect of the β'' and higher order terms in (2.32). So let us return again to (2.31) and use the first three terms in (2.32) for $\beta(\omega)$ to get

$$f_0(t) = \exp(-i\phi_c)\int_{-\infty}^{\infty} S(\omega)\ \exp(-i\beta'' z\omega^2)\ \exp(i\omega t_d)\ d\omega/2\pi \tag{2.35}$$

where $\phi_c = \omega_c t - \beta(\omega_c)z$, $t_d = t - \beta'(\omega_c)z$ is the delayed time and $\beta'' \equiv \beta''(\omega_c)$.

To get a feeling for the effect of the term β'' on the output signal waveform, let us assume that $\beta'' z\omega^2$ is sufficiently less than unity in the significant frequency range of $S(\omega)$ so that $\exp(-i\beta'' z\omega^2) \simeq 1 - i\beta'' z\omega^2$. Using this approximation in (2.35), we get

$$f_0(T) \simeq \exp(-i\phi_c)\{s(t_d) + i\beta'' z[d^2 s(t_d)/dt_d^2]\} \tag{2.36}$$

To interpret this result, remember that the output waveform is actually the real part of (2.36), which is

$$s(t_d)\ \cos(\phi_c) - \{\beta'' z d^2 s(t_d)/dt_d^2\}\ \sin(\phi_c) \equiv s_0(t_d)\ \cos(\phi_c + \theta(t_d)) \tag{2.37}$$

where

$$s_0(t_d) = \{s^2(t_d) + (\beta'' z d^2 s(t_d)/dt_d^2)^2\}^{1/2} \tag{2.38}$$

$$\theta(t_d) = \tan^{-1}\{\beta'' z[d^2 s(t_d)/dt_d^2]/s(t_d)\} \tag{2.39}$$

Thus, besides the delay, the output envelope $s_0\ (t_d)$ has been distorted by the presence of the second derivative of s in (2.38). One aspect of this distortion is the broadening of the output envelope (see prob. 2.14) which limits the rate of transmission in telecommunication systems. In addition to this amplitude distortion, there is an introduced phase modulation which has not been present at the input; this is expressed by the θ term in (2.39). The generation of phase, or frequency, modulation is usually referred to as AM to PM or FM conversion.

At this point, a word of caution about the group velocity is due. If the signal transmitted is sufficiently broadband, it can be badly distorted and dispersed over time to the extent that a signal velocity may be hard to define. In such cases the group velocity will lose its meaning. While the group velocity, as defined in (2.34), is less than the velocity of light c in the case of a hollow closed waveguide, it can exceed c in other cases. Actually, in a more general context, one can verify that v_g is less than c if the phase velocity is a decreasing function of frequency, and vice versa (see prob. 2.12). Thus the group velocity can exceed c in certain cases, such as propagation in a conducting medium; whence it cannot be associated with the signal velocity. To clarify this issue, Stratton (1941, sec. 5.18) presents an illuminating analysis which dates back to Sommerfeld and Brillouin in 1914. These authors treated the propagation of a step function in a conducting medium by the application of the Laplace transform. They conclude that the leading edge of the propagating disturbance does not appear at a location of distance z from the source before the elapse of time equal to z/c, which means that the velocity of the leading edge, or the wavefront, can never exceed the velocity of light irrespective of the values of the phase or group velocities. The arrival of the first wavefront is associated with a zero amplitude which is the result of a cancellation between free and forced oscillations in the medium. At a later time of arrival of the wavefront, the forced oscillation starts to dominate over the free oscillation which damps out with time. Brillouin associates the signal velocity with the time at which the forced oscillation, or the steady state response, starts to take over. At this time, a noticeable increase of the signal occurs. Because this time comes later than z/c, the signal velocity is always less than c. This picture is also clear in the more recent work of Wait (e.g. Collin and Zucker, 1969b, chapter 24 and Wait, 1982, chapter 7).

2.7 Radial waveguides

So far, we have considered modes which travel along the axial direction z of the cylindrical structure. It follows that the surfaces of equal phase are planar surfaces orthogonal to z. Another class of modes is that in which the waves travel radially and therefore the equiphase surfaces are cylindrical in shape. As an example, consider the sectoral horn having perfect electric walls and shown in Fig. 2.3. The modes in this horn travel along the ρ direction and are standing waves in the z and ϕ directions.

We can still classify modes into *TE to z* and *TM to z* modes. Considering *TE to z* modes, an appropriate expression for H_z is

$$H_z = \sin(\beta z)\,\cos(\nu\phi) H_\nu^{(2)}(\lambda\rho) \tag{2.40}$$

The transverse fields (to z) can still be given by (1.20) and (1.23) but the factor $-i\beta$ should now be interpreted as the operator $\partial/\partial z$. In order to satisfy the boundary conditions of vanishing E_ρ and E_ϕ at the top and bottom walls $z=0$ and a, the wavenumber β must be limited to any of the discrete values:

$$\beta_m = m\pi/a, \quad m = 1, 2, \ldots \tag{2.41}$$

The other boundary conditions are $E_\rho = 0$ at the side walls $\phi = 0$ and $\phi = \phi_0$. Since E_ρ is proportional to $\partial H_z/\partial\phi$, it follows that:

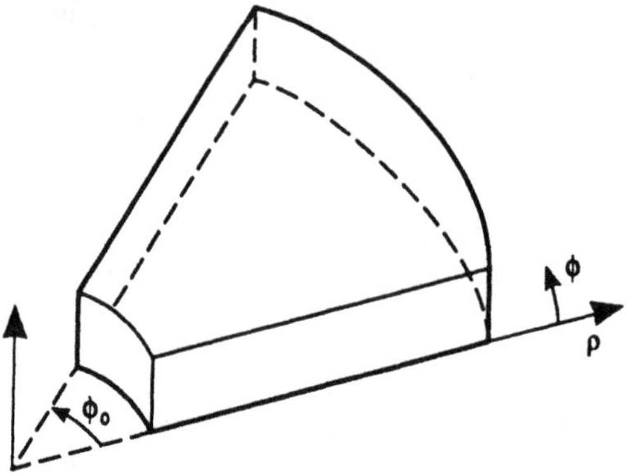

Fig. 2.3 Sectoral horn with perfect electric walls

$$\nu_0 = n\pi, \quad \text{or,} \quad \nu_n = n\pi/\phi_0, \; n = 0, 1, 2 \ldots \tag{2.42}$$

Modes are thus labelled by pairs of integers (m, n), with the following relation between the wavenumbers:

$$(m\pi/a)^2 + \lambda^2 = k^2 = \omega^2\mu\varepsilon \tag{2.43}$$

and H_z is given by:

$$H_{z,\,mn} = A \sin(\beta_m z) \cos(\nu_n \phi) H_{\nu_n}^{(2)}(\lambda_m \rho) \tag{2.44}$$

Similar treatment for *TM to z* modes gives E_z as:

$$E_{z,\,mn} = B \cos(\beta_m z) \sin(\nu_n \phi) H_{\nu_n}^{(2)}(\lambda_m \rho) \tag{2.45}$$

where $m = 0, 1, 2 \ldots$ and $n = 1, 2 \ldots$.

Obviously, a mode is cutoff if λ_m is imaginary, i.e. if $k < m\pi/a$. It is thus clear that the TM_{01} mode is the dominant mode of the structure. However, even when λ_m is real, the mode does not behave as a propagating wave at all values of the radial distance ρ. This is so because of the special character of the radial wave function $H_\nu^{(2)}(\lambda\rho)$ as discussed below. The rate of phase change of radial waves is not constant, but is a function of the radial distance ρ, and changes gradually into attenuation at sufficiently small values of ρ. This behaviour is, obviously, in contrast with plane wave functions which have constant rates of phase change. To provide a quantitative discussion of this point, let us define the phase constant $\beta_\nu(\rho)$ of the radial wave function. Taking a clue from the planar wave function $e^{-i\beta z}$, we define $\beta_\nu(\rho)$ as:

$$\begin{aligned} \beta_\nu(\rho) &= i[\partial H_\nu^{(2)}(\lambda\rho)/\partial\rho]/H_\nu^{(2)}(\lambda\rho) \\ &= i\lambda H_\nu^{(2)\prime}(\lambda\rho)/H_\nu^{(2)}(\lambda\rho) \end{aligned} \tag{2.46}$$

which is generally complex. The real part is the rate of phase change with ρ and the imaginary part is the rate of attenuation. Rewriting $\beta_\nu(\rho)$ as $k_\nu(\rho) - i\alpha_\nu(\rho)$, we deduce the explicit expressions:

$$k_\nu(\rho) = (2\lambda/\pi x)/[J_\nu^2(x) + N_\nu^2(x)] \tag{2.47}$$

$$\alpha_\nu(\rho) = \lambda[J_\nu(x)J_{\nu+1}(x) + N_\nu(x)N_{\nu+1}(x)]/[J_\nu^2(x) + N_\nu^2(x)] - \nu\lambda/x \tag{2.48}$$

where $x = \lambda\rho$, and λ is assumed real. To study the behaviour of the phase and attenuation factors in the above two equations, consider the two extreme cases $x \gg \nu$ and $x \ll \nu$. In the first case, use of asymptotic formulae of the Bessel functions in Appendix 2 in either (2.46) or (2.47)–(2.48) leads to:

$$k_\nu \overset{x \gg \nu}{\Rightarrow} \lambda[1 - (4\nu^2 - 1)/8\lambda^2\rho^2] \tag{2.49}$$

$$\alpha_n \overset{x \gg \nu}{\Rightarrow} 1/2\rho \tag{2.50}$$

which shows that k_ν approaches λ and α_ν approaches zero as ρ tends to ∞, as we should expect. In the other extreme case $x \ll \nu$, the small argument approximation for the Bessel functions given in Appendix 2 leads to:

$$k_\nu \overset{x \ll \nu}{\Rightarrow} (2\pi/\rho) \,.\, (\lambda\rho/2)^{2\nu}/\Gamma^2(\nu) \tag{2.51}$$

$$\alpha_\nu \overset{x \ll \nu}{\Rightarrow} \lambda N_{\nu+1}(x)/N_\nu(x) - \nu\lambda/x \Rightarrow \nu/\rho \tag{2.52}$$

where $\Gamma(.)$ is the gamma function. These two equations show that the rate of attenuation dominates the rate of phase change as ρ tends to zero. Thus the radial wave function changes character as ρ decreases from large values to small ones; behaving as a propagating wave for sufficiently large values of ρ and as an evanescent wave for sufficiently low values of ρ. The change from one character to the other occurs gradually. This behaviour is known as the gradual cutoff of radial waves. To illustrate this behaviour more clearly, the functions k_ν and α_ν are plotted versus $x \equiv \lambda\rho$ for $\nu = 1, 2, 3$ and in Figs 2.4 and 2.5.

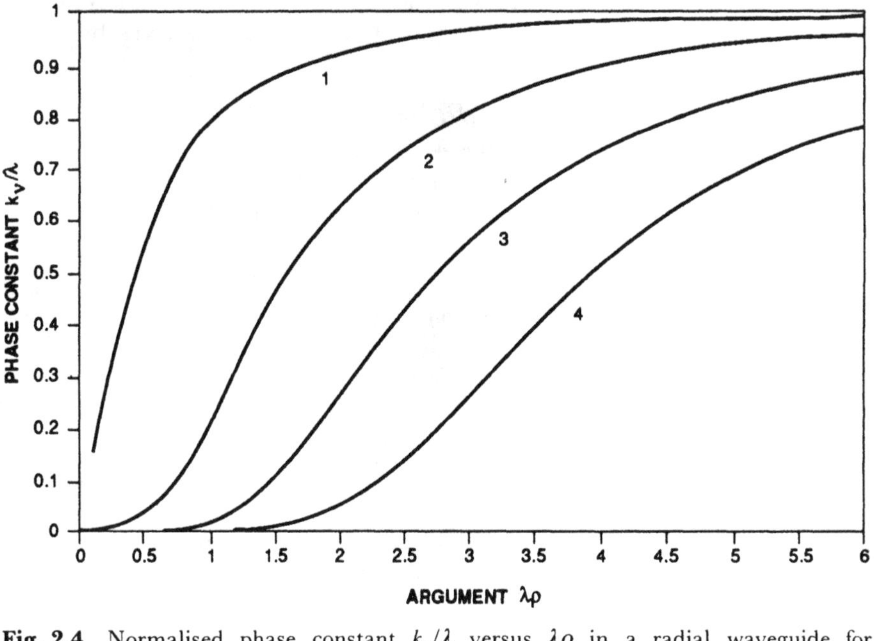

Fig. 2.4 Normalised phase constant k_ν/λ versus $\lambda\rho$ in a radial waveguide for $\nu = 1, 2, 3, 4$

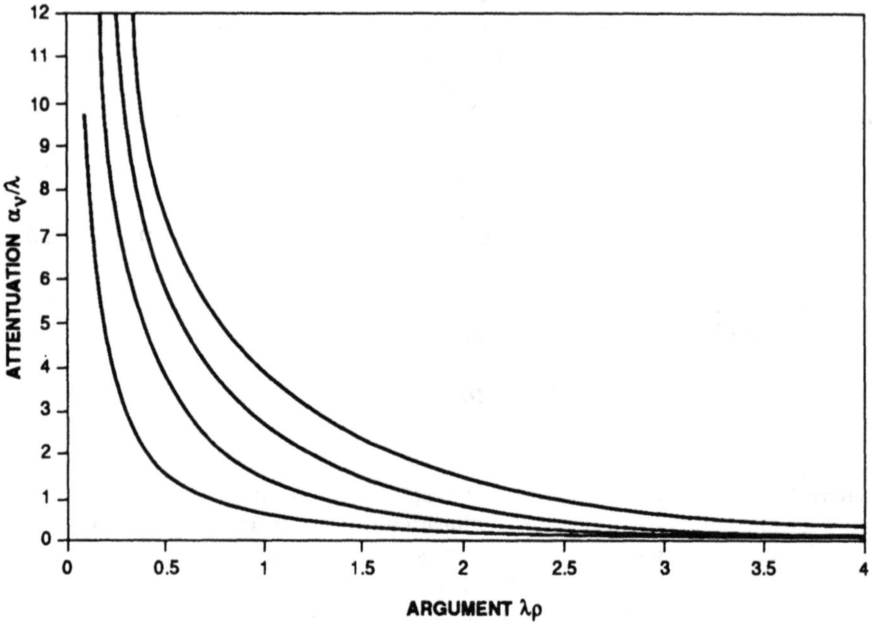

Fig. 2.5 Normalised attenuation constant α_ν/l versus $\lambda\rho$ in a radial waveguide for $\nu = 1, 2, 3, 4$

The radial wave impedance is now discussed. Starting with *TE to z* modes, the wave impedance of the outwardly travelling TE_{mn} mode is given by:

$$Z_{TE}^{+} = E_\phi / H_z$$
$$= (i\omega\mu/\lambda_m)[H_{\nu_n}^{(2)\prime}(\lambda_m\rho)/H_{\nu_n}^{(2)}(\lambda_m\rho)] \tag{2.53}$$

where (1.23) has been used and the + superscript signifies that the wave is travelling outwardly. The mode impedance is closely related to the complex phase constant β_ν defined by (2.46). Explicitly, Z_{TE}^{+} is given in terms of β_ν as

$$Z_{TE}^{+} = \omega\mu\beta_\nu(\rho)/\lambda_n^2 \tag{2.54}$$

For inwardly travelling mode (in the $-\rho$ direction), the mode impedance is given by $-E_\phi/H_z$ and the radial function $H_\nu^{(1)}(.)$ replaces $H_\nu^{(2)}(.)$, hence it follows that for real λ_n

$$Z_{TE}^{-} = \omega\mu\beta_\nu^{*}(\rho)/\lambda^2 = (Z_{TE}^{+})^{*} \tag{2.55}$$

where the * superscript stands for the complex conjugate operation. So, the mode impedances for outwardly and inwardly travelling modes are not equal except when ρ tends to ∞ and both of them tend to a real value equal to that of a plane wave.

A similar treatment for *TM to z* modes leads to wave admittances:

$$Y_{TM}^{+} = -H_\phi/E_z = \omega\varepsilon\beta_\nu(\rho)/\lambda^2 \tag{2.56}$$

and

$$Y_{TM}^{-} = (Y_{TM}^{+})^{*} \tag{2.57}$$

2.8 Curved waveguides

Interest in curved waveguides has existed for a long period of time owing to their application in microwave transmission. In this section, modal solutions in a curved waveguide of rectangular cross section are considered. The geometry of the problem fits into a circular cylindrical coordinate system, as shown in Fig. 2.6. The cross sections lie in the planes $\phi=$constant, and the z axis coincides with the waveguide height. The waveguide walls are assumed to behave as perfect electric conductors.

We seek modal solutions which behave as travelling waves along the ϕ direction. Hence the ϕ dependence of all fields of a given mode is taken as $\exp(-i\nu\phi)$, where ν is an eigenvalue to be determined for each mode from the boundary conditions. Such waves which travel along ϕ are often called circulating waves.

Among the many authors who have contributed to the formulatin and analysis of bent waveguides are Lewin (1955), Waldron (1957), Quinn (1965) and Cochran and Pecina (1966). The latter authors have produced extensive numerical results for the phase factors of the dominant modes of the guide as well as closed form expressions valid at high frequencies and/or large radii of curvature. In the following we derive the modal fields and phase factors and discuss the main features of modes in a curved waveguide.

Considering *TE to z* modes, an appropriate form for H_z, which satisfies Maxwell's equations and behaves as a circulating wave, is:

$$H_z=\{AJ_\nu(\lambda\rho)+BN_\nu(\lambda\rho)\}\sin(\beta z)\ e^{-i\nu\phi} \tag{2.58}$$

where $\lambda=(k^2-\beta^2)^{1/2}$. Since H_z must vanish at both walls $z=0$ and $z=a$, then β can only take the discrete values $m\pi/a$, m being an integer. At the other two walls $\rho=\rho_1=R_0-b/2$ and $\rho=\rho_2-R_0+b/2$, the field component E_ϕ must be

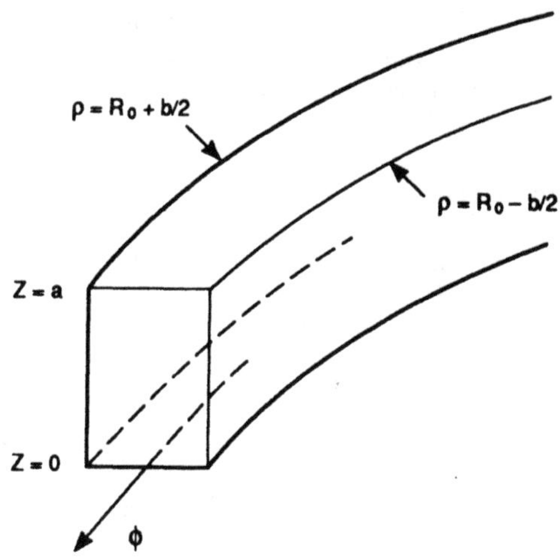

Fig. 2.6 A curved waveguide with rectangular cross section $a\times b$ and a mean radius of cruvature R_0

zero. Since E_ϕ is proportional to $\partial H_z/\partial \rho$, the applications of these two boundary conditions to (2.58) leads to the following two equations:

$$AJ'_\nu(\lambda\rho_l) + BN'_\nu(\lambda\rho_l) = 0$$

with $l=1$ and 2, and the prime superscript stands for the process of differentiation of the Bessel functions with respect to the argument. Eliminating A and B in the two equations results in:

$$J'_\nu(\lambda\rho_1)/N'_\nu(\lambda\rho_1) = J'_\nu(\lambda\rho_2)/N'_\nu(\lambda\rho_2) \tag{2.59}$$

This is the modal equation for *TE to z* modes and the only unknown in the equation is ν. The azimuthal phase constant may be defined by ν/R_0, which is equal to the rate of phase change along the mean circular path $\rho=R_0$. Solution of (2.59) results in a set of eigenvalues ν_n. Modes are thus labelled by pairs of integers (m, n), such that for TE_{mn} modes, $m=1, 2, \ldots$ and $n=0, 1, 2, \ldots$, $\beta_m = m\pi/a$ and ν_n is the nth root of (2.59). The choice of n starting from zero makes the mode labelling compatible with modes of a straight waveguide, in which the dominant mode is the TE_{10}. A similar treatment for *TM to z* modes reveals that E_z can still be given by (2.58) with the sin function replaced by cos in order to satisfy the boundary conditions at the walls $z=0$ and $z=a$. On the two curved walls $\rho=\rho_1$ and $\rho=\rho_2$, E_z must vanish, and this leads to:

$$J_\nu(\lambda\rho_1)/N_\nu(\lambda\rho_1) = J_\nu(\lambda\rho_2)/N_\nu(\lambda\rho_\nu) \tag{2.60}$$

which is the modal equation for the *TM to z* modes, whose solution determines the set of eigenvalues ν_n. The *TM* to *z* modes are thus labelled by pairs of integers (m, n) with $m\geqslant 0$ and $n\geqslant 1$. Hence the dominant *TM to z* mode in a curved waveguide is the TM_{01} mode. In fact, this corresponds to the TE_{01} mode in the straight rectangular waveguide with transverse dimensions a and b.

The modal equations (2.59) and (2.60) have been numerically solved by Cochran and Pecina (1966) for typical dimensions of waveguides in the *S* to *X* bands of frequency (2–10 GHz). Approximate formulae for the azimuthal phase factor have also been given by these authors. The formulae are valid either at high frequencies ($\lambda R_0 \gg 1$) or for gradual bends ($R_0 \gg b$). The cutoff frequencies of modes in bent waveguides are also obtainable from the modal equations (2.59) and (2.60) by inserting $\nu=0$ in the equations and solving for values of λ. The cutoff wavenumbers are then given by $k_c = \{(m\pi/a)^2 + \lambda^2)\}^{1/2}$. Results obtained by Cochran and Pacina (1966) reveals that the guide curvature tends to increase the cutoff frequencies of *TE* modes and to reduce them for *TM* modes. Exceptions are the TE_{m0} modes whose cutoff frequencies are independent of curvature. Actually the cutoff condition for these modes corresponds to $\lambda=0$ and the cutoff wavenumbers are therefore given by $k_c = m\pi/a$.

In order to get insight into the mode characteristics in curved waveguides, let us consider gentle curvatures such that $R_0 \gg b$. Furthermore, we limit our discussion to lower order modes at frequencies sufficiently higher than their cutoff frequencies so that the condition $\lambda R_0 \gg 1$ is also valid. Under these conditions, the azimuthal phase factor ν/ρ will be close to λ so that the following inequality holds:

$$|\nu/\lambda\rho - 1/ \ll 1 \tag{2.61}$$

A ρ-dependent radial wavenumber may be defined as $k_\rho = (\lambda^2 - \nu^2/\rho^2)^{1/2}$. Now recalling the differential equation for the radial wave function $R(\rho)$ given by (1.38) and repeated here for convenience:

$$(\rho^2\partial^2/\partial\rho^2 + \rho\partial/\partial\rho - \nu^2 + \lambda^2\rho^2)R(\rho) = 0 \tag{2.62}$$

let us define a dimensionless variable t by:

$$t = -ck_\rho^2/\lambda^2 = c[(\nu/\lambda\rho)^2 - 1] \tag{2.63}$$

where c is assumed to be a slowly varying function of ρ. From the inequality (2.61), it is clear that $t \ll c$. Using (2.63) to change the variable ρ into t in (2.62) we find that:

$$\rho\partial/\partial\rho \simeq [-2c\nu^2/(\lambda\rho)^2]\partial/\partial t = -2(c+t)\partial/\partial t \simeq -2c\partial/\partial t$$

and

$$\rho^2\partial^2/\partial\rho^2 \simeq 4c^2\partial^2/\partial t^2$$

Anticipating that c will be much greater than unity, we ignore the term $\rho\partial/\partial\rho$ in comparison with $\rho^2\partial^2/\partial\rho^2$ in (2.62); hence we get:

$$(4c^2\partial^2/\partial t^2 - \lambda^2\rho^2 t/c)R(\rho) = 0$$

This equation is reduced to the Airy differential equation:

$$(\partial^2/\partial t^2 - t)R = 0 \tag{2.64}$$

if c is chosen such that $4c^3 = (\lambda\rho)^2$, i.e.

$$c = (\lambda\rho/2)^{2/3} \tag{2.65}$$

One can then verify that c is readily $\gg 1$, as we have already anticipated. In addition, the derivative of c with respect to ρ is proportional to $\rho^{-1/3}$ which means that c is slowly varying with ρ.

The Airy differential equation can be regarded as a one dimensional wave equation in which the wavenumber squared is a linear function of the distance. The solutions are the Airy functions $Ai(t)$ and $Bi(t)$ as defined, for example, by Abramowitz and Stegun (1965). The main properties of these functions are reviewed in Appendix 2. Both functions are oscillatory for negative values of t, while for positive t, the $Ai(t)$ has the nature of an evanescent wave and the $Bi(t)$ the nature of a growing wave. Travelling wave solutions are made up of proper combinations of these two solutions and are termed $w_1(t)$ and $w_2(t)$, as defined in Appendix 2.

Now we write down approximate modal fields for the curved waveguide of Fig. 2.6 by using Airy functions. The approximation is valid at sufficiently large radii of curvature and high frequencies. Starting with TE to z modes:

$$H_z = \{fAi(t) + gBi(t)\}\ \sin(m\pi z/a)\ e^{-i\nu\phi} \tag{2.66}$$

where:

$$t = (\lambda\rho/2)^{2/3}(\nu^2/\lambda^2\rho^2 - 1) \tag{2.67}$$

$$\lambda = (\omega^2\mu\varepsilon - (m\pi/a)^2)^{1/2} \tag{2.68}$$

Application of the boundary conditions $E_\phi \alpha\ \partial H_z/\partial\rho = 0$ at $\rho = \rho_1$ and $\rho = \rho_2$ provides two homogeneous equations for the two constants f and g. The

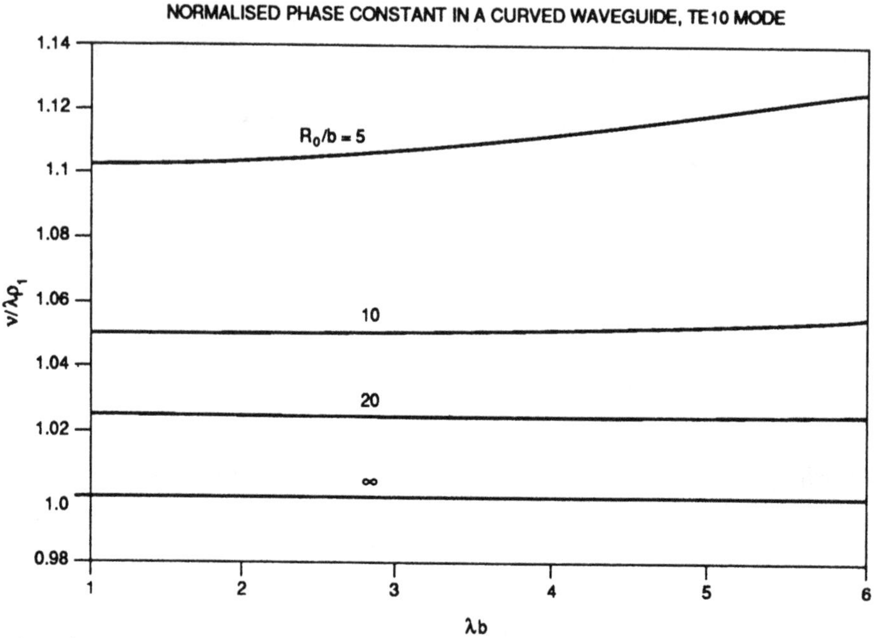

Fig. 2.7 Normalised azimuthal phase constant $\nu/k\rho_1$ versus λb in the curved guide of Fig. 2.6 for the TE_{10} mode. R_0/b is a parameter

determinant of the coefficients of these two equations should therefore vanish. Noting that $\partial/\partial\rho \simeq (-2c/\rho)\partial/\partial t = -(2\lambda^2\rho)^{1/3}\partial/\partial t$, we get the modal equation as:

$$Ai'(t_1)Bi'(t_2) - Ai'(t_2)Bi'(t_1) = 0 \qquad (2.69)$$

where t_1 and t_2 correspond to $\rho=\rho_1$ and ρ_2 respectively. In getting to the above equation, we have neglected the deviation of the ratio $(\rho_1/\rho_2)^{1/3}$ from unity. This obviously conforms with our assumption of large radius of curvature ($R_0 \gg b$). It is worth mentioning that the above equation can also be obtained from (2.59) by invoking the approximate relationships between the Airy and Bessel functions displayed in Appendix 2.

For TM to z modes, E_z is expressed by:

$$E_z = \{fAi(t) + gBi(t)\} \cos(m\pi z/a)\, e^{-i\nu\phi} \qquad (2.70)$$

E_z is required to vanish at $\rho=\rho_1$ and $\rho=\rho_2$; hence the modal equation becomes:

$$Ai(t_i)B_i(t_2) - Ai(t_2)Bi(t_1) = 0 \qquad (2.71)$$

Equations (2.69) and (2.71) are solved numerically for the first mode of each; that is the TE_{10} and TM_{01} modes respectively. The results are shown in Figs. 2.7 and 2.8 by plotting $(\nu/\lambda\rho_1)$ versus λb with R_0/b as a parameter. The transition to the straight waveguide is shown by the curves having R_0/b tending to ∞. In this case, $\nu/\rho_1\lambda$ tends to β_l/λ; β_l being the longitudinal phase constant which is equal to λ for the TE_{10} mode and to $(\lambda^2 - \pi^2/b^2)^{1/2}$ for the TM_{01} mode.

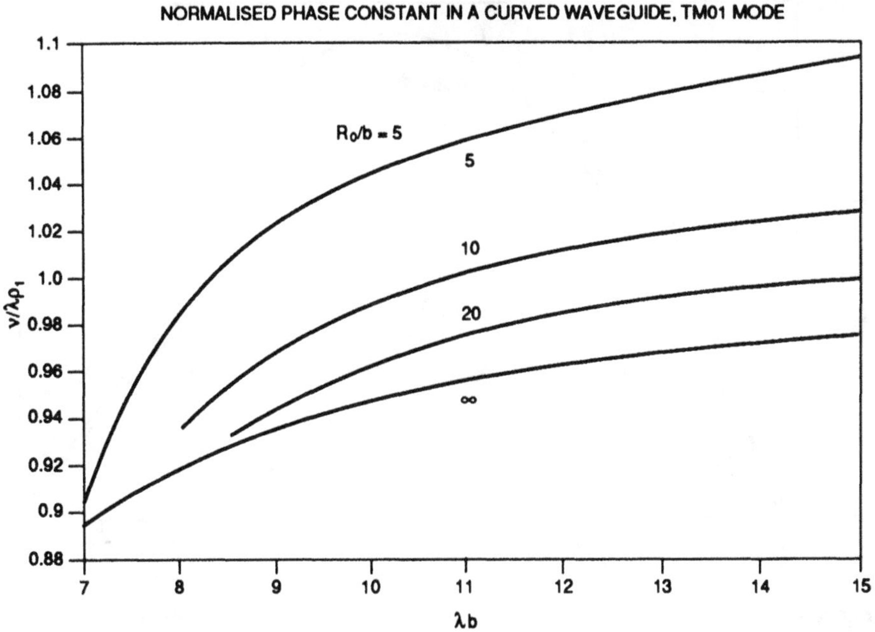

Fig. 2.8 Normalised azimuthal phase constant $\nu/k\rho_1$ versus λb in the curved guide of Fig. 2.6 for the TM_{01} mode. R_0/b is a parameter

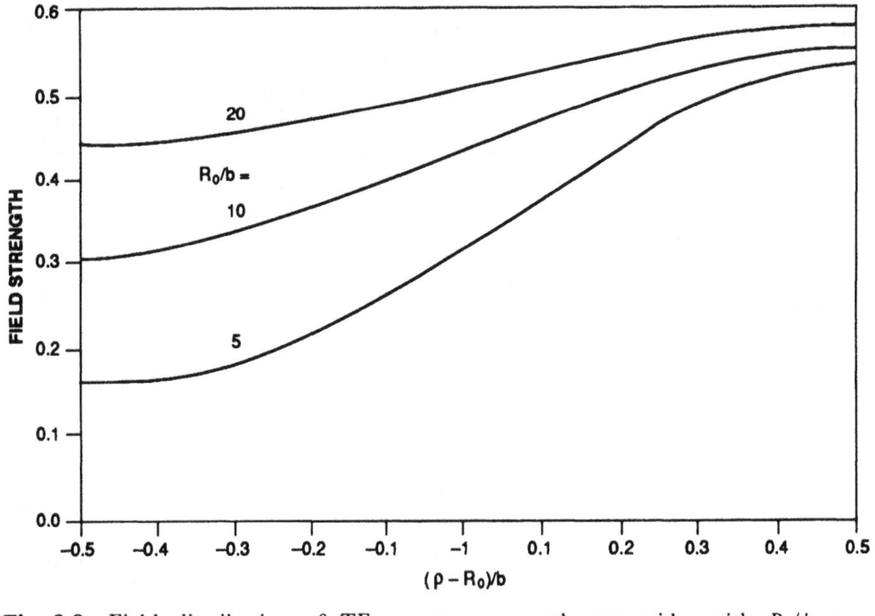

Fig. 2.9 Field distribution of TE_{10} across a curved waveguide, with R_0/b as a parameter and $\lambda b = 6{\cdot}0$

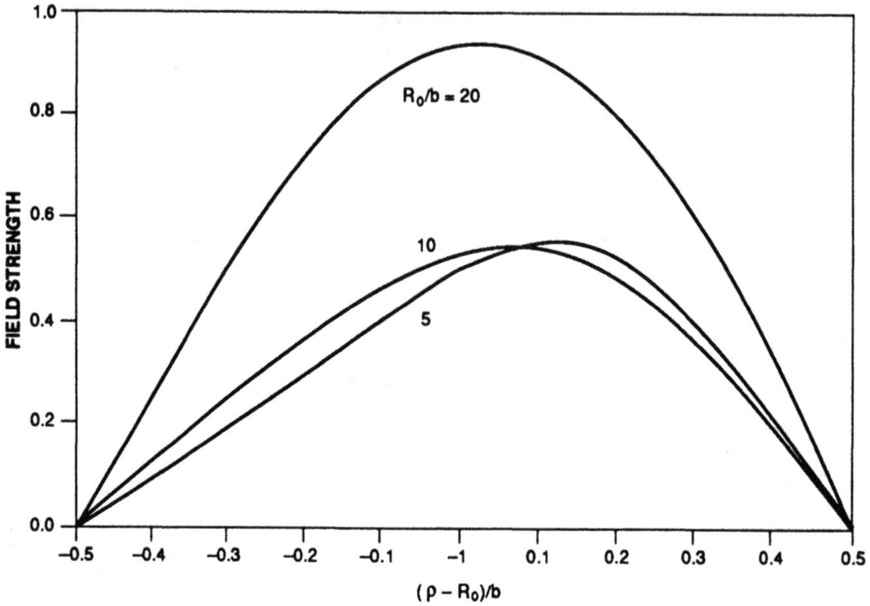

Fig. 2.10 Field distribution of TM_{01} mode across a curved waveguide, with R_0/b as a parameter and $\lambda b = 12{\cdot}0$

It is seen from Figs. 2.7 and 2.8 that the azimuthal wavenumber ν/ρ_1 increases with curvature. This means that the phase velocity slows down with increased curvature. This becomes clear when we recall that the phase velocity is equal to $\omega R_0/\nu$. Meanwhile, as $\nu/\lambda\rho_1$ increases and exceeds unity, the fields become evanescent near ρ_1, since then t_1 becomes positive (see (2.67)). This is clearly seen in the field distributions across the waveguide as given in Figs. 2.9 and 2.10 which show that, with increased curvature, the fields tend to become detached from the convex surface ρ_1 and cling to the concave surface ρ_2. This phenomenon is related to that of the whispering gallery modes which were observed firstly by Lord Rayleigh and discussed extensively by Wait (1967, 1968).

2.9 Mode orthogonality

An extremely useful property of modes in waveguides is the mode orthogonality property. This property plays a key role in solving problems of excitation and scattering in waveguides.

In a limited sense, orthogonality between modes means that each mode carries its own power independently of other modes. Mathematically stated, if (e_n, h_n) are the transverse (to the direction of propagation) vector fields of mode number n, in an arbitrary ordered scheme, and (e_m, h_m) are those of mode m, then, unless $m = n$,

$$\int_S (e_n \times h_m^*) \cdot \hat{z}\, dS = 0 \tag{2.72}$$

where the * denotes the conjugate operation, S is the cross-sectional area and the unit vector $\hat{z}$ is normal to it. So there is no power carried between fields of different modes.

The power orthogonality property, as stated above, is found to be valid only in a lossless guide, i.e. a guide with lossless materials (see prob. 2.17). A more general definition of orthogonality is given in terms of the volt-ampere rather than active power. Symbolically stated, this orthogonality is expressed by:

$$\int_S (e_n \times h_m) \cdot \hat{z}\, dS = 0 \tag{2.73}$$

In this sense the orthogonality is valid even if the waveguide includes lossy materials. Relation (2.73), however, requires that the filling material be isotropic, as will be proved below. In the case of anisotropic materials, the orthogonality relationship has still another form which involves the use of the concept of the adjoint waveguide. This, however, lies outside the scope of this text and the reader is referred, on this point, to the excellent paper of Bressler, Joshi and Marcuvitz (1958).

Now, we prove the orthogonality property (2.73) for a cylindrical waveguide with perfectly conducting walls (whether electric or magnetic) and isotropic filling materials. So, both ε and μ are scalar quantities, although they can be complex signifying the presence of loss. To this end, let (E_n, H_n) and (E_m, H_m) be the total vector fields of modes number n and m, of which (e_n, h_n) and (e_m, h_m) are the transverse fields. Obviously, the fields of each mode satisfy the source free Maxwell's equations (1.1) and (1.2). Now, we form the quantity:

$$\nabla \cdot (E_n \times H_m - E_m \times H_n)$$

by using the identity:

$$\nabla \cdot (A \times B) = B \cdot \nabla \times A - A \cdot \nabla \times B$$

and the source free Maxwell's equations for the fields of the two modes. The result is simply given by

$$\nabla \cdot (E_n \times H_m - E_m \times H_n) = 0 \tag{2.74}$$

which is known as the Lorentz reciprocity theorem.

Integrating over a volume bounded by two cross sections $z = z_1$ and $z = z_1 + dz$ and the perfectly conducting side walls of the guide, and by virtue of the divergence theorem of integration (see A1.18)

$$\int_{vol} \nabla \cdot (E_n \times H_m - E_m \times H_n) = \int_A (E_n \times H_m - E_m \times H_n) \cdot \hat{n}\, dA = 0 \tag{2.75}$$

where A is the closed surface enclosing the chosen volume and $\hat{n}$ is the outward unit normal at any point on this surface, (see Fig. 2.11). On the side walls, either $E \times \hat{n}$ or $H \times \hat{n}$ is zero for any of the two modes. In other words, the tangential electric/magnetic fields vanish over the perfectly electric/magnetic

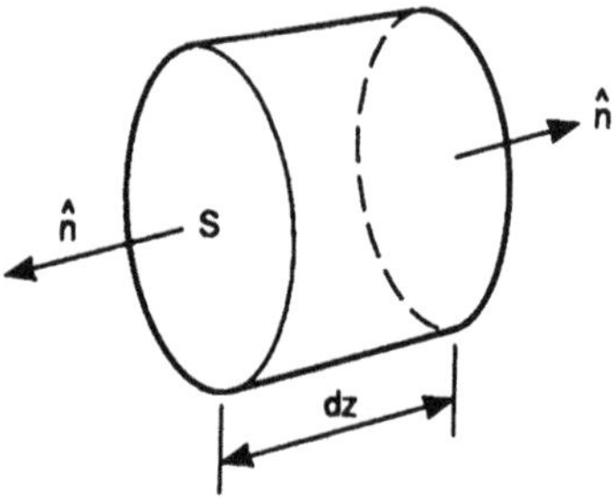

Fig. 2.11 A small length of a cylindrical waveguide with arbitrary cross section

side walls. Hence, the integration over these walls is equal to zero. On the cross sectional surfaces $z=z_1$ and $z=z_1+dz$, $\hat{n}=-\hat{z}$ and $\hat{n}=\hat{z}$ respectively; hence, with dz tending to zero, (2.75) reduces to

$$(\partial/\partial z)\int_S (\mathrm{E}_n \times \mathrm{H}_m - \mathrm{E}_m \times \mathrm{H}_n)\,.\,\hat{z}\,\mathrm{d}S=0 \tag{2.76}$$

where the integration is taken over the cross sectional area denoted by S. While the fields in (2.76) are the total modal fields, we notice that the longitudinal or z components do not contribute to the integrand. Hence it is admissible to substitute the transverse fields (e_n, h_n) and (e_m, h_m) for their respective total fields. Noticing also that the differential operator $\partial/\partial z$ is equivalent to a multiplication by $-i\beta_n$ for mode n and $-\mathrm{i}\beta_m$ for mode m, (2.76) takes the form:

$$(\beta_n+\beta_m)\int_S (\mathrm{e}_n \times \mathrm{h}_m - \mathrm{e}_m \times \mathrm{h}_n)\,.\,\hat{z}\,\mathrm{d}S=0$$

So, unless $\beta_n=-\beta_m$,

$$\int_S (\mathrm{e}_n \times \mathrm{h}_m - \mathrm{e}_m \times \mathrm{h}_n)\,.\,\hat{z}\,\mathrm{d}S=0 \tag{2.77}$$

Now if mode m is left unchanged, while mode n is taken to be a reflected mode, that is β_n is replaced by $-\beta_n$, then h_n changes sign if e_n is kept unaltered. Under this new situation, (2.76) becomes:

$$(\beta_m-\beta_n)\int_S (\mathrm{e}_n \times \mathrm{h}_m + \mathrm{e}_m \times \mathrm{h}_n)\,.\,\hat{z}\,\mathrm{d}S=0$$

which shows that, unless $\beta_n=\beta_m$,

$$\int_S (\mathrm{e}_n \times \mathrm{h}_m + \mathrm{e}_m \times \mathrm{h}_n)\,.\,\hat{z}\,\mathrm{d}S=0 \tag{2.78}$$

Adding (2.77) and (2.78), we conclude that, if $\beta_n \neq \pm\beta_m$, then:

$$\int_S (\mathrm{e}_n \times \mathrm{h}_m)\,.\,\hat{z}\,\mathrm{d}S=0 \tag{2.79}$$

Equations (2.78) and (2.79) are used extensively in excitation and scattering problems. In slightly different forms, they can be recast as:

$$\int_S (\mathbf{e}_n \times \mathbf{h}_m + \mathbf{e}_m \times \mathbf{h}_n) \cdot \hat{\mathbf{z}} \, dS = 2N_n\delta_{n,m} \tag{2.80}$$

$$\int_S (\mathbf{e}_n \times \mathbf{h}_m) \cdot \hat{\mathbf{z}} \, dS = N_n\delta_{n,m} \tag{2.81}$$

where

$$N_n = \int_S (\mathbf{e}_n \times \mathbf{h}_n) \cdot \hat{\mathbf{z}} \, dS \tag{2.82}$$

and $\delta_{n,m}$ is equal to unity only if $n=m$; otherwise it is equal to zero.

2.10 Modal expansion of fields

Now that the orthogonality relations among the modal fields have been established, we can expand an arbitrary field distribution over the cross section of the guide in terms of the modal fields.

So assume that we are given the transverse electric field vector $\mathbf{E}_t(0)$ over the cross section $z=0$. Modal fields will then be excited and travel down the guide. If the guide extends uniformly along z, then no reflection will occur. The transverse electric field at any cross section $z>0$ can then be expanded as a sum of modal fields. This is based on the completeness of the infinite set of modes inside a closed waveguide. Thus, we write:

$$\mathbf{E}_t(z) = \sum_n A_n \mathbf{e}_n \exp(-i\beta_n z) \tag{2.83}$$

where the dependence of $\mathbf{E}_t$ and $\mathbf{e}_n$ on the transverse coordinates is implicit and the sum is over all possible modes. The coefficients A_n are excitation coefficients of modes. Since $\mathbf{E}_t(0)$ is known, then we have:

$$\mathbf{E}_t(0) = \sum_n A_n \mathbf{e}_n$$

and if we utilise the orthogonality relation (2.81) by post-multiplying both sides of the last equation by $\times \mathbf{h}_m$ and integrating over the cross section, we get A_n as:

$$A_n = \int_S (\mathbf{E}_t(0) \times \mathbf{h}_n) \cdot \hat{\mathbf{z}} \, dS / N_n \tag{2.84}$$

The transverse magnetic field $\mathbf{H}_t(z)$ is given by a modal expansion similar to that of the electric field in (2.83). In particular, the input transverse magnetic field at $z=0$ is given by:

$$\mathbf{H}_t(0) = \sum_n A_n \mathbf{h}_n = \sum_n \left[\int_S (\mathbf{E}_t(0) \times \mathbf{h}_n) \cdot \hat{\mathbf{z}} \, dS \right] \mathbf{h}_n / N_n$$

which is seen to be completely determined by $E_t(0)$. In an exactly similar fashion, the modal excitation coefficients A_n due to an input transverse magnetic field $H_t(0)$ are given by:

$$A_n = \int_S (e_n \times H_t(0)) \cdot \hat{z} \, dS/N_n \tag{2.85}$$

and the input transverse electric field is then completely determined by $H_t(0)$. So, to sum up this part, we conclude that either E_t or H_t at $z=0$ is required to resolve the modal amplitudes in the region $z>0$, provided that this region is matched so that no reflected modes exist.

Conversely, if both incident and reflected modes exist in the $z>0$ region because of the presence of any type of discontinuity down the guide, then both E_t and H_t at $z=0$ are needed to completely determine the waves down the guide as will be shown by the following argument. Appropriate field expansions in this case are:

$$E_t(z) = \sum_n A_n e_n \exp(-i\beta_n z) + B_n e_n \exp(i\beta_n z)$$

$$H_t(z) = \sum_n A_n h_n \exp(-i\beta_n z) - B_n h_n \exp(i\beta_n z)$$

Putting $z=0$, and post-multiplying the first equation by $\times h_m$ and pre-multiplying the second by $e_m \times$, we get after integrating over the cross section and using the orthogonality relationship (2.81):

$$\int_S (E_t(0) \times h_m) \cdot \hat{z} \, dS = (A_m + B_m) N_m$$

$$\int_S (e_m \times H_t(0)) \cdot \hat{z} \, dS = (A_m - B_m) N_m$$

Adding and subtracting these two equations results in the modal amplitudes as:

$$A_m = \int_S (E_t(0) \times h_m + e_m \times H_t(0)) \cdot \hat{z} \, dS/2N_m \tag{2.86}$$

$$B_m = \int_S (E_t(0) \times h_m - e_m \times H_t(0)) \cdot \hat{z} \, dS/2N_m \tag{2.87}$$

This is the most general result for modal excitation by known aperture fields.

2.11 Mode excitation by transverse current sheets

Now we consider waveguide excitation by means of current sheets. To simplify the analysis, let us restrict attention to transverse current sheets. So, assume that j (ampere/m) is a given electric current distribution lying totally in the cross sectional area $z=0$ of a cylindrical waveguide. The question is to find the amplitudes of excited modes in both regions $z>0$ and $z<0$. The current sheet j produces a discontinuity in the magnetic field vector H, equal in magnitude to the surface current j itself. This follows directly from the second equation (1.2) of Maxwell, and is expressed by:

$$\mathrm{j}=\hat{\mathrm{z}}\times(\mathrm{H}_t(0^+)-\mathrm{H}_t(0^-)) \tag{2.88}$$

where $\mathrm{H}_t(0^\pm)$ is the tangential magnetic field just to the right and to the left of the current, respectively. On the other hand j does not produce any discontinuity in E_t, so that

$$\mathrm{E}_t(0^+)=\mathrm{E}_t(0^-) \tag{2.89}$$

Now we may express the transverse fields in the regions $z \gtrless 0$ by the following modal expansions:

$$\mathrm{E}_t(z)=\sum_n A_n^\pm \mathrm{e}_n \exp(\mp \mathrm{i}\beta_n z) \tag{2.90}$$

$$\mathrm{H}_t(z)=\sum_n \pm A_n^\pm \mathrm{h}_n \exp(\mp \mathrm{i}\beta_n z) \tag{2.91}$$

where the upper/lower signs correspond to $z \gtrless 0$. Invoking relation (2.89) in the first of the above two equations, we immediately deduce that $A_n^+=A_n^-$ for all n. Using (2.88) in the second equation, we get

$$\mathrm{j}\times\hat{\mathrm{z}}=\sum_n (A_n^+ + A_n^-)\mathrm{h}_n = 2\mathrm{H}_t(0^+) = -2\mathrm{H}_t(0^-)$$

which simply means that the magnetic field on both sides of the current in equal in magnitude to half the surface current density. At this point, it is most straightforward to use (2.85) to obtain the modal amplitudes in terms of $\mathrm{H}_t(0^+)$ distribution. Thus we get:

$$A_m^+ = \int_S \mathrm{e}_m \times (\mathrm{j}\times\hat{\mathrm{z}}) \cdot \hat{\mathrm{z}}\ \mathrm{d}S/2N_m$$

which, on using identity (A.19) and the assumption that j is transverse to $\hat{\mathrm{z}}$, reduces to:

$$A_m^+ = A_m^- = -\int_S \mathrm{j}\cdot\mathrm{e}_m\ \mathrm{d}S/2N_m \tag{2.92}$$

Next, consider the case of excitation by a transverse magnetic current distribution which lies totally in the cross section $z=0$. The current sheet, denoted by m volt/m produces a discontinuity in the transverse E across the plane $z=0$ while the transverse magnetic field remains continuous. So, in place of (2.88) and (2.89), we now have:

$$\mathrm{m}=\hat{\mathrm{z}}\times(\mathrm{E}_t(0^-)-\mathrm{E}_t(0^+)), \quad \text{and}$$

$$\mathrm{H}_t(0^+)=\mathrm{H}_t(0^-)$$

So, following a similar argument to the previous case, we deduce that the exicted modal amplitudes on both sides of the magnetic current sheet are related by $A_m^+ = -A_m^-$ and that

$$\mathrm{E}_t(0^+) = -\mathrm{E}_t(0^-) = -\mathrm{m} \times \hat{\mathrm{z}}/2$$

Now, applying (2.84) we get the modal amplitudes as:

$$A_m^+ = -A_m^- = \int_S \mathrm{m} \cdot \mathrm{h}_m \, \mathrm{d}S/2N_m \tag{2.93}$$

This completes the problem of waveguide excitation by transverse current sheets.

2.12 Scattering at a longitudinal discontinuity

In a microwave circuit, junctions between dissimilar waveguides will inevitably exist as interfaces between different circuit components. Certain discontinuities may be inserted longitudinally down the guide for the purpose of matching between two unequal levels of impedance. Discontinuities occur also in natural waveguides; a well known example is the earth–ionosphere guide for very low frequency (VLF) radio waves. The height of the effective reflecting layer of the ionosphere changes as the guide crosses the shadow region between day and night. The effective height of the ionosphere for VLF waves changes from about 65 to 95 km from day to night regions, causing a discontinuity in the earth–ionosphere waveguide.

A canonical problem of interest in this respect is a sharp discontinuity between two dissimilar waveguides. A gradual transition, or a taper, from one waveguide to the other may then be treated as a series of small sharp discontinuities in cascade. The analysis to follow is based on the technique of modal matching at the plane of discontinuity. This technique is the most straightforward when complete eigenmodes of the waveguides involved are known. The geometry of the problem is depicted in Fig. 2.12. The two

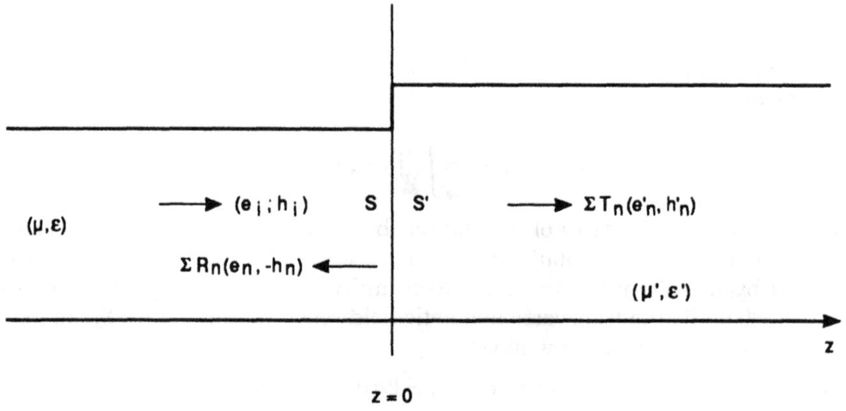

Fig. 2.12 A longitudinal discontinuity

waveguides on the left hand and right hand sides of the discontinuity plane $z=0$ may differ in cross sectional area or shape, in the guide walls, or in the properties of materials filling the guides. In any case, we require that a complete set of orthogonal modes be known for each of the two waveguides on the two sides of the discontinuity. So let us assume that the transverse modal fields on the left waveguide are $(\mathrm{e}_n, \mathrm{h}_n)\exp(-i\beta_n z)$, $n=1, 2 \ldots$ and those on the right waveguide are $(\mathrm{e}'_n, \mathrm{h}'_n)\exp(-i\beta'_n z)$. Here the prime is used to distinguish between similar quantities on the two waveguides. Let the ith mode be incident from left with unit amplitude. In general, reflected modes and transmitted modes of all orders will be scattered away from the discontinuity plane $z=0$. Our purpose is to determine the amplitudes (generally complex) of these scattered modes. To this end, let $R_{n,i}$ be the complex amplitude of the nth reflected mode and $T_{n,i}$ be the complex amplitude of the transmitted nth mode. Thus the total transverse fields $\mathrm{E}_t(0)$ and $\mathrm{H}_t(0)$ at $z=0$ can be expanded in terms of modal fields on both left and right waveguides as:

$$\mathrm{E}_t(0)=\mathrm{e}_i+\sum_n R_{n,i}\,\mathrm{e}_n=\sum_n T_{n,i}\,\mathrm{e}'_n \tag{2.94}$$

$$\mathrm{H}_t(0)=\mathrm{h}_i-\sum_n R_{n,i}\,\mathrm{h}_n=\sum_n T_{n,i}\,\mathrm{h}'_n \tag{2.95}$$

where the dependence of all fields on the transverse coordinates is implicit. The summations are taken over the infinite set of modes of either waveguide. To proceed, post-multiply the first equation by $\times\mathrm{h}'_m$ and pre-multiply the second by $\mathrm{e}'_m\times$ and integrate over the cross section S' of the right hand side waveguide. Then, on using the orthogonality relation (2.81), we get

$$\langle \mathrm{e}_i, \mathrm{h}'_m\rangle+\sum_n R_{n,i}\langle \mathrm{e}_n, \mathrm{h}'_m\rangle=T_{m,i}N'_m \tag{2.96}$$

$$\langle \mathrm{e}'_m, \mathrm{h}_i\rangle-\sum_n R_{n,i}\langle \mathrm{e}'_m, \mathrm{h}_n\rangle=T_{m,i}N'_m \tag{2.97}$$

where the abbreviation $\langle \mathrm{e}_n, \mathrm{h}'_m\rangle$ is used to stand for the inner product $\int_{S'}(\mathrm{e}_n\times\mathrm{h}'_m)\cdot\hat{z}\,\mathrm{d}S'$. It is worth noting that when the cross sections S and S' are different the integration is performed over their common area. The normalisation factor N'_m is equivalent to $\langle \mathrm{e}'_m, \mathrm{h}'_m\rangle$ with the integration over S'. Subtracting and adding the last two equations results in

$$\sum_n R_{n,i}\{\langle \mathrm{e}_n, \mathrm{h}'_m\rangle+\langle \mathrm{e}'_m, \mathrm{h}_n\rangle\}=\langle \mathrm{e}'_m, \mathrm{h}_i\rangle-\langle \mathrm{e}_i, \mathrm{h}'_m\rangle \tag{2.98}$$

$$\sum_n R_{n,i}\{\langle \mathrm{e}_n, \mathrm{h}'_m\rangle-\langle \mathrm{e}'_m, \mathrm{h}_n\rangle\}+\langle \mathrm{e}_i, \mathrm{h}'_m\rangle+\langle \mathrm{e}'_m, \mathrm{h}_i\rangle=2T_{m,i}N'_m \tag{2.99}$$

with $m=1, 2, 3, \ldots$.

These equations can be solved numerically if the sum terms are suitably truncated at, say, N reflected modes and M transmitted modes. We will then have $2M$ equations in $N+M$ unknowns which can be solved for if $M \geqslant N$.

2.12.1 An approximate solution

Alternatively, an approximate closed form solution is obtainable if the discontinuity is sufficiently weak such that the inner products $\langle e_n, h'_m \rangle$ are considered significant only when $n=m$. Thus retaining only the term $n=m$ in the summation of (2.98) $R_{m,i}$ is immediately determined:

$$R_{m,i} \simeq \frac{\langle e'_m, h_i \rangle - \langle e_i, h'_m \rangle}{\langle e_m, h'_m \rangle + \langle e'_m, h_m \rangle} \tag{2.100}$$

In (2.99) the sum term is of second order smallness, and hence can be totally neglected. Thus

$$T_{m,i} \simeq \{\langle e_i, h'_m \rangle + \langle e'_m, h_i \rangle\}/2N'_m \tag{2.101}$$

where $m=1, 2, \ldots$.

The results in (2.100) and (2.101) have been obtained by Dragone (1977) in the context of evaluating mode conversion on a nonuniform corrugated waveguide. It is usually convenient to adopt a mode normalisation such that N_m and N'_m are equal to unity. If these are not unity, the normalisation can still be effected in (2.101) by replacing N'_m by $(N_i N'_m)^{1/2}$. In this latter form, the reciprocity relationship clearly holds; that is $T_{m,i} = T_{i,m}$.

2.12.2 An exact solution

Let us now return to the infinite set of equations (2.98) and (2.99) to show that under certain conditions an exact solution can be obtained (Dragone, 1984). To this end, we note that in many cases the inner products $\langle e_n, h'_m \rangle$ and $\langle e'_m, h_n \rangle$ are expressible in the special form

$$\langle e_n, h'_m \rangle \pm \langle e'_m, h_n \rangle = \frac{C_n C'_m}{\beta_n \mp \beta'_m} \tag{2.102}$$

By invoking (2.102) in (2.98)–(2.99), the latter take the forms

$$\sum_{n=1}^{\infty} \frac{R_{n,i} C_n}{\beta_n - \beta'_m} + \frac{C_i}{\beta_i + \beta'_m} = 0 \tag{2.103}$$

$$\sum_{n=1}^{\infty} \frac{R_{n,i} C_n}{\beta_n + \beta'_m} + \frac{C_i}{\beta_i - \beta'_m} = 2T_{m,i} N'_m / C'_m \tag{2.104}$$

for $m=1, 2, \ldots$.

In these forms the two infinite sets of equations ($m=1, 2, \ldots \infty$) can be solved by the so called method of residue calculus (e.g. Mittra and Lee, 1971). This relies on forming a residue series that resembles the infinite set of equations to be solved and the unknown coefficients are then determined by comparing terms. To apply this method to the set of equations (2.103), consider the complex integral

$$\frac{1}{2\pi i}\oint \frac{f(w)}{w-\beta'_m}\,\mathrm{d}w \tag{2.105}$$

where the integration is taken over an infinitely large circle in the complex w plane. Keeping in mind that we aim at expressing (2.105) as an infinite residue series which resembles the left hand side of (2.103), we impose the following conditions on $f(w)$:

(i) $f(w)$ has simple poles at $w=\beta_n$, $n=1, 2 \ldots \infty$ and at $w=-\beta_i$.
(ii) $f(w)$ has simple zeros at $w=\beta'_m$, $m=1, 2 \ldots \infty$ in order to cancel the poles in (2.105) at these locations.
(iii) The residue of $f(w)$ at the pole $w=-\beta_i$ is equal to $-C_i$.
(iv) Finally $f(w)$ should approach zero as $|w|$ approaches ∞ in such a way that the integral in (2.105) vanishes over the infinite circle.

The form of $f(w)$ that satisfies conditions (i) and (ii) is

$$f(w)=\frac{f_0}{w+\beta_i}\prod_{r=1}^{\infty}\frac{w/\beta'_r-1}{w/\beta_r-1} \tag{2.106}$$

where f_0 is a constant that is to be determined from condition (iii) above. The result of applying this condition is that

$$f_0=-C_i\prod_{r=1}^{\infty}\frac{1+\beta_i/\beta_r}{1+\beta_i/\beta'_r} \tag{2.107}$$

Therefore $f(w)$ is completely determined by

$$f(w)=\frac{-C_i}{w+\beta_i}\prod_{r=1}^{\infty}\frac{(w/\beta'_r-1)(1+\beta_i/\beta_r)}{(w/\beta_r-1)(1+\beta_i/\beta'_r)} \tag{2.108}$$

It can be shown that $f(w)$ behaves as $w^{-3/2}$ as $|w|$ tends to ∞; hence condition (iv) is satisfied. Now writing the residue series of the integral (2.105) and equating to zero, we get a set of equations of the form (2.103). Upon comparing terms $R_{n,i}C_n$ is found to be equal to the residue of $f(w)$ at the pole β_n, or

$$R_{n,i}=-(C_i/C_n)\,\frac{\beta_n-\beta'_n}{\beta_i+\beta'_n}\prod_{r\neq n}\frac{(\beta_n/\beta'_r-1)(1+\beta_i/\beta_r)}{(\beta_n/\beta_r-1)(1+\beta_i/\beta'_r)} \tag{2.109}$$

which is an exact result. The product in (2.109) is over all values of r other than n. In particular, the reflection coefficient of mode i, the incident mode, is

$$R_{i,i}=\frac{\beta'_i-\beta_i}{\beta'_i+\beta_i}\prod_{r\neq i}\frac{(\beta_i-\beta'_r)(\beta_i+\beta_r)}{(\beta_i+\beta'_r)(\beta_i-\beta_r)} \tag{2.110}$$

If mode i is the only propagating mode in each of the two guides, then only β_i and β'_i are real while all other β and β' are purely imaginary (in lossless guides). Hence the product term is equal to unity in magnitude and $R_{i,i}$ is given in magnitude merely by the first term in (2.110). This remarkably simple, yet exact result, seems to have been first derived by Dragone (1984). Let us now turn attention to the infinite set of equations (2.104). With the same $f(w)$ in (2.108) consider the complex integral

$$\frac{1}{2\pi i}\oint \frac{f(w)}{w+\beta'_m} \tag{2.111}$$

The integration is taken over an infinite circle in the w plane. Under condition (iv) stated earlier, the integral vanishes. The poles of the integrand lie at $w=\beta_n$, $n=1, 2 \ldots \infty$, $w=-\beta_i$ and $w=-\beta'_m$, all being simple poles. Expanding the integral in terms of its residue series and equating to zero,

$$\sum_{n=1}^{\infty} \frac{\mathrm{Res}(f(\beta_n))}{\beta_n+\beta'_m}+\frac{\mathrm{Res}(f(-\beta_i)}{-\beta_i+\beta'_m}+f(-\beta'_m)=0 \tag{2.112}$$

where $\mathrm{Res}(f(p))$ stands for the residue of f at the pole p. Comparison of (2.112) with (2.104) determines $T_{m,\,i}$ as

$$T_{m,\,i}=(-C'_m/2N'_m)f(-\beta'_m)$$

$$=\frac{(C_iC'_m/2N'_m)}{\beta_i-\beta'_m}\prod_{n=1}^{\infty}\frac{(1+\beta'_m/\beta'_n)(1+\beta_i/\beta_n)}{(1+\beta'_m/\beta_n)(1+\beta_i/\beta'_n)} \tag{2.113}$$

which is the exact result for the transmission coefficient of mode m when mode i is incident. At this point it is interesting to compare the approximate expressions (2.100) and (2.101) with the exact ones (2.109) and (2.113). First, by invoking condition (2.102) in (2.100) and (2.101), these take the forms

$$R_{m,\,i}\simeq-(C_i/C_m)(\beta_m-\beta'_m)/(\beta_i+\beta'_m) \tag{2.114}$$

$$T_{m,\,i}\simeq(C_iC'_m/2N'_m)/(\beta_i-\beta'_m) \tag{2.115}$$

which are seen to be the same as the exact expressions (2.109) and (2.113) except for the product terms. Of particular interest is that, in the case when the incident mode is the only propagating mode in both guides, the magnitude of its reflection coefficient $R_{i,\,i}$ is given exactly by the approximate formula (2.114).

As an example of using the above results let us consider a very simple model of a discontinuity in the earth–ionosphere waveguide; namely, an abrupt change of the ionospheric height is assumed to occur at the shadow line separating the day and night regions. The geometry of the problem is given in Fig. 2.13. The earth is assumed to behave as a perfect electric conductor and the ionosphere as a perfect magnetic conductor. These assumpitons are reasonably valid in the VLF band; 3–30 kHz (Wait, 1970*b*). With the usual vertically polarised excitation, the magnetic field is totally y directed and the electric field has components along the x and z axes. The transverse modal fields in the *day* waveguide (e_n, h_n) are, therefore:

$$\mathrm{h}_n=\hat{y}\cos((n-1/2)\pi x/d)\exp(-i\beta_n z) \tag{2.116}$$

$$\mathrm{e}_n=\hat{x}(\beta_n/\omega\varepsilon)(\mathrm{h}_n\,.\,\hat{y}) \tag{2.117}$$

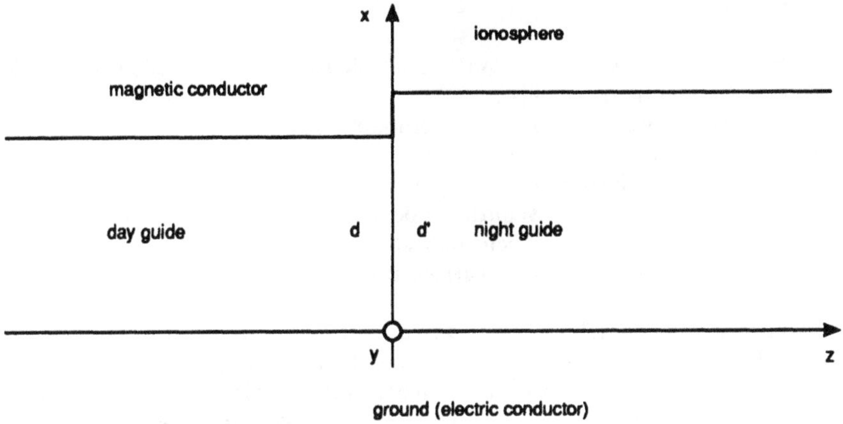

Fig. 2.13 A planar model of the earth–ionosphere waveguide with a day/night discontinuity

with

$$\beta_n = (\omega^2 \mu \varepsilon - (n - 1/2)^2 \pi^2 / d^2)^{1/2} \tag{2.118}$$

Similar expressions apply to modes (e'_n, h'_n) in the *night* waveguide with d replaced by d'. It is then an elementary exercise to derive expressions for N_n, N'_n and the inner products $\langle e_n, h'_m \rangle$ and $\langle e_m, h'_n \rangle$. The results are:

$$N_n = \int_0^d e_x h_y \, dx = \beta_n d / 2\omega\varepsilon$$

$$N'_n = \int_0^{d'} e'_x h'_y \, dx = \beta' d' / 2\omega\varepsilon \tag{2.119}$$

$$\langle e_n, h'_m \rangle \pm \langle e'_m, h_n \rangle = C_n C'_m / (\beta_n \mp \beta'_m)$$

where:

$$C_n = (-)^n \pi (n - 1/2) / d(\omega\varepsilon)^{1/2} \tag{2.120}$$

$$C'_m = \cos((m - 1/2)\pi d / d') / (\omega\varepsilon)^{1/2} \tag{2.121}$$

A direct substitution from (2.119)–(2.121) into (2.109) and (2.113) results in the scattered mode amplitudes due to the change in the ionospheric height. For instance, if mode $i = 1$ is incident from the day guide, the amplitude $T_{m,1}$ of the transmitted mth mode is:

$$T_{m,1} = \frac{-(\pi/2)\cos((m - 1/2)\pi d/d')}{\beta'_m d' (\beta_1 - \beta'_m) d} \prod_{n=1}^{\infty} \frac{(1 + \beta'_m/\beta'_n)(1 + \beta_1/\beta_n)}{(1 + \beta'_m/\beta_n)(1 + \beta_1/\beta'_n)} \tag{2.122}$$

While in the above example we have assumed lossless boundaries, we shall see in the next chapter how to introduce more realistic boundaries by using the concept of constant impedance surfaces.

2.13 Problems

2.1 Show that a *TEM* mode is not a possible solution in a hollow waveguide with perfectly conducting walls.
2.2 Find the frequency range of monomode propagation in an empty rectangular waveguide of dimensions $a+b$, $a=2b$, such that the mode impedance does not exceed $2\eta_0$, where η_0 is the free space impedance.
2.3 Find the first five lowest order modes in a rectangular waveguide with perfect magnetic walls and dimensions $a \times a/2$.
2.4 Show that a cuttoff mode carries no real power down the cylindrical waveguide.
2.5 Write down all field components of a *TE to x* modes in a rectangular waveguide with perfect electric walls. Find the mode impedance $-E_y/H_x$ and the power flow in terms of this impedance and the maximum value of H_x in the cross section. Repeat for *TM to x* modes noting that the mode admittance is equal to H_y/E_x.
2.6 Find the cutoff wavelengths of the first five lowest order modes in a circular cylindrical waveguide with perfectly conducting walls in terms of the perimeter $P \equiv 2\pi a$.
2.7 Show that the power carried by a propagating TE_{mn} mode in a circular cylindrical waveguide with a perfect electric wall and a radius a is given by the formula:

$$P_f = H_0^2(\omega\mu\beta\pi/2k_c^4)J_m^2(k_c a)\,[(k_c a)^2 - m^2]$$

where H_z has the expression:

$$H_z = H_0 J_m(k_c\rho)\ \sin m\phi\ \exp(-\mathrm{i}\beta z)$$

Hint: Use the recurrence relation (A2.18) for the Bessel functions and the following identity: $\int_0^x J_m^2(z) z\, dz \equiv \frac{1}{2}x^2[J_m^2(x) - J_{m-1}(x)J_{m+1}(x)]$. Use also the modal equation $J_m'(k_c a) = 0$ for TE_{mn} modes.
2.8 Show that the power carried by a propagating TM_{mn} mode in a circular waveguide with a perfect electric wall and a radius a is given by:

$$P_f = E_0^2(\pi\omega\varepsilon\beta/2k_c^4)\,[k_c a J_m'(k_c a)]^2$$

where E_z has the expression:

$$E_z = E_0 J_m(k_c\rho)\ \cos m\phi\ \exp(-\mathrm{i}\beta z)$$

Use the hint of problem 2.7 but note that $J_m(k_c a) = 0$ for TM_{mn} modes.
2.9 Derive an explicit expression for the attenuation rates of *TE* and *TM* modes in a circular waveguide with a slightly imperfect electric wall. (Use the results of problems 2.7 and 2.8.)
2.10 Show that the following radial wave functions in elliptical coordinates can be approximated by:

$$R_{e1}^{(1)}(q, u) \simeq A_1 J_1(2q^{1/2} \cosh u) - A_3 J_3(2q^{1/2} \cosh u)$$
$$R_{01}^{(1)}(q, u) \simeq B_1 J_1(2q^{1/2} \sinh u) - A_3 J_3(2q^{1/2} \sinh u)$$

for small q, where $A_3/A_1 = B_3/B_1 \simeq -q/8$.
2.11 Find the cutoff frequencies of the three lowest order modes in an elliptical

waveguide with perfect electric wall in terms of v/a, where v is the unbounded wave velocity and a is the length of the semimajor axis, under the following conditions:
(*i*) $b/a = 0{\cdot}98$, (*ii*) $b/a = 0{\cdot}707$, (*iii*) $b/a = 0{\cdot}5$
Hint: Use Fig. 2.2 or eqns. (2.21) and the following equation for the cutoff q of TM_{c01} *mode* (Kretzschmar, 1970):

$$q_{c01} = -0{\cdot}0016e + 1{\cdot}448e^2 - 0{\cdot}314e^3 + 1{\cdot}425e^4, \qquad 0{\cdot}05 \leqslant e \leqslant 0{\cdot}5$$
$$= -0{\cdot}222 - 0{\cdot}278e + 1{\cdot}308e^2 + 0{\cdot}341/(1-e), \quad 0{\cdot}5 \leqslant e \leqslant 0{\cdot}95$$

2.12 Show that the group velocity v_g and the phase velocity v_p are related by:

$$v_g = v_p^2/(v_p - \omega \; dv_p/d\omega)$$

Hence v_g is less than v_p only if v_p is a decreasing function of frequency.
2.13 Calculate the spread of time delay of a signal having a 20% bandwidth around a carrier frequency equal to ω_0 when it propagates for a distance of 150 m in a monomode waveguide with a mode cutoff frequency equal to $\omega_0/2$.
2.14 Consider a gaussian pulse $s(t) = \exp(-t^2/\tau^2)$ modulating a carrier of frequency ω_0 which is propagating for a distance z in a monomode waveguide. By using the first three terms in (2.32) show that the output pulse is spread out in time according to the form

$$\exp[-(t-t_d)/\tau_1^2] \text{ where } \tau_1^2 = \tau^2 + 2i(d^2\beta/d\omega^2)z$$

The reader is referred to Wait (1969a) and (1970a) for further discussion on pulse broadening and distortion in dispersive guides.
2.15 Derive relations (2.47) and (2.48) for the phase and attenuation rates in a radial waveguide. Derive also their limiting forms (2.49)–(2.52) under the conditions stated in the text.
2.16 An atmospheric surface duct is composed of a highly conducting earth and an atmospheric refractive index profile given by: $n^2(z) = n_0^2 - Kz$, where z is the height above the earth, K is a constant (m^{-1}), and $n = (\varepsilon/\varepsilon_0)^{1/2}$. Considering a *TE* wave having E solely in the horizontal y direction and no field variation with y, show that valid solutions of E_y are given by: $E_y = E_0 Ai(t) \exp(-i\beta x)$, where

$$t = -(k_0/K)^{2/3}(Kz + n_0^2 - \beta^2/k_0^2), \; k_0 = \omega(\mu_0\varepsilon_0)^{1/2},$$

and Ai is the Airy function. By applying the boundary condition at the earth's surface, show that the modal values of β are:

$$(\beta_m/k_0)^2 = t_m(k_0/K)^{-2/3} + n_0^2$$

$m = 1, 2, \ldots$ and t_m are the roots of $Ai(t) = 0$.
Repeat for *TM* modes showing that the above equation applies if t_m are the roots of $Ai'(t) = 0$.
2.17 Starting from Maxwell's equations in lossless media, prove the following reciprocity relationship between two source free electromagnetic field distributions (E, H) and (E′, H′)

$$\nabla \cdot (\mathbf{E} \times \mathbf{H}'^* - \mathbf{E}'^* \times \mathbf{H}) = 0$$

where * denotes the complex conjugate operation. Now, by taking (E, H) =

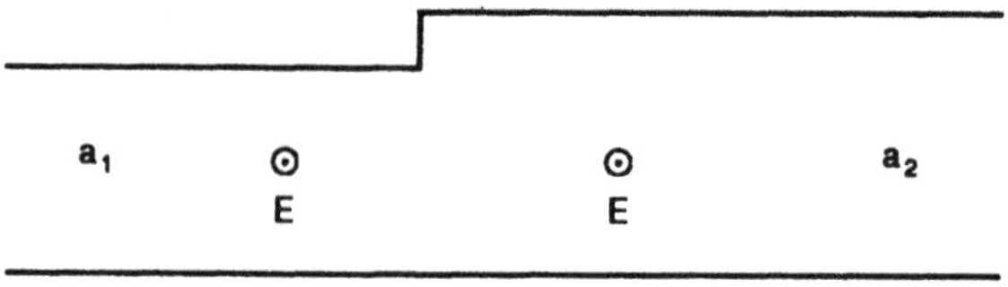

Fig. pr. 2.18 A junction between two rectangular waveguides with the same b dimension

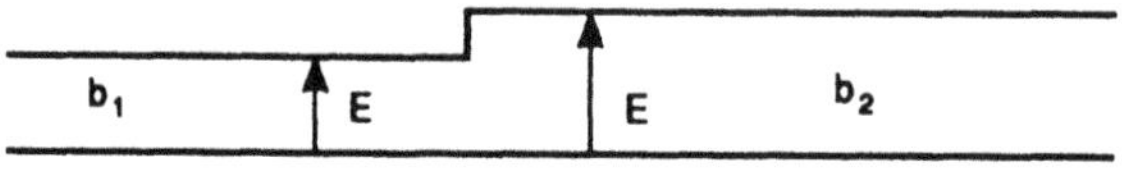

Fig. pr. 2.19 A junction between two rectangular waveguides with the same a dimension.

$(\mathrm{E}_n, \mathrm{H}_n)$ and $(\mathrm{E}', \mathrm{H}') = (\mathrm{E}_m, \mathrm{H}_m)$, the fields of the nth and mth modes in a cylindrical guide with perfect electric or magnetic walls, derive the orthogonality relationship (2.72).

2.18 Consider a junction between two rectangular waveguides with different a dimensions but the same b dimension as shown in Fig. pr.2.18. Assuming that the TE_{10} mode is incident from the left waveguide and neglecting reflections in this guide, find the relative amplitudes of the transmitted modes in the other guide. (Note that the transmitted modes are of TE_{m0} type only.)

2.19 Reconsider the junction in problem 2.18, but now take a to be the same on both sides of the junction and b to be different (see Fig. pr.2.19). With the TE_{10} mode incident from the left, find the relative amplitudes of the transmitted modes to the right. Note that the transmitted modes can be of the types TE_{1n} and TM_{1n} (to z). Alternately they are of the type TE_{1n} to x only.

2.20 Consider the excitation of a rectangular waveguide by a y directed electric current probe of length $l < b$ inserted from the bottom broad wall at $x = x_0$. Find the power delivered to the TE_{10} in terms of the probe current. Assuming that the TE_{10} is the only propagating mode, show that the input resistance seen by the probe is given by:

$$R_{in}/\eta_0 = (k_0/\beta_{10})\,(l^2/4ab)\,\sin^2(\pi x_0/a)$$

Find also the amplitudes of excitation of the evanescent modes, and hence the probe input reactance.

2.21 A circular cylindrical waveguide is excited by a small current loop of area dA centered at the point $(\rho_s, \phi_s, 0)$ and oriented normal to the direction of local transverse magnetic field of the TE_{11} mode. Find the amplitude of excitation of the dominant TE_{11} mode. Assuming that this is the only propagating mode, find the input resistance seen by the current source feeding the probe. (The magnetic moment of an electrically small loop is equal to $i\omega\mu_0 I\, dA$.)

2.22 Reconsider the junction of Fig.pr.2.18. Assuming unit incident power in the TE_{10} mode from the left, find the reflection and transmission coefficients of the two lowest order modes on the left and right sides respectively of the junction. Take $k_0 a = 1{\cdot}5\,\pi$, $a'/a = 1{\cdot}2$ and $b = b' = a/2$.

2.23 Show that the inner products $\langle e_n, h'_m \rangle$ and $\langle e'_m, h_n \rangle$ for the junctions of Figs. pr.2.18 and pr.2.19 satisfy the special relation (2.102).
2.24 Prove the results given by (2.119)–(2.121).
2.25 Write a computer code to compute $T_{m,1}$ in equation (2.122) for the following input data: the frequency f in kHz, the ionospheric day height h_d in km and the ionospheric night height h_n in km. Find $T_{m,1}$ for the propagating transmitted modes for the following set of data: $f = 15$, $h_d = 65$, $h_n = 85$.

2.14 References

ABELE, T.A., ALSBERG, D.A., and HUTCHISON, P.T. (1975): 'A higher capacity digital communication system using TE_{01} transmission in circular waveguide', *IEEE Trans.* **MTT–23,** pp. 326–333

ABRAMOWITZ, M., and STEGUN, I.A. (1965): 'Handbook of mathematical functions' (Dover Publications, New York)

BHARTIA, P., and BAHL, I.J. (1984): 'Millimeter wave engineering and applicatons' (Wiley-Interscience Publications, NY) Chapters 5, 6

BOYD, R.J., *et al.* (1977): 'Waveguide design and fabrication', *Bell Syst. Tech. J.*, **56,** pp. 1873–1897

BRESSELER, A.D., JOSHI, G.H., and MARCUVITZ, N. (1958): 'Orthogonality properties for modes in passive and active uniform waveguides', *J. Apply. Phys.*, **29,** pp. 794–799

CARLIN, J.W., and D'AUGUSTINO, P. (1973): 'Normal modes in overmoded dielectric lined circular waveguide', *Bell. Syst. Tech. J.*, **52,** pp. 453–456

CARLIN, J.W., and MOORTHY, S.C. (1977): 'TE_{01} transmission in waveguide with axial curvature', *Bell Syst. Tech. J.*, **56,** pp. 1849–1872

CLARRICOATS, P.J.B., and SLINN, R. (1967): 'Numerical solutions of waveguide discontinuity problem', *Proc. IEE*, **114,** pp. 878, 886

COCHRAN, J.A., and PECINA, R.G. (1966): 'Mode propagation in continuously curved waveguide', *Radio Science*, **1** (new series), pp. 679–696

COLLIN, R.E. (1966): 'Foundations for microwave engineering' (McGraw-Hill)

DRAGONE, C. (1977): 'Reflection, transmission and mode conversion in a corrugated feed', *Bell Syst. Tech.*, **56,** pp. 835–867

DRAGONE, C. (1984): 'Scattering at a junction of two waveguides with different surface impedances', *IEEE Trans. Microwave Theory Tech.*, vol. **MTT–32**, pp. 1319–1327

FALCIASECCA, G., and ROGAI, S. (1977): 'Random discrete imperfections in millimeter waveguide systems', *IEEE Trans.* **MTT–25,** pp. 911–915

HARRINGTON, R.F. (1961): 'Time harmonic electromagnetic fields' (McGraw-Hill, New York, Toronto, London)

KINZER, J.P., and WILSON, I.G. (1947): 'Some results on cylindrical cavity resonators', *Bell Syst. Tech. J.*, **26,** pp. 423–431

KRETZSCHMAR, J.G. (1970): 'Wave propagation in hollow conducting elliptical waveguides', *IEEE Trans.*, **MTT–18,** pp. 547–554

LEWIN, L. (1955): 'Propagation in curved and twisted waveguides of rectangular cross-section', *Proc. IEE*, Pt. B, **102,** pp. 75–80

MARCUSE, D. [1958]: 'Attenuation of the TE_{01} wave within the curved helix waveguide', *Bell Syst. Tech. J.*, **37,** p. 1649

MITTRA, R. and LEE, S.W. (1971): 'Analytical techniques in the theory of guided waves', Macmillan series in electrical science (Macmillan Co., New York, London)

MORGAN, S.P., and YOUNG, J.A. (1956): 'Helix waveguides', *Bell Syst. Tech. J.*, **35,** pp. 1347–1384

NAGELBERG, E.R., and SHEFER, J. (1965): 'Mode conversion in circular waveguides', *Bell Syst. Tech. J.*, **44,** 1321–1339

QUINN, J.P. (1965): '*E* and *H* plane bands for high power oversized rectangular waveguide', *IEEE Trans.*, **MTT–13,** pp. 54–63

STRATTON, J.A. (1941): 'Electromagnetic theory' (McGraw-Hill)

UNGER, H.G. (1961): 'Normal modes and mode conversion in helix waveguides', *Bell Syst. Tech. J.*, **40,** pp. 255–280

WAIT, J.R. (1960): 'On the excitation of electromagnetic surface waves on a curved surface', *IRE Trans.*, **AP–8,** pp. 445–449

WAIT, J.R. (1967): 'Electromagnetic whispering gallery modes in a dielectric rod', *Radio Science,* **2,** pp. 1005–1117

WAIT, J.R. (1968): 'The whispering gallery nature of the earth ionosphere waveguide at VLF', *IEEE Trans.* **AP–16**, 147

WAIT, J.R. (1969*a*): 'On the optimum bandwidth for propagated pulsed signals', *Proc. IEEE,* **57,** no. 10, pp. 1784–1785

WAIT, J.R. (1969*b*): 'Electromagnetic fields of sources in lossy media', *in* COLLIN and ZUCKER (Eds.): 'Antenna theory' (McGraw-Hill) pp. 438–513

WAIT, J.R. (1970*a*): 'Distortion of pulsed signals when the group delay is a nonlinear function of frequency', *Proc. IEEE,* **58,** pp. 1292–1294

WAIT, J.R. (1970*b*): 'Electromagnetic waves in stratified media', Pergamon Press.

WAIT, J.R. (1982): 'Geo-electromagnetism' (Academic Press) Chap. 7

WALDRON, R.A. (1957): 'Theory of the helical waveguide of rectangular cross section', *J. Brit. Inst. Radio Engrs.,* **17,** pp. 577–592

Additional references

KRETZSCHMAR, J.G. (1972): 'Attenuation characteristics of hollow conducting elliptical waveguides', *IEEE Trans.* **MTT–20,** pp. 280–284

LEWIN, L., and AL-HARIRI (1974): 'The effect of cross section curvature on attenuation in elliptic waveguides and a basic correction to previous formulas', *IEEE Trans.,* **MTT–22,** pp. 504–509

LEWIN, L., CHANG, D.C., and KUESTER, E.F. (1977): 'Electromagnetic waves and curved structures', *IEE* Electromagnetic wave series 2 (Peter Peregrinus, England)

Chapter 3

Waveguides with constant impedance walls

3.1 Introduction

In the previous chapter waveguides with perfectly reflecting walls have been studied. A perfectly reflecting wall has been defined as one having either a zero impedance (a perfect electric wall) or a zero admittance (a perfect magnetic wall). Strictly speaking, such a wall does not exist and some reflection loss, however small, is necessarily present. Power lost upon reflection from a wall is consumed either in the form of heat due to ohmic losses or in the form of radiation in the outer medium of the waveguide. A physically realisable wall will therefore have a finite surface resistance or conductance. In addition a waveguide wall will generally store magnetic or electric energy which can be represented by a finite surface reactance or susceptance. Some surfaces are intentionally made highly reactive, such as corrugated metal surfaces, in order to guide surface waves as will be described later in this chapter. Thus a waveguide wall is generally characterised by a complex surface impedance or admittance which relates the electric and magnetic fields tangential to the wall. An anisotropic wall, as in most cases of interest, has a tensor, rather than a scalar, surface impedance or admittance.

Now we introduce the concept of a *constant* surface impedance wall. The surface impedance of a wall is necessarily a function of the angle of incidence of a plane wave on the wall, or equivalently the tangential wavenumber to the wall. However, the surface impedance of a constant impedance wall is assumed to be independent of the tangential wavenumber of the incident wave. If this assumption is sufficiently accurate over the significant range of tangential wavenumbers, the use of the constant surface impedance concept will be well justified. In such cases, application of this concept will not only reduce much of the mathematical complications, but also lead to sufficiently accurate results. An example in which the concept is quite valid is a highly conducting metal wall. The reflection coefficient from such a wall is hardly dependent on the angle of incidence of a plane wave from the air region, for all real values of this angle as well as an extensive range of complex angles. Other notable examples include the earth's surface [e.g. Wait, 1970 and King and Wait, 1976] and many types of corrugated surfaces.

In the next section we consider a variety of cases in which a boundary interface between two regions can be effectively considered as a constant impedance surface. Corrugated surfaces and dielectric coated metallic surfaces will also be characterised as constant impedance surfaces. Mode characterisation in waveguides with constant impedance walls is treated in sections 3.3, 3.4 and 3.5. Mode orthogonality in waveguides with constant impedance walls is proved in section 3.6 and is followed by mode conversion and reflection at a longitudinal discontinuity.

3.2 Examples of constant impedance surfaces

Interfaces between media with distinct permittivity and conductivity constants can, under certain conditions, be considered as constant impedance surfaces. Examples of such interfaces with different geometries are shown in Fig. 3.1. Another class of possible constant impedance surfaces are corrugated and dielectric coated conducting surfaces. In this section, the impedance parameters of these surfaces will be derived along with conditions for their constancy with the tangential wavenumber.

3.2.1 The planar interface

A planar interface between an air region and a lossy dielectric region with permittivity ε, conductivity σ and permeability μ is shown in Fig. 3.1*a*. Let a single spectral plane wave be incident from the air region (ε_0, μ_0) and have longitudinal field components along z given by

$$(E_{zi}, H_{zi}) = (a_0, b_0)\exp(-i\beta z - i\lambda y - ux) \tag{3.1}$$

where

$$\beta^2 + \lambda^2 - u^2 = k_0^2 \equiv \omega^2\mu_0\varepsilon_0 \tag{3.2}$$

Reflected fields in the air will be set up and will have the form

$$(E_{zr}, H_{zr}) = (a_r, b_r)\exp(-i\beta z - i\lambda y + ux) \tag{3.3}$$

The form of the transmitted fields in the dielectric medium, is

$$(E_{zt}, H_{zt}) = (a_t, b_t)\exp(-i\beta z - i\lambda y - vx) \tag{3.4}$$

where

$$\beta^2 + \lambda^2 - v^2 = k^2 \equiv -i\omega\mu(\sigma + i\omega\varepsilon) \tag{3.5}$$

In the above equations a_0 and b_0 are arbitrary amplitudes of the incident *TM* to z and *TE* to z waves respectively. The corresponding reflected amplitudes are a_r and b_r and the transmitted amplitudes are a_t and b_t. The wavenumbers β and λ tangential to the interface surface are the same in both media in accordance with Snell's law so as to satisfy the boundary conditions all over the interface.

The other field components are obtainable from (1.27)—(1.28). In particular the transmitted field components E_{yt} and H_{yt} are given by:

$$E_{yt} = pE_{zt} + ZH_{zt} \tag{3.6}$$

$$H_{yt} = -YE_{zt} + pH_{zt} \tag{3.7}$$

where

$$Z = -i\omega\mu v/(k^2 - \beta^2) \tag{3.8}$$

$$Y = -(i\omega\varepsilon + \sigma)v(k^2 - \beta^2) \tag{3.9}$$

$$p = -\beta\lambda/(k^2 - \beta^2) \tag{3.10}$$

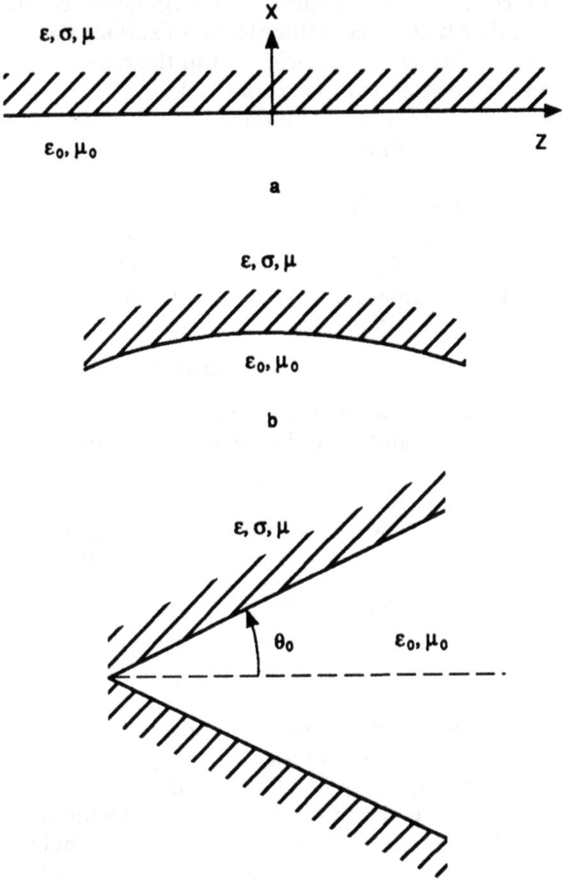

Fig. 3.1 Examples of constant impedance interface surfaces

a Planar interface
b Circular cylindrical interface
c Conical interface

Equations (3.6) and (3.7) are valid at all points in the half space $x \geqslant 0$, including the boundary interface $x=0$. Therefore they can also be considered as the boundary conditions to be satisfied by the fields in the other half space $x \leqslant 0$. Therefore, (3.6) and (3.7) may be applied to the interface $x=0$ and rewritten as:

$$E_y = pE_z + ZH_z \tag{3.11}$$

$$H_y = -YH_z + pH_z \tag{3.12}$$

Obviously Z is a surface impedance, Y a surface admittance and p a dimensionless factor. The three parameters fully characterise the interface surface $x=0$ as seen by the medium $x \leqslant 0$. In other words, if the half space $x \geqslant 0$

is source free it suffices to know Z, Y and p to determine the conditions imposed on the fields in the other half space. Other forms of surface parameters, such as the impedance tensor relating the electric field to the magnetic field tangential to the surface, can also be defined (see probl. 3.1).

So far, (3.11) and (3.12) are exact boundary conditions and the values of Z, Y and p are dependent on the tangential wavenumbers β and λ of the incident wave. They have to be used as such if exact solutions are to be found. However, great simplification is obtained if the surface parameters Z, Y and p can be assumed constant, in the sense that they are independent of the wavenumbers β and λ. In the following we find conditions under which this assumption is reasonably valid. A key parameter in (3.8)—(3.10) is the wavenumber v which may be written as:

$$v=(k_0^2-k^2+u^2)^{1/2} \text{ with Real } (v)>0.$$

which is a function of the tangential wavenumbers β and λ through its dependence on u. However, under the following condition:

$$|k^2-k_0^2| \gg |u^2=|\beta|^2+\lambda^2-k_0^2| \tag{3.13}$$

the wavenumber v can be considered constant; $v \simeq i(k^2-k_0^2)^{1/2}$. Furthermore if the product $\beta\lambda \ll |k^2-k_0^2|$, then Z, Y and p in (3.8)—(3.10) take the forms:

$$Z \simeq i\omega\mu/v \simeq \omega\mu/(k^2-k_0^2)^{1/2} \tag{3.14}$$

$$Y \simeq (i\omega\varepsilon+\sigma)/v \simeq (\omega\varepsilon-i\sigma)/(k^2-k_0^2)^{1/2} \tag{3.15}$$

$$|p|=|\lambda\beta/(k^2-k_0^2)| \ll 1 \tag{3.16}$$

These are the constant surface impedance and admittance forms for the planar interface between two media. One can identify two possible situations in which condition (3.13), and consequently (3.14)—(3.16), are valid.

(i) $|k| \gg |k_0|$, i.e. medium 1 is much more dense than medium 0

(ii) $(\beta^2+\lambda^2)^{1/2} \simeq k_0$, i.e. the wave in the air region is grazingly incident at the interface. The condition $\beta\lambda \ll |k^2-k_0^2|$ signifies that the fields should not be rapidly varying on the surface. Thus the constant impedance concept is not expected to work near sharp discontinuities along the surface such as sharp corners or bends, or sudden changes of surface parameters.

Now applying (3.14)—(3.16), the surface boundary conditions (3.11)—(3.12) reduce to

$$E_y=ZH_z \tag{3.17}$$

$$H_y=-YE_z \tag{3.18}$$

These two boundary conditions may be used to find the reflected and transmitted amplitudes in (3.3) and (3.4). To this end we write down the total tangential fields to the $x=0$ plane in the air (incident + reflected):

$$\left.\begin{aligned} E_z&=a_0+a_r \\ H_z&=b_0+b_r \\ E_y&=A(a_0+a_r)-B\eta_0(b_0-b_r) \\ H_y&=B\eta_0^{-1}(a_0-a_r)+A(b_0+b_r) \end{aligned}\right\} \tag{3.19}$$

where

$$A=-\beta\lambda/(k_0^2-\beta^2),\ B=ik_0u/(k_0^2-\beta^2) \text{ and } \eta_0=(\mu_0/\varepsilon_0)^{1/2}.$$

Substituting from (3.19) into (3.17)—(3.18) and performing some algebra, we arrive at the following relations between the reflected and incident wave amplitudes:

$$a_r=R_e a_0+\chi\eta_0 b_0 \tag{3.20}$$

$$b_r=-\chi\eta_0^{-1}a_0+R_h b_0 \tag{3.21}$$

where

$$R_e=\{[i\omega\varepsilon_0 u+Y(k_0^2-\beta^2)][i\omega\mu_0 u-Z(k_0^2-\beta^2)]-(\beta\lambda)^2\}/D \tag{3.22}$$

$$R_h=\{[i\omega\varepsilon_0 u-Y(k_0^2-\beta^2)][i\omega\mu_0 u+Z(k_0^2-\beta^2)]-(\beta\lambda)^2\}/D \tag{3.23}$$

$$\chi=-2ik_0\beta\lambda u/D \tag{3.24}$$

$$D=[i\omega\varepsilon_0 u-Y(k_0^2-\beta^2)][i\omega\mu_0 u-Z(k_0^2-\beta^2)]+(\beta\lambda)^2 \tag{3.25}$$

Equation (3.20)—(3.21) reveal that *TE* and *TM* to z waves are, in general, coupled by the interface of the two media. In other words, if none of the wavenumbers β, λ and u is zero, an incident *TE* to z wave produces a reflected *TE* and *TM* waves and vice-versa. On the other hand, if $\lambda=0$, then $\chi=0$ and the formulae for R_e and R_h reduce to:

$$R_e=(i\omega\varepsilon_0/u-Y)/(i\omega\varepsilon_0/u+Y) \tag{3.26}$$

$$R_h=(i\omega\mu_0/u-Z)/(i\omega\mu_0/u+Z) \tag{3.27}$$

which are the well known reflection coefficients for *TM* to z (and to x) and *TE* to z (and to x) waves respectively.

3.2.2 Cylindrical interface

Referring to Fig. 3.1*b*, it is required to find the surface impedance and admittance reflected by the outer medium (1) to the inner cylindrical region (0). An appropriate expression for the total longitudinal fields in the inner region is:

$$(E_{z0},\ H_{z0})=(a_0,\ b_0)I_m(u\rho)\exp(-i\beta z-im\phi) \tag{3.28}$$

where $I_m(.)$ is the modified Bessel function of first kind and order m and

$$\beta^2-u^2=k_0^2\equiv\omega^2\mu_0\varepsilon_0$$

The corresponding field components in the outer region $\rho\geqslant a$ should have the form:

$$(E_{z1},\ H_{z1})=(a_t,\ b_t)K_m(v\rho)\exp(-i\beta z-im\phi) \tag{3.29}$$

where

$$\beta^2-v^2=k^2\equiv-i\omega\mu(\sigma+i\omega\varepsilon)$$

$K_m(.)$ is the modified Bessel function of the second kind and is chosen to ensure field decay as $\rho\Rightarrow\infty$. The constants $(a_0,\ b_0)$ in (3.28) are *TM* and *TE* wave amplitudes respectively, and (a_t, b_t) are the corresponding transmitted amplitudes. Now using (1.27)—(1.28) to derive the transmitted field components E_ϕ and H_ϕ at the interface $\rho = a$, we get:

$$E_\phi = pE_z + ZH_z \tag{3.30}$$

$$H_\phi = -YE_z + pH_z \tag{3.31}$$

where

$$Z = (-i\omega\mu/v)K'_m(va)/K_m(va) \tag{3.32}$$

$$Y = -[(i\omega\varepsilon + \sigma)/v]K'_m(va)/K_m(va) \tag{3.33}$$

$$p = \beta m/v^2 a \tag{3.34}$$

and the prime means differentiation with respect to the argument.

As in the case of a planar interface the wavenumber v can be considered constant under the condition:

$$|k^2 - k_0^2| \gg |u^2|$$

which requires that either (i) $|k| \gg |k_0|$, β or (ii) $\beta \simeq k_0$ (grazing incidence). Under this condition v is approximated by $v \simeq i\ (k^2 - k_0^2)^{1/2}$, leading to surface impedance and admittance independent of β. However they are not exactly constant because of their dependance on the azimuthal wavenumber m, as is clear from (3.32)—(3.33). Nevertheless, for low values of m and large values of $|va|$ relative to unity, the ratio K'_m/K_m is well approximated by (see Appendix 2):

$$K'_m(va)/K_m(va) \simeq -(1 + 1/2va)$$

and Z and Y take the constant values:

$$Z = [\omega\mu/(k^2 - k_0^2)^{1/2}](1 + 1/2i(k^2 - k_0^2)^{1/2}a) \tag{3.35}$$

$$Y = [(\omega\varepsilon - i\sigma)/(k^2 - k_0^2)^{1/2}](1 + 1/2i(k^2 - k_0^2)^{1/2}a) \tag{3.36}$$

On comparison with (3.14) and (3.15) for the planar surface, it is clear that these are the limiting forms of (3.35) and (3.36) when a tends to infinity as expected. It should be emphasised that, under the same conditions leading to (3.35)—(3.36), the parameter p is $\ll 1$.

3.2.3 Conical interface

Finally we consider a conical interface separating two different media. The geometry is given in Fig. 3.1*c*. The vector fields E and H in the outer region; $\theta > \theta_0$ and $r > r_0$, can be written in terms of the spherical wave functions given in section 1.4. We can choose:

$$\mathrm{E} = A\mathrm{n}_{emn} + B\mathrm{m}_{omn} \tag{3.37}$$

$$-i\eta\mathrm{H} = A\mathrm{m}_{emn} + B\mathrm{n}_{omn} \tag{3.38}$$

where

$$\eta = \omega\mu/k,\ k^2 = -i\omega\mu(i\omega\varepsilon + \sigma)$$

Using the explicit expressions (1.139)—(1.140) for the m and n vector functions, we deduce the relations:

$$E_\phi = pE_r + ZH_r \tag{3.39}$$

$$H_\phi = -YE_r + pH_r \tag{3.40}$$

which are valid everywhere in the outer region, including the interface surface $\theta = \theta_0$. The parameters Z, Y and p are:

$$Z = \eta(ikr/v(v+1))\,[\partial P_v^m(\cos\theta)/\partial\theta]/P_v^m(\cos\theta) \tag{3.41}$$

$$Y = \eta^{-1}(ikr/v(v+1))\,[\partial P_v^m(\cos\theta)/\partial\theta]/|P_v^m(\cos\theta) \tag{3.42}$$

$$p = (-m/\sin\theta)\,(ikr/v(v+1))\,(\hat{H}_v'^{(2)}(kr)/\hat{H}_v^{(2)}(kr)] \tag{3.43}$$

Fields in the inner region $\theta < \theta_0$ have similar forms to (3.37)—(3.38) with k and v replaced by k_0 and v_0. Now, in matching the tangential fields at the boundary $\theta = \theta_0$ we encounter a basic difficulty; the field dependence on r is not the same on both sides of the boundary, being $\hat{H}_{v0}^{(2)}(k_0 r)$ on one side and $\hat{H}_v^{(2)}(kr)$ on the other. To deal rigorously with this difficulty one should use a summation over v in the outer region for every v_0 in the inner region. This approach has been used by Yeh (1964) and discussed by Wait (1969) in a similar situation. However, since we are only looking for approximate boundary conditions, we shall use, in what follows, a rather simplified approach. First, we define a radial (complex) phase constant in the inner and outer regions; namely:

$$\beta_0(r) = ik_0 \hat{H}_{v0}'^{(2)}(k_0)/\hat{H}_{v0}^{(2)}(k_0 r)$$

for the inner region $\theta \leqslant \theta_0$, and

$$\beta(r) = ik\hat{H}_v'^{(2)}(kr)/\hat{H}_v^{(2)}(kr)$$

for the outer region $\theta \leqslant \theta_0$. The boundary conditions at $\theta = \theta_0$ require that $\beta_0(r) = \beta(r)$ at all values of r. Now in order to simplify the analysis we require that this equality holds only for large values of r, $k_0 r \gg 1$, whence upon using the result in problem 1.15

$$\beta_0(r) \simeq k_0[1 - v_0(v_0+1)/2k_0^2 r^2]^{1/2} \tag{3.44}$$

In the outer medium we argue that under the assumption that $|k| \gg |k_0|$, v is expected to be of the same order as $|kr|$, with both being much greater than unity. In this case $\hat{H}_v^{(2)}(kr) \equiv (\pi kr/2)^{1/2} H_{v+1/2}^{(2)}(kr)$ is well approximated by the Airy function $w_1(t)$, [see (A2.52)]; hence:

$$\beta(r) \simeq -ik(2/kr)^{1/3} w_1'(t)/w_1(t)$$

where $t = (kr/2)^{2/3}[(v+1/2)^2/(kr)^2 - 1]$. For $|t| \gg 1$, $\beta(r)$ is further reduced to [see (A2.48)]:

$$\beta(r) \simeq k[1 - (v+1/2)^2/(kr)^2]^{1/2} \tag{3.45}$$

On equating (3.44) with (3.45), we get

$$v + 1/2 = [(k^2 - k_0^2)r^2 - v_0(v_0+1)/2]^{1/2} \simeq (k^2 - k_0^2)^{1/2} r \tag{3.46}$$

which shows that v is readily of the order of $|kr|$ as we have previously anticipated. Next, the Legendre function appearing in (3.41)—(3.42) can be approximated, for large v, by (e.g. Abramowitz and Stegun, 1965)

$$P_v^m(\cos\theta) \simeq \frac{\Gamma(v+m+1)}{\Gamma(v+3/2)}\left(\frac{\pi}{2}\sin\theta\right)^{-1/2} \cos\left[\left(v+\frac{1}{2}\right)\theta + \left(m-\frac{1}{2}\right)\pi/2\right]$$

hence

$$\frac{\partial P_v^m(\cos\theta_0)/\partial\theta_0}{P_v^m(\cos\theta_0)} \simeq -i(v+1/2) - 1/(2\sin\theta_0) \tag{3.47}$$

where it has been assumed that $|\text{Imag.}(v\theta_0)| \gg 1$ and $\theta_0 \ll 1$. Substituting from (3.46)—(3.47) in (3.41)—(3.42) we finally get the approximate constant surface impedance and admittance as:

$$Z \simeq \frac{\omega\mu_0}{(k^2-k_0^2)^{1/2}}\left(1 - \frac{i}{2(k^2-k_0^2)^{1/2} r \sin\theta_0}\right) \tag{3.48}$$

$$Y \simeq \frac{\omega\varepsilon - i\sigma}{(k^2-k_0^2)^{1/2}}\left(1 - \frac{i}{2(k^2-k_0^2)^{1/2} r \sin\theta_0}\right) \tag{3.49}$$

Comparing these with the corresponding expressions for the cylindrical interface, we find that they are identical provided that $r \sin\theta_0$ replaces the cylindrical radius a. This should not be surprising if we recall that (3.48)—(3.49) are derived by using the approximations $kr \gg 1$ and $\theta_0 \ll 1$; hence the conical surface approaches the cylindrical surface. It is worth noting also that the parameter p in (3.43) approaches the value $[-m\beta_0/[(k^2-k_0^2) r \sin\theta_0]$ in accordance again with the corresponding value of p for the cylindrical interface.

3.2.4 Corrugated surfaces

An important class of constant impedance surfaces is the class of corrugated surfaces. Two examples of cylindrical corrugated surfaces are shown in Fig. 3.2*a*, *b*. These are the transversely corrugated and longitudinally corrugated surfaces. Waveguides having these types of surfaces have been designed as low crosspolar waveguiding structures as will be discussed in detail in chapter 5. The planar corrugated surfaces can be treated as special cases of cylindrical surfaces with infinitely large radii.

First consider the transversely corrugated cylindrical surface of Fig. 3.2*a*. Usually the surface has many corrugations per wavelength. Therefore the fields within any one slot can be assumed uniform along the longitudinal direction z, although they can vary from one slot to the next according to the longitudinal wavenumber in the guide. The slot can only support TM to z fields with field components E_z, H_ϕ and H_ρ within the slot. The ridges will present a zero impedance to the field component E_ϕ, while the circumferential currents flowing in the ridges can support a finite magnetic field H_z at $\rho = a$. Therefore the impedance Z presented by the corrugations at the surface $\rho = a$ is:

$$Z(\rho = a) = E_\phi / H_z = 0 \tag{3.50}$$

The admittance Y at $\rho = a$ is determined from the fields inside the slots; namely;

$$Y(\rho = a) = -H_\phi / E_z = -(\partial E_z/\partial\rho)/i\omega\mu_0 E_z|_{\rho=a}$$

An appropriate expression for E_z which must vanish at the wall $\rho = b$ within a slot and is independent of z is

$$E_z(\rho, \phi) = [J_m(k_0\rho)N_m(k_0b) - J_m(k_0b)N_m(k_0\rho)] \cos m\phi \tag{3.51}$$

Hence the admittance Y at $\rho = a$ is:

$$Y = i\eta_0^{-1} \frac{J'_m(k_0a)N_m(k_0b) - J_m(k_0b)N'_m(k_0a)}{J_m(k_0a)N_m(k_0b) - J_m(k_0b)N_m(k_0a)} \tag{3.52}$$

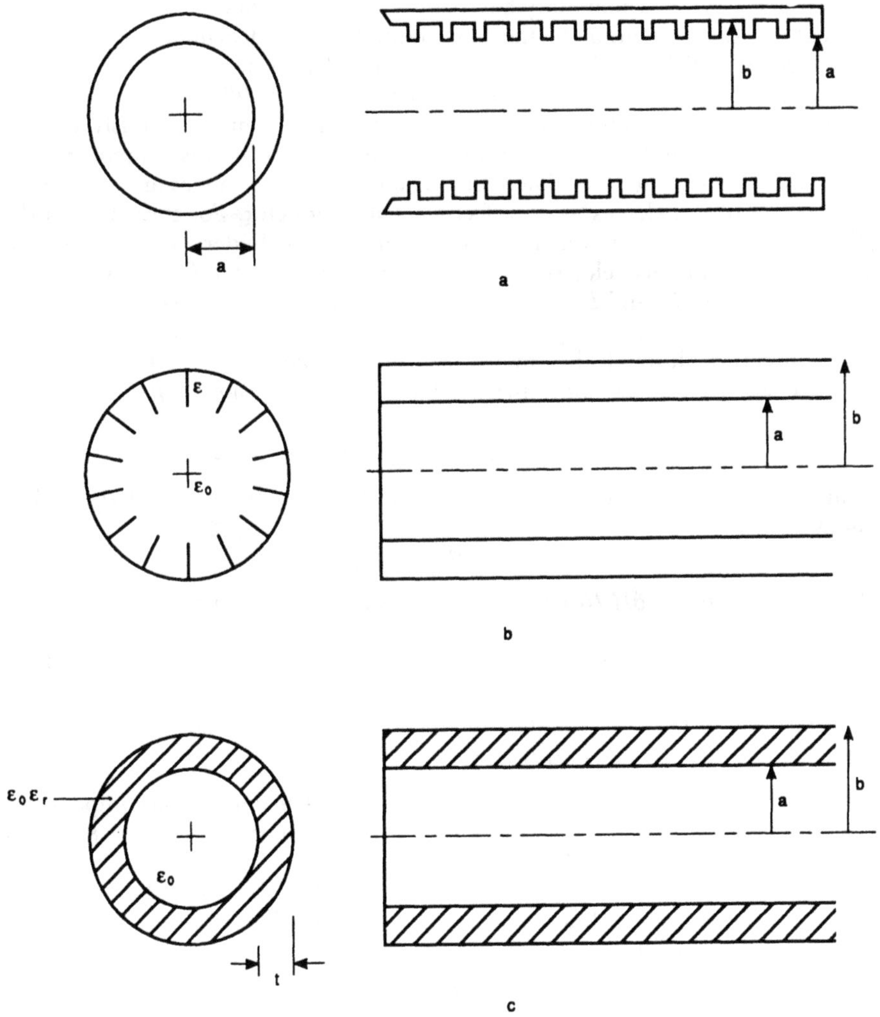

Fig. 3.2 Constant impedance cyindrical surfaces

a Transversely corrugated surface
b Longitudinally corrugated surface
c Dielectric coated surface

Except for the dependence on the azimuthal wavenumber m, the surface admittance can be considered constant. However, for small m and $k_0a \gg 1$, the above expression is reduced to

$$Y \simeq -i\eta_0^{-1}(\cot k_0(b-a) + 1/2k_0a) \tag{3.53}$$

where the large argument approximation for the Bessel functions have been utilized [see (A2.8)—(A2.11)]. Now expression (3.53) readily represents a constant surface admittance. This expression, along with (3.50), characterise the surface $\rho = a$ as a boundary surface to the inner region of the guide.

Next we turn attention to the longitudinally slotted cylindrical surface of Fig. 3.2*b*. It will be evident shortly that, in order to have an approximately constant impedance surface, the slotted region must be filled with a dielectric, i.e. $\varepsilon > \varepsilon_0$. The longitudinal electric field at the ridges must obviously vanish, and for a large number of slots around the circumference, this component will also vanish within the slots. Therefore, as far as TM to z modes are concerned, the surface $\rho = a$ behaves as a perfect electric conductor. For these modes the slots have no effect on the inner fields and the cylinder acts as a closed waveguide with perfect electric walls. On the other hand, for TE to z modes, the slots support H_z, E_ϕ and E_ρ fields and will present a finite surface impedance at the surface $\rho = a$. Within any one slot the fields may be assumed uniform along the ϕ direction although they may vary from one slot to the next according to the field variation in the inner region. Thus, inside any slot $a \leqslant \rho \leqslant b$, we write for H_z:

$$H_z(\rho, z) = [J_0(v\rho)N_1(vb) - J_1(vb)N_0(v\rho)] \exp(-i\beta z) \tag{3.54}$$

which readily satisfies the boundary condition $E_\phi \propto \partial H_z/\partial\rho = 0$ at $\rho = b$. The radial wavenumber v is governed by:

$$v^2 + \beta^2 = k_0^2\varepsilon/\varepsilon_0 \tag{3.55}$$

Since $E_\phi = (i\omega\mu_0/v^2)\partial H_z/\partial\rho$, we derive the surface impedance at $\rho = a$ as:

$$Z = (E_\phi/H_z)_{\rho=a} = \frac{-i\omega\mu_0}{v} \frac{J_1(va)N_1(vb) - J_1(vb)N_1(va)}{J_0(va)N_1(vb) - J_1(vb)N_0(va)} \tag{3.56}$$

In order that Z be approximately independent of β, we impose the condition:

$$k_0(\varepsilon/\varepsilon_0 - 1)^{1/2} \gg (k_0^2 - \beta^2)^{1/2}$$

which renders $v \simeq k_0(\varepsilon/\varepsilon_0 - 1)^{1/2}$, and Z in (3.56) a constant surface impedance. The above condition implies that either (1)$\varepsilon \gg \varepsilon_0$ or (ii) $\beta \simeq k_0$ and ε is only moderately higher than ε_0.

The case of a planar corrugated surface is obtainable from the cylindrical surface by taking the limit a tends to infinity. Therefore (3.53) for the transversely corrugated surface becomes:

$$Y = -i\eta_0^{-1} \cot k_0d \tag{3.57}$$

where d is the slot depth. For the longitudinally corrugated surface (i.e. the propagation occurs along the slot length) equation (3.56) yields:

$$Z = [i\eta_0/(\varepsilon/\varepsilon_0 - 1)^{1/2}] \tan[k_0d(\varepsilon/\varepsilon_0 - 1)^{1/2}] \tag{3.58}$$

The above two equations ae the expected results based on simple transmission line theory.

3.2.5 Dielectric lined surfaces

A reactive surface can be realised by coating a metallic wall with a dielectric layer. To show this, consider a cylindrical metallic wall of radius b which is assumed, for simplicity, to be a perfect electric conductor. The wall is coated on the inside by a dielectric layer of relative permittivity ε_r and uniform thickness t, as depicted in Fig. 3.2*c*. We wish to find the effective surface impedance Z and admittance Y as seen at the dielectric-air interface $\rho = a$. To this end, we write the following for the longitudinal field components in the region $a \leqslant \rho \leqslant b$:

$$E_z = A[J_m(v\rho)N_m(vb) - J_m(vb)N_m(v\rho)] \equiv Af(v\rho) \quad (3.59)$$

$$H_z = B[J_m(v\rho)N'_m(vb) - J'_m(vb)N_m(v\rho)] \equiv Bg(v\rho) \quad (3.60)$$

apart from a common factor $\exp(-im\phi - i\beta z)$, and

$$\beta^2 + v^2 = k_0^2\varepsilon_r$$

The above expressions readily satisfy the boundary conditions at the perfect electric wall $\rho = b$. Now, using (1.27)—(1.28) for the transverse components at $\rho = a$, we get

$$E_\phi = pE_z + ZH_z$$

$$H_\phi = -YE_z + pH_z$$

where

$$Z = (i\omega\mu_0/v)\,[g'(va)/g(va)] \quad (3.61)$$

$$Y = (i\omega\varepsilon_0\varepsilon_r/v)\,[f'(va)/f(va)] \quad (3.62)$$

$$p = -\beta m/v^2 a \quad (3.63)$$

where the functions f and g are defined by (3.59)—(3.60) and the prime stands for differentiation with respect to argument.

As long as $v \equiv (k_0^2\varepsilon_r - \beta^2)^{1/2} \gg (k_0^2 - \beta^2)^{1/2}$, we can make the approximation $v \simeq k_0(\varepsilon_r - 1)^{1/2}$, which renders Z and Y in the above equations to be constants with respect to β. However, they remain dependent on the azimuthal number m, but if m is small and va is much greater than unity, Z and Y are further reduced to

$$Z \simeq (i\omega\mu_0/v)\,\frac{\tan vt}{1 - \tan vt/2va} \quad (3.64)$$

$$Y \simeq (-i\omega\varepsilon_0\varepsilon_r/v)\,(\cot vt + 1/2va) \quad (3.65)$$

where t, the dielectric thickness, is assumed $\ll a$. Within the same approximations, the surface parameter p is $\ll 1$. Clearly Imag(Z) or Y in the above expressions can have any positive or negative values; hence, any capacitive or inductive surface can be realised. Furthermore, since Z and Y^{-1} are generally not equal, the air-dielectric interface is anisotropic. Finally we note that in the limit of infinite radius a, (3.64)—(3.65) yield the surface, impedance and admittance of a dielectric coated conducting plane.

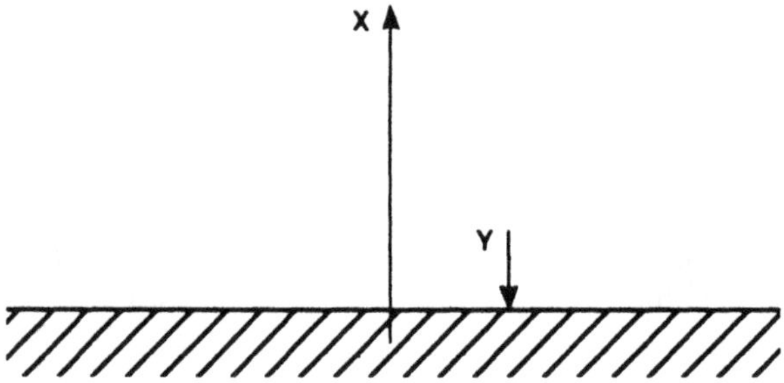

Fig. 3.3 A planar constant impedance surface

3.3 Guided modes on constant impedance plane surfaces

Now we study the main properties of guided modes on structures that are characterised by constant impedance walls. The structures chosen for our study in this section are the single planar constant impedance surface and the two parallel plate guide. The circular cylindrical guide with a constant impedance wall is studied in section 3.4 and, in section 3.5, noncircular cylindrical guides are considered. The practical significance of every one of these guides will be clear during the course of study.

3.3.1 A single planar constant impedance surface

In the following we show that a planar surface with a finite reactive surface impedance or admittance will support a guided mode whose fields decay away from the surface. Such a mode is called a surface wave mode (e.g. Barlow and Brown, 1962). Consider first a TM to z mode on the planar surface of Fig. 3.3. In order to have a guided mode the fields must decay in the direction normal to the surface (x axis). Taking the fields to be uniform along the y axis we can write E_z in the form:

$$E_z = \exp(-ux)\exp(-i\beta z) \tag{3.66}$$

where we choose $\mathrm{Real}(u) > 0$ to have a proper guided mode. Meanwhile,

$$\beta^2 - u^2 = k_0^2$$

The other nonzero field components are E_x and H_y, where from (1.25)

$$H_y = (i\omega\varepsilon_0/u^2)\partial E_z/\partial x$$

The boundary condition to be satisfied at the constant admittance surface $x = 0$ is:

$$Y = H_y/E_z|_{x=0} = -i\omega\varepsilon_0/u$$

or

$$u = -i\omega\varepsilon_0/Y \tag{3.67}$$

At this point, we note that for $\mathrm{Real}(u)$ to be positive, we should have $\mathrm{Imag}(Y)$ negative, i.e., in order to have a guided TM surface wave mode, the surface should have an inductive impedance. Defining a normalised surface impedance by $\Delta \equiv \eta_0 Y^{-1}$, (3.67) takes the more convenient form (Wait, 1970):

$$u = -ik_0\Delta \tag{3.68}$$

The longitudinal wavenumber β is then given by:

$$\beta = k_0(1-\Delta^2)^{1/2} \tag{3.69}$$

from which we deduce that if the surface is purely inductive, i.e. $\Delta = i|\Delta|$, then $\beta > k_0$ and the phase velocity is less than c. A surface wave of this type is therefore called a slow wave in contrast with modes in hollow guides which are fast modes. A surface wave mode can be realised on a corrugated metallic surface or a dielectric coated surface.

In the general case of a lossy surface Δ is complex, say:

$$\Delta = \delta_1 + i\delta_2$$

where both δ_1 and δ_2 are positive real. The longitudinal wavenumber is then also complex and given by:

$$\beta/k_0 = (1-(\delta_1+i\delta_2)^2)^{1/2} \simeq (1-\delta_1^2+\delta_2^2)^{1/2}(1-i\delta_1\delta_2/(1-\delta_1^2+\delta_2^2))$$

where $\delta_1\delta_2$ is assumed $\ll 1-\delta_1^2+\delta_2^2$. The phase velocity v and the longitudinal attenuation constant α are then given by:

$$v/c = k_0/\mathrm{Real}(\beta) \simeq (1-\delta_1^2+\delta_2^2)^{-1/2} \tag{3.70}$$

$$\alpha(\mathrm{neper/meter}) = k_0\delta_1\delta_2 v/c \tag{3.71}$$

It is seen from (3.70) that the surface mode can be fast or slow depending on whether $\delta_1 > \delta_2$ or $\delta_2 > \delta_1$ respectively.

A notable example of a fast *TM* surface wave is that guided by a homogeneous earth and known as the Zenneck wave (Wait, 1964, 1970). The earth's admittance is given by (3.15) yielding Δ as:

$$\Delta = k_0(k^2-k_0^2)^{1/2}/k^2 = (\varepsilon_r - 1 - i\sigma/\omega\varepsilon_0)^{1/2}/(\varepsilon_r - i\sigma/\omega\varepsilon_0) \tag{3.72}$$

where ε_r is the relative dielectric constant and σ is the conductivity of the earth. It can be verified that the Zenneck wave on a homogeneous earth is a fast wave (see problem 3.10); namely in the low frequency limit,

$$v/c = (1-x^2(\varepsilon_r+1)/2)^{-1/2} \tag{3.73}$$

$$\alpha/k_0 = (x/2)\,(1-x^2(\varepsilon_r+1)^2/4)v/c \tag{3.74}$$

where $x = \omega\varepsilon_0/\sigma \ll 1$.

In the high frequency limit $x \gg 1$, both v/c and α tend to constant values:

$$v/c = (1-(\varepsilon_r-1)/\varepsilon_r^2)^{-1/2} \tag{3.75}$$

$$\alpha = \sigma(\mu_0/\varepsilon_0)^{1/2}[(\varepsilon_r-2)/2\varepsilon_r^3]v/c \tag{3.76}$$

The velocity and attenuation of the Zenneck wave are given in Fig. 3.4a and 3.4b. Although the Zenneck wave enjoys a considerably low rate of attenuation along the earth's surface it is hard to excite as demonstrated by Wait and Hill (1979) and Hill and Wait (1980). This is due to the slow decay of the field in the direction normal to the surface; hence the requirement of a large aperture source for efficient excitation of this wave.

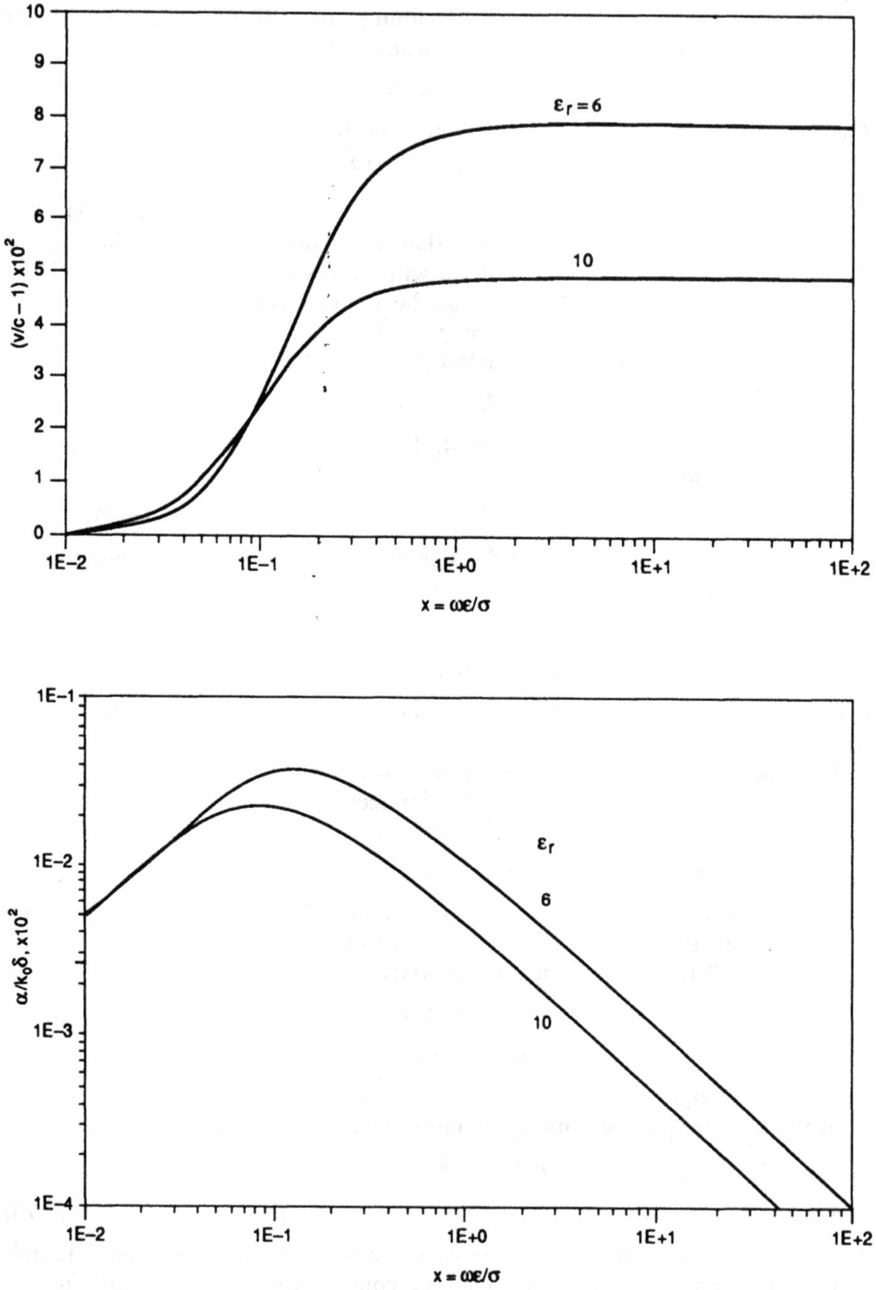

Fig. 3.4 Propagation parameters of the Zenneck wave over earth versus $x = \omega\varepsilon/\sigma$

a PHase velocity
b Attenuation

A similar treatment of *TE* modes on planar surfaces shows that a *TE* surface wave will be supported by a capacitive surface. Namely if H_z is expressed by the form (3.66), then $u = -i\omega\mu_0/Z$ where $Z = -E_y/H_z$ is the surface impedance of the planar surface $x=0$. Defining a normalized surface admittance $\Delta = Z^{-1}\eta_0$, u is given by $-ik_0\Delta$ in accordance with (3.68). It then follows that equation (3.69) for β and the next equations for v/c and α are still valid.

3.3.2 Constant impedance parallel plate waveguide

A prototype for several guiding structures is a parallel plate waveguide with finite constant surface impedance walls. Examples include waveguides with corrugated or dielectric coated walls used as feeds for microwave horn antennas, and natural waveguides such as tunnels and the earth-ionosphere guide. A general theory for guided waves in a parallel plate waveguide with finite impedance has been introduced by Wait (1967b). In the following we review his work, but with a slight change of notation; namely, we derive the modal equation for the longitudinal wavenumbers, and identify the allowable modes. Next, we consider the excitation problem for the structure.

Considering *TM* to z waves and looking for the natural modes of the structure as depicted in Fig. 3.5., we write for E_z:

$$E_z = [\exp(ux) + A\exp(-ux)]\exp(-i\beta z) \tag{3.77}$$

where $\beta^2 - u^2 = k_0^2$. With $\partial/\partial y = 0$, the other nonzero field components are E_x and H_y. Using (1.24)—(1.25), we get:

$$H_y = (i\omega\varepsilon_0/u)(\exp(ux) - A\exp(-ux)]\exp(-i\beta z) \tag{3.78}$$

$$E_x = (\beta/\omega\varepsilon_0)H_y \tag{3.79}$$

The boundaries at $x=0$ and $x=d$ are characterised by constant surface admittances Y_1 and Y_2 respectively. The boundary conditions are then given as:

$$Y_1 = (H_y/E_z)_{x=0} = (i\omega\varepsilon_0/u)[(1-A)/(1+A)] \tag{3.80}$$

$$Y_2 = (-H_y/E_z)_{x=d} = (-i\omega\varepsilon_0/u)[(1-Ae^{-2ud})/(1+Ae^{-2ud})] \tag{3.81}$$

These two boundary conditions determine a discrete infinite set of eigenvalues for u and their associated A^s. Eliminating A between these two equations, one gets:

$$\left(\frac{ik_0 - u\hat{Y}_1}{ik_0 + u\hat{Y}_1}\right)\left(\frac{ik_0 - u\hat{Y}_2}{ik_0 + u\hat{Y}_2}\right)\exp(-2ud) = 1 \tag{3.82}$$

where $\hat{Y}_{1,2} = Y_{1,2}\eta_0$. Now one can identify two types of eigenmodes for the structure. The first type consists of two surface wave modes which are the perturbed versions of those supported individually by each of the two planar (inductive) surfaces. The other type is the infinite waveguide modes which correspond to the modes of a two parallel plate waveguide with perfectly conducting walls. In what follows we expose the main features of each of these two types of modes.

Surface wave modes

To identify the two surface wave modes of the structure, we take the limit d tending to ∞. Solutions of (3.82) are then $u = -ik_0/\hat{Y}_1$ and $u = -ik_0/\hat{Y}_2$ corresponding to the two surface wave modes of the isolated plates in

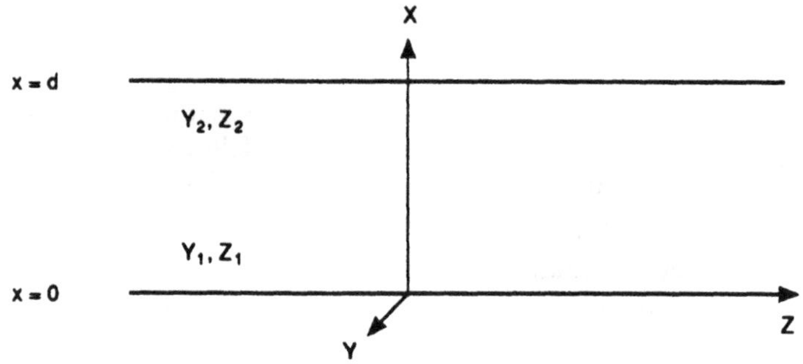

Fig. 3.5 Constant impedance (admittance) parallel plane waveguide

accordance with (3.67). Next, as d takes on finite values, the surface wave mode associated with each plate is perturbed by the proximity of the other plate. Using a simple perturbation analysis the perturbed eigenmodes are approximately given by:

$$u_{1,2} \simeq \frac{-ik_0}{\hat{Y}_{1,2}}\left[1 \pm 2\,\frac{\hat{Y}_2+\hat{Y}_1}{\hat{Y}_2-\hat{Y}_1}\exp(i2k_0d/\hat{Y}_{1,2})\right] \tag{3.83}$$

which are sufficiently accurate for large values of d provided that Y_1 is not close to Y_2, while for moderate values of d they can serve as initial solutions of (3.82) in any root finding algorithm. For the case of two identical plates $Y_1 = Y_2 = Y$, there are still two surface wave modes characterised by even and odd field patterns about the central axis of the guide. The modal equation (3.82) in this case reduces to:

$$\left(\frac{ik_0 - u\hat{Y}}{ik_0 + u\hat{Y}}\right)\exp(-ud) = \pm 1 \tag{3.84}$$

whose approximate solutions based on large d are

$$u \simeq (-ik_0/\hat{Y})\,(1 \pm 2\exp(ik_0d/\hat{Y}) \tag{3.85}$$

Waveguide modes

To solve for the waveguide modes it is more convenient to define $u = ik_0C$ and $\beta = k_0S$, with $C^2 + S^2 = 1$, whence (3.82) takes the form:

$$\left(\frac{1-C\hat{Y}_1}{1+C\hat{Y}_1}\right)\left(\frac{1-C\hat{Y}_2}{1+C\hat{Y}_2}\right)\exp(-2ik_0Cd) = \exp(-2in\pi) \tag{3.86}$$

where n is any positive integer $\geqslant 1$. In general this equation can only be solved numerically. However closed form solutions of sufficient accuracy can be obtained in few special cases of which we mention the following two cases.

Case 1: Both $|\hat{Y}_1^{-1}|$ and $|\hat{Y}_2^{-1}| \leqslant |C|$, corresponding to well conducting plates. In this case the first two terms on the LHS of (3.86) may be approximated by exponential functions so that:

$$\exp(-2/C\hat{Y}_1 - 2/C\hat{Y}_2)\ \exp(-2ik_0Cd) \simeq \exp(-2in\pi)$$

hence

$$\Delta/C + ik_0Cd \simeq in\pi \tag{3.87}$$

where Δ, a normalised surface impedance, is defined by

$$\Delta = 1/\hat{Y}_1 + 1/\hat{Y}_2$$

Equation (3.87) is solved as a second degree algebric equation in C to give:

$$C_n = n\pi/2k_0d \pm [(n\pi/2k_0d)^2 + i\Delta/k_0d]^{1/2}$$

Only the positive sign before the radical can be chosen in order that C_n reduces to $n\pi/k_0d$ for the case $\Delta = 0$. Now for $|k_0d\Delta| \ll (n\pi/2)^2$, the radical is expanded to give:

$$C_n \simeq n\pi/k_0d + i\Delta/n\pi + \Delta^2k_0d/(n\pi)^3 + \ldots \tag{3.88}$$

The corresponding characteristic longitudinal wavenumbers β_n are obtained from:

$$S_n = \beta_n/k_0 = (b_n^2 - 2i\Delta/k_0d - \Delta^2/(n\pi)^2 \ldots)^{1/2} \tag{3.89}$$

where $b_n = (1-(n\pi/k_0d)^2)^{1/2}$.

Away from mode cutoff, that is when $n\pi \ll k_0d$, and for $\Delta \ll k_0d$, (3.89) can be approximated by:

$$\beta_n/k_0 \simeq b_n - (i\Delta/k_0d + (\Delta/n\pi)^2/2)/b_n \tag{3.90}$$

from which the attenuation α_n is

$$\alpha_n = -\text{Imag}(\beta_n) \simeq \text{Real}(\Delta)/b_nd + (k_0/b_n)\ \text{Imag}(\Delta^2)/2n^2\pi^2 \tag{3.91}$$

It can be shown that the first term on the RHS of (3.91) agrees with the perturbation analysis of section 2.5. It is worth noting that (3.91) applies whether the attenuation is due to ohmic losses or to radiation losses, such as when Δ is purely real, corresponding to a lossless outside medium. It is seen from (3.91) that the attenuation of *TM* modes (H is parallel to the impedance planes) is inversely proportional to the spacing d between the impedance planes. A similar treatment of *TE* modes reveals that the attenuation of these modes is inversely proportional to d^3 (see problem 3.12). Finally, we note that near to the mode cutoff ($b_n \ll 1$), (3.90) fails to predict the attenuation rate, and in this case one should resort to the more exact formula (3.89).

Case 2: $CY_1 \gg 1$ and $CY_2 \ll 1$, corresponding to a highly conducting lower surface and a poorly conducting upper surface. As an example, these conditions are satisfied for a planar model of the earth-ionosphere waveguide at very low frequencies (VLF) (Wait, 1970, 1986). The modal equation (3.86) for *TM* (or vertically polarised) waves can then be approximated by:

$$-\exp(-2/C\hat{Y}_1)\ \exp(-2C\hat{Y}_2)\ \exp(-2ik_0Cd) = \exp(-2in\pi) \tag{3.92}$$

or

$$1/C\hat{Y}_1 + C\hat{Y}_2 + ik_0Cd = i\pi(n-1/2),\ n = 1, 2, \ldots$$

Solving for C_n and using the condition $\hat{Y}_1^{-1}k_0 d \ll \pi^2$

$$C_n \simeq (n-1/2)/(k_0 d - i\hat{Y}_2) + i\hat{Y}_1^{-1}/\pi(n-1/2) \tag{3.93}$$

We can then proceed to obtain $S_n = \beta_n/k_0$ from which the modal phase velocity v_n and the modal attenuation rate α_n are deduced; namely, far above cutoff $(n-\frac{1}{2} \ll k_0 d)$, these quantities are given by:

$$v_n/c \simeq \left[b_n + \mathrm{Imag}(\hat{Y}_2)\,\frac{\pi^2(n-1/2)^2}{(k_0 d)^3 b_n} + \mathrm{Imag}(\hat{Y}_1^{-1})/k_0 d_n\right]^{-1} \tag{3.94}$$

$$\alpha_n/k_0 \simeq \mathrm{Real}(\hat{Y}_2)\,\frac{\pi^2(n-1/2)^2}{(k_0 d)^3 b_n} + \mathrm{Real}(\hat{Y}_1^{-1})/k_0 d b_n \tag{3.95}$$

where

$$b_n = [1 - ((n-\tfrac{1}{2})/k_0 d)^2)^{1/2}$$

Applying these results to the planar model of the earth-ionosphere waveguide and taking the earth to have a conductivity $\sigma_1 \gg \omega\varepsilon$

$$Y_1^{-1} \simeq (1+i)\,(\omega\mu_0/2\sigma_1)^{1/2} \tag{3.96}$$

At VLF, the ionosphere acts as a lossy dielectric with $\varepsilon \simeq \varepsilon_0$ and $\sigma/\omega\varepsilon_0 \simeq \omega_0^2/\omega\nu$, where ω_0 is the plasma frequency and ν the electron collision frequency ($\gg \omega$). Denoting $\sigma/\omega\varepsilon_0$ by x, and using (3.15)

$$\hat{Y}_2 = (1-ix)/(-ix)^{1/2} = 2^{-1/2}[(x^{-1/2} + x^{1/2}) + i(x^{-1/2} - x^{1/2})] \tag{3.97}$$

From (3.97) it is clear that $\mathrm{Real}(\hat{Y}_2)$, and hence the attenuation caused by the ionosphere, has a broad minimum around $x=1$, i.e. as a function of the ionospheric conductivity the attenuation is minimum when $\sigma = \omega\varepsilon_0$; a result which has been reported by Wait [1970, chap. 6].

Mode excitation

Now we consider the modal excitation by a specified source. To start with, the source is taken to be a y oriented magnetic line source lying at $x = x_s$ and $z = 0$ (Fig. 3.5). This may represent a narrow aperture of x oriented E field in the $z=0$ plane. One can derive the amplitudes of the excited modes by application of the reciprocity theorem and the mode orthogonality property (to be proved in section 3.6) as demonstrated in section 2.11. However, there is much to be learned by using the more direct approach described below.

Firstly, the magnetic line source may be considered as a magnetic surface current sheet m on the plane $x = x_s$. Thus

$$\mathrm{m(volt/m)} = V\delta(z)\hat{y} = (V/2\pi)\int_{-\infty}^{\infty} \exp(\pm i\beta z)\, d\beta\hat{y} \tag{3.98}$$

where the last equality stems from the definition of the delta function (e.g., Clemmow, 1966, chap. 1).

The excited fields can be expressed, in the most general form, as a Fourier integral over the longitudinal wavenumber β:

$$E_{z1} = \int_{-\infty}^{\infty} V_1(\beta)\,[e^{ux} + R_1(u)e^{-ux}]\ e^{-i\beta z}\ d\beta,\ 0 \leqslant x \leqslant x_s \tag{3.99}$$

$$E_{z2} = \int_{-\infty}^{\infty} V_2(\beta)\,[e^{-u(x-d)} + R_2(u)\ e^{u(x-d)}]\ e^{-i\beta z}\ d\beta,\ x_s \leqslant x \leqslant d \tag{3.100}$$

where $R_1(u)$ and $R_2(u)$ are identified as the spectral reflection coefficients at the lower and upper boundaries $x=0$ and $x=d$ respectively and $u^2=\beta^2-k_0^2$. The y component of the magnetic field is given in terms of E_z by (1.25). Therefore

$$H_{y1} = i\omega\varepsilon_0 \int_{-\infty}^{\infty} V_1(\beta)\,[e^{ux} - R_1(u)\ e^{-ux}]\,(e^{-i\beta z}/u)\ d\beta,\ 0 \leqslant x \leqslant x_s \tag{3.101}$$

$$H_{y2} = -i\omega\varepsilon_0 \int_{-\infty}^{\infty} V_2(\beta)\,[e^{-u(x-d)} - R_2(u)\ e^{u(x-d)}]\,(e^{-i\beta z}/u)\ d\beta,\ x_s \leqslant x \leqslant d \tag{3.102}$$

Application of the boundary conditions (3.80) and (3.81) at the boundary surfaces for each spectral component yields immediately

$$R_{1,2}(u) = (ik_0 - u\hat{Y}_{1,2})/(ik_0 + u\hat{Y}_{1,2}) \tag{3.103}$$

The boundary conditions at the source plane $x=x_s$ require a discontinuity of E_z by an amount equal to the source magnetic current; namely

$$E_{z1}(x_s, z) - E_{z2}(x_s, z) = V\delta(z) \tag{3.104}$$

Meanwhile, the magnetic field H_y is continuous across the $x=x_s$ plane.

$$H_{y1}(x_s, z) = H_{y2}(x_s, z) \tag{3.105}$$

Using the Fourier transform of the Kronecker delta function in (3.104) and equating the spectral components of both sides of the equation, one gets

$$V_1(\beta)\,[\exp(ux_s) + R_1(u)\ \exp(-ux_s)] - V_2(\beta)$$
$$[\exp(-u(x_s-d)) + R_2(u)\ \exp(u(x_s-d))] = V/2\pi \tag{3.106}$$

Similarly, using (3.101)—(3.102) to equate the spectral components on both sides of (3.105) one gets

$$V_1(\beta)\,[\exp(ux_s) - R_1(u)\ \exp(-ux_s)] = -V_2(\beta)$$
$$[\exp(-u(x_s-d)) - R_2(u)\ \exp(u(x_s-d))] \tag{3.107}$$

The last two equations are solved simultaneously to yield the unknown functions $V_{1,2}(\beta)$ in terms of the source strength V. Thus

$$V_1(\beta) = Vg(x_s)\ \exp(-ud)/4\pi D(\beta) \tag{3.108}$$

$$V_2(\beta) = -Vf(x_s)\ \exp(-ud)/4\pi D(\beta) \tag{3.109}$$

where $f(x)$, $g(x)$ and $D(\beta)$ are defined by:

$$f(x) \equiv \exp(ux) - R_1(u)\ \exp(-ux) \tag{3.110}$$

$$g(x) \equiv \exp(-u(x-d)) - R_2(u)\ \exp(u(x-d)] \tag{3.111}$$

$$D(\beta) = 1 - R_1(u)R_2(u)\ \exp(-2ud) \tag{3.112}$$

Thus the complete integral representation (3.99)—(3.102) of the fields is determined. For example, the magnetic field H_y in the region $0 \leqslant x \leqslant x_s$ is expressed by:

$$H_{y1} = (i\omega\varepsilon_0 V/4\pi) \int_{-\infty}^{\infty} [g(x_s)f(x)\ e^{-ud}\ e^{-i\beta z}/uD(\beta)]\ d\beta \qquad (3.113)$$

and in the region $x_s \leqslant x \leqslant d$,

$$H_{y2} = (i\omega\varepsilon_0 V/4\pi) \int_{-\infty}^{\infty} [f(x_s)g(x)\ e^{-ud}\ e^{-i\beta z}/uD(\beta)]\ d\beta \qquad (3.114)$$

At this point we observe that the zeros of $D(\beta)$ are the modal eigenvalues of the guide. Thus the integrands in (3.113)—(3.114) have simple poles at all values of u (or β) which are solutions to the modal equation (3.82). It is then possible to express the integrals as sums of residues at these poles. To this end, deform the contour of integration to an infinite semicircle in the lower complex plane of β and encircle all poles to get

$$H_{y1}(x;\ x_s) = (\omega\varepsilon_0 V/2) \sum_p \frac{g_p(x_s)f_p(x)\ \exp(-u_p d - i\beta_p z)}{u_p(dD(\beta)/d\beta)|_{\beta=\beta_p}} \qquad (3.115)$$

where the integer p runs over all the natural modes of the structure. It is important to observe that under the condition $D(\beta)=0$, satisfied by each mode, the functions f_p and g_p are related by

$$g_p(x) = -f_p(x)R_2(u_p)\ \exp(-u_p d)$$

as can be verified from (3.110)—(3.112). When this relation is used in (3.115) to substitute for $g_p(x_s)$, the reciprocity property is revealed, since then an interchange of x and x_s leaves the expression unaltered. It can also be verified that (3.115) is valid for all values of x, $0 \leqslant x \leqslant d$. Now it is possible to recast this equation into the more convenient form

$$H_y(x;\ x_s) = \sum_p A_p f_p(x)\ \exp(-i\beta_p z) \qquad (3.116)$$

where A_p is the amplitude of excitation of the p th mode

$$A_p = \frac{-(\omega\varepsilon_0 V/2)f_p(x_s)}{u_p R_1(u_p)\ (dD/d\beta)_{\beta=\beta_p}} \qquad (3.117)$$

We can then infer the expression of A_p for a general distribution of magnetic line sources (or E_x aperture) in the $z=0$ plane by integrating over x_s. Therefore, for a general $V(x)$,

$$A_p = \frac{-\omega\varepsilon_0}{2u_p R_1(u_p)\ (dD/d\beta)_{\beta=\beta_p}} \int_0^d V(x)f_p(x)\ dx \qquad (3.118)$$

Obviously (3.118) reduces to (3.117) when $V(x)$ is a delta function at $x=x_s$. We have thus obtained the fields excited by an arbitrary source distribution in the waveguide cross section. The fields are expressed as a discrete sum of the natural modes which form a complete set of eigenfunctions. It is worth noting that, in the process of deforming the contour of integration of (3.113) in the complex β plane, we encounter a branch point at $\beta = k_0$ due to the double value of the wavenumber $u = (\beta^2 - k_0^2)^{1/2}$. An integration around this point, such that Real$(u) = 0$ must then be performed. However, owing to the fact that the integrand is an even function of u, as can be verified in this case, the branch cut integration is equal to zero. Actually, a branch cut contribution would account for a continuous spectrum of radiated field which does not exist for the closed guiding structure in hand. In the next chapter on open waveguides, it will be shown that both discrete and continuous sets of modes are needed to represent an arbitrary field supported by the structure.

3.4 Circular cylindrical waveguide with constant impedance wall

A circular cylindrical waveguide with a constant impedance wall is depicted in Fig. 3.6. The wall at $\rho = a$ is characterised by a constant impedance Z and a constant admittance Y defined by:

$$Z = E_\phi / H_z \tag{3.119}$$

$$Y = -H_\phi / E_z \tag{3.120}$$

A few examples of surfaces that can be described by (119)—(120) have already been given in section 3.2. Now we wish to identify all possible modes which are supported by the guide. For finite values of Z and Y, we must anticipate that the modes are generally hybrid, i.e. having both TE and TM (to z) fields. For a given mode, the longitudinal field components in the region $0 \leqslant \rho \leqslant a$ can be expressed by:

$$E_z = iJ_m(u\rho/a) \cos m\phi \tag{3.121}$$

$$\eta_0 H_z = i\Lambda J_m(u\rho/a) \sin m\phi \tag{3.122}$$

apart from a common factor $e^{i\omega t - i\beta z}$. η_0 is the unbounded wave impedance of the medium filling the guide, Λ is the hybrid factor relating TE to TM wave parts, and

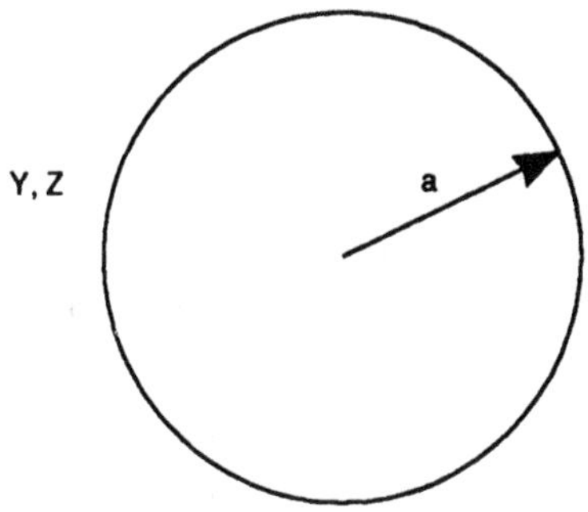

Fig. 3.6 A circular cyulindrical waveguide with constant impedance wall

$$(u/a)^2+\beta^2=k_0^2\equiv\omega^2\mu_0\varepsilon_0 \tag{3.123}$$

The transverse field components are derived from the longitudinal components by using (1.27)—(1.28). For convenience, they are written down explicitly:

$$E_\phi=-(v/u)\,[m\hat{\beta}\{J_m(u\rho/a)/(u\rho/a)\}+\Lambda J'_m(u\rho/a)]\ \sin m\phi \tag{3.124}$$

$$E_\rho=(v/u)\,[\hat{\beta}J'_m(u\rho/a)+m\Lambda\{J_m(u\rho/a)/(u\rho/a)\}]\ \cos m\phi \tag{3.125}$$

$$\eta_0H_\phi=(v/u)\,[J'_m(u\rho/a)+m\hat{\beta}\Lambda(J_m(u\rho/a)/(u\hat{\rho}/a))]\ \cos m\phi \tag{3.126}$$

$$\eta_0H_\rho=(v/u)\,[m\{J_m(u\rho/a)/(u\rho/a)\}+\hat{\beta}\Lambda J'_m(u\rho/a)]\ \sin m\phi \tag{3.127}$$

where $v=k_0a$, $\hat{\beta}=\beta/k_0$ and the prime on the Bessel function denotes differentiation with respect to the argument. In order to determine the allowable eigenvalues of the modes of the structure, we first impose the boundary conditions (3.119)—(3.120) at $\rho=a$ to get:

$$F_m(u)+m\hat{\beta}\Lambda=-i\eta_0Yu^2/v \tag{3.128}$$

$$F_m(u)+m\hat{\beta}/\Lambda=-i\eta_0^{-1}Zu^2/v \tag{3.129}$$

where

$$F_m(u)\equiv uJ'_m(u)/J_m(u) \tag{3.130}$$

At this point, let us define a normalised surface susceptance B and reactance X by

$$Y\eta_0=iB$$

$$Z/\eta_0=iX$$

Substituting in (3.128)—(3.129) and rearranging

$$F_m(u)-Bu^2/v=-m\hat{\beta}\Lambda \tag{3.131}$$

$$F_m(u)-Xu^2/v=-m\hat{\beta}/\Lambda \tag{3.132}$$

Subtracting (3.131) from (3.132)

$$\Lambda-\Lambda^{-1}=(B-X)u^2/m\hat{\beta}v \tag{3.133}$$

and multiplying the two equations

$$[F_m(u)-Bu^2/v]\,[F_m(u)-Xu^2/v]=(m\hat{\beta})^2 \tag{3.134}$$

The latter is the modal equation for the eigenvalues β or u, while (3.133) is an equation for the mode hybrid factor Λ, which can be solved as a second degree algebric equation in Λ yielding

$$\Lambda=(B-X)u^2/2m\hat{\beta}v\pm[1+((B-X)u^2/2m\hat{\beta}v)^2]^{1/2} \tag{3.135}$$

This is an explicit expression for the hybrid factor in terms of the surface susceptance and reactance as well as the modal eigenvalue. The two possible solutions in (3.135) correspond to the two types of hybrid modes HE_{mn} and EH_{mn} (Clarricoats, 1971). At the cutoff condition $\beta=0$, these two types of modes reduce to E and H modes respectively (for which $\Lambda=0$ and ∞).

In order to gain insight into the mode characteristics in a circular guide with finite impedance wall, we consider a few special cases of the modal equation (3.134).

3.4.1 Azimuthly symmetric modes, $m=0$

In this case the modal equation decouples into two equations for *TM* (or *E*) modes and *TE* (or *H*) modes; namely

$$F_0(u) = -uJ_1(u)/J_0(u) = Bu^2/v \tag{3.136}$$

$$F_0(u) = -uJ_1(u)/J_0(u) = Xu^2/v \tag{3.137}$$

Each of these two equations has an infinite set of solutions giving rise to TM_{0n} and TE_{0n} modes. In the special, and most familiar, case of a perfectly conducting wall, $X=0$ and $B=\infty$, the modal solutions reduce to the well known solutions: $u=u_{0n}$ for *TM* modes and $u=u'_{0n}$ for *TE* modes. Here u_{0n} and u'_{0n} are the n th zeros of $J_0(u)$ and $J'_0(u)$ respectively.

Next it is interesting to find the surface wave mode associated with a reactive wall. For such a mode the fields must be evanescent in the interior region of the guide. Hence it is appropriate to write $u=iw$, where we expect w to be purely real if X or B is real. The modal equations (3.136)—(3.137) now take the form

$$F_0(iw) = wI_1(w)/I_0(w) = -w^2B/v \tag{3.138}$$

for the TM_0 surface wave mode. For the TE_0 mode B is replaced by X. Observing that the LHS of (3.138) is always positive (for real w), we deduce immediately that a surface wave mode is supported only if B (or X) is negative; that is, a *TM* surface wave mode is supported by an inductive surface and a *TE* mode is supported by a capacitive surface. This agrees with the results of section 3.3 for a planar surface. Next a graphical solution for (3.138) is presented. To this end, it is more convenient to recast this equation in the form

$$wI_0(w)/I_1(w) = -k_0a\Delta \tag{3.139}$$

where $\Delta = 1/B$ or $1/X$. A plot of the LHS of (3.139) against w is shown in Fig. 3.7. The solution for w is obtained by the point of intersection of a horizontal line at a height of $-k_0a\Delta$ with the plotted curve. It is clear that a cutoff condition for the surface wave mode exists if

$$-k_0a\Delta \leqslant 2 \tag{3.140}$$

This can also be interpreted in terms of a cutoff frequency, a cutoff surface curvature or a cutoff surface reactance A lower frequency than the cutoff value, a higher curvature or a lower reactance leads to the cutoff of the surface wave mode. In the context of our problem this cutoff means the conversion of the mode to an ordinary waveguide mode (with real value of u). Finally the limit of a planar surface is approached as a tends to ∞, hence w tends to ∞ and (3.139) reduces to $w/a = -k_0\Delta$. Apart from a slight difference of meanings of variables, this result agrees with (3.68) as expected.

3.4.2 A low impedance wall

A particulary interesting special case of the circular cylindrical waveguide with finite impedance wall is that in which $|X| \ll 1$ and $|B^{-1}| \ll 1$. A finite wall loss is allowed for by taking X and/or B as complex quantities. These conditions may represent a highly conducting metallic wall or a dielectric lined metallic wall where the lining is very thin relative to a wavelength. The modes can then be considered to be perturbed versions of *TE* and *TM* modes in a guide with

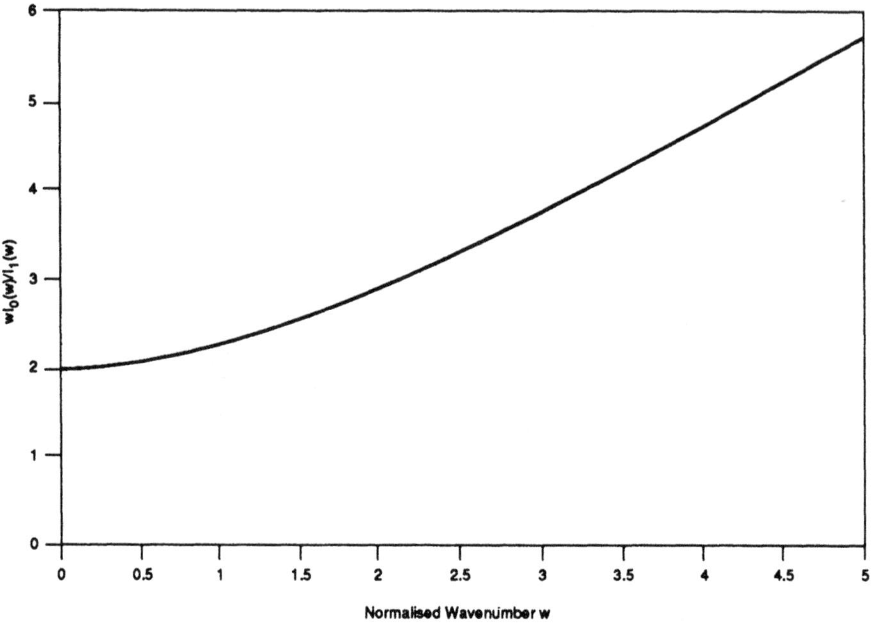

Fig. 3.7 Plot of the LHS of (3.139) versus the normalised wavenumber w

perfectly conducting wall. In what follows we find the modified modal phase and attenuation constants of these perturbed modes.

Remembering that all modes in a guide with imperfectly conducting walls are generally hybrid (unless $m=0$ or $\beta=0$), we argue that for the perturbed *TE* modes there is a small longitudinal electric field. This is manifested by a large, but finite, hybrid factor Λ. From (3.133) and since $\Lambda^{-1} \ll \Lambda$ and $X \ll B$, then

$$\Lambda|_{TE} \simeq Bu^2/mv\hat{\beta} \tag{3.141}$$

Using this in (3.132)

$$F_m(u) \simeq Xu^2/v - v(m\hat{\beta})^2/Bu^2 \tag{3.142}$$

In the ideal case of a perfectly conducting wall, both X and B^{-1} are zero; so is the LHS of (3.142). In this ideal case the solution to the modal equation is $u=u'_{mn}=n$ th zero of $J'_m(u)$, n any integer. In the case under consideration the RHS of (3.142) is small relative to unity; hence a perturbation analysis can be applied. So, writing $u=u'_{mn}+\delta u$, we get

$$\delta u \simeq \frac{X(u_{mn})^2/v - B^{-1}v(mb_{mn}/u'_{mn})^2}{[dF_m(u)/du]_{u=u'_{mn}}} \tag{3.143}$$

where $b_{mn}=(1-(u'_{mn}/v)^2)^{1/2}$. Evaluation of the denominator shows that it is equal to $[-u'_{mn}+m^2/u'_{mn}]$. The corresponding value of β (possibly complex) is obtained from (3.123) yielding:

$$\beta_{mn}a = [v^2b^2_{mn} - 2u'_{mn}\delta u - (\delta u)^2]^{1/2} \tag{3.144}$$

A good feature of this formula is that, unlike the perturbation analysis of section 2.5, it remains valid down to the cutoff frequency; namely, when $v = u'_{mn}$ (or $b_{mn} = 0$):

$$\beta_{mn} a \simeq (-2u'_{mn}\delta u)^{1/2} = u'_{mn}[2X/v(1 - m^2/u'^2_{mn})]^{1/2} \tag{3.145}$$

whose real and imaginary parts give the phase and attenuation constants at cutoff. Here cutoff refers to the waveguide with perfect walls. Next, at frequencies sufficiently higher than the cutoff, (3.144) can be simplified by noticing that $v^2 b^2_{mn} \gg 2u'_{mn}\delta u$; hence:

$$\beta_{mn} a \simeq v b_{mn} - u'_{mn}\delta u / v b_{mn}$$

Using (3.143) and taking the imaginary part, one gets the attenuation constant α due to the wall losses for TE modes:

$$\alpha|_{TE} \simeq \frac{u'^2_{mn}}{a^2 k_0 b_{mn}(u'^2_{mn} - m^2)} (\Delta_1 u'^2_{mn}/k_0 a + \Delta_2 k_0 a (m b_{mn}/u'_{mn})^2) \tag{3.146}$$

where

$$\Delta_1 = -\mathrm{Imag}(X) = \mathrm{Real}(Z/\eta_0),$$

and

$$\Delta_2 = \mathrm{Imag}(B^{-1}) = \mathrm{Real}(Y\eta_0)^{-1}$$

A comment on equation (3.146) is now in order. The term involving Δ_1 accounts for the attenuation due to the circumferential currents on the wall. The term involving Δ_2 accounts for the attenuation due to the longitudinal currents. For the case of a highly conducting wall with conductivity σ, both Δ_1 and Δ_2 are equal to $(\omega\varepsilon/2\sigma)^{1/2}$. Hence, the first term in (3.146) is proportional to $\omega^{-3/2}$, and the second term is proportional to $\omega^{1/2}$ away from cutoff. This is in agreement with the perturbation analysis of section 2.5. Clearly, for TE_{0n} modes $(m = 0)$ the attenuation decreases with frequency according to $\omega^{-3/2}$.

Next consider perturbed TM modes in a circular guide with small X and B^{-1}. For these modes the hybrid factor Λ is $\ll 1$; i.e. there is a small longitudinal component of magnetic field. Using (3.133) Λ is approximately given by:

$$\Lambda^{-1} \simeq -(B - X)u^2/m\beta v$$

Substituting in (3.132)

$$F_m(u) = uJ'_m(u)/J_m(u) \simeq Bu^2/v \tag{3.147}$$

Effectively, the longitudinal magnetic field is so small that the associated circumferential currents on the wall are negligible. As a result the reactance X does not show up in the above modal equation. To solve (3.147) under the condition $B^{-1} \ll 1$, we set $u = u_{mn} + \delta u$, where u_{mn} is the n th zero of $J_m(u)$, and δu is obtained as

$$\delta u \simeq (B^{-1}v/u^2_{mn})/(dF^{-1}_m(u)/du)$$

where the denominator, evaluated at $u = u_{mn}$, ie equal to $1/u_{mn}$. Therefore

$$u \simeq u_{mn} + B^{-1}v/u_{mn} \tag{3.148}$$

The corresponding βa is

$$\beta a = [v^2 b_{mn}^2 - 2B^{-1}v - B^{-2}v^2/u_{mn}^2]^{1/2} \tag{3.149}$$

which gives the complex propagation constant at all frequencies including those near cutoff.

3.4.3 Balanced hybrid mode condition

Another special case of practical significance is when both X and B are $\ll k_0 a$. This is often achieved in practice by using corrugated surfaces or dielectric coated surfaces, as will be explained in detail in chapter 5. The lower order modes, in this case, are characterised by reduced wall losses compared to the conventional smooth wall guides, as demonstrated by Clarricoats *et. al.* (1975*a*, *b*). Besides, some of the lower order modes exhibit a high polarisation purity. To present a simplified discussion on modes in this case, let us first assume that $X = B$. Referring to (3.135), it then follows that the hybrid factor $\Lambda = \pm 1$; hence the name *balanced hybrid mode* given to modes under this condition. Furthermore if we let $X = B = 0$, the modal equation (3.134) reduces to the pair:

$$F_m(u) \pm m\hat{\beta} = 0 \tag{3.150}$$

corresponding to $\Lambda = \pm 1$ respectively, as can be verified from (3.128) or (3.129). Referring to equations (3.124)—(3.127) for the modal field components, we observe that the tangential components E_ϕ and H_ϕ vanish at the wall when (3.150) is satisfied. This at once explains the reduced attenuation of modes at the balanced hybrid condition. To continue we present solutions to (3.150) in the limit of large v, whence $\hat{\beta} \simeq 1$. Therefore

$$J_m'(u) \pm mJ_m(u)/u = 0 \tag{3.151}$$

Using identity (A2.26) this reduces to

$$J_{m-1}(u) = 0, \text{ or } J_{m+1}(u) = 0 \tag{3.152}$$

corresponding to HE_{mn} and EH_{mn} modes for which $\Lambda = \pm 1$ respectively. The dominant mode, having the least value of u, is the HE_{11} mode. For this mode $u = 2{\cdot}4048$, which is the first zero of $J_0(u)$. The next higher order modes with $m = 1$ are the EH_{12} and the HE_{12} modes for which $u = 5{\cdot}1362$ and $5{\cdot}5201$ respectively. The designations *HE* and *EH* for hybrid modes follow Clarricoats and Saha (1971). The mode EH_{11} is found to correspond to a surface wave mode for a finite value of the wall admittance.

Turning attention again to the modal equation (3.134), we now give approximate solutions valid near the hybrid balanced condition, namely $X \ll 1$ and $B \ll 1$, and in addition $v \gg 1$. So using a first order perturbation analysis around $u = u_{mn}$, the roots of (3.152), and $\hat{\beta} \simeq 1$, we get

$$u = u_{mn} + \delta u,$$

where

$$2F_m(u_{mn})F_m'(u_{mn})\delta u - F_m(u_{mn})(B + X)u^2/v \simeq -m^2u^2/v^2$$

Noting that $F_m = \mp m$ for *HE* and *EH* modes respectively and the derivative $F_m' = -u_{mn}$, then

$$u \simeq u_{mn} - [(B+X)/2v \pm m//2v^2]/u_{mn} \tag{3.153}$$

The corresponding β, for $v \gg u$ is

$$\beta = b_{mn} + (u_{mn}^2/v^2 b_{mn})\,[(B(+X)/2v \pm m/2v^2] \tag{3.154}$$

where

$$b_{mn} = (1 - u^2/v^2)^{1/2}$$

The attenuation due to wall losses is then equal to

$$\alpha \simeq -\mathrm{Imag}((B+X)/2)\,[u_{mn}^2/a^3 k_0^2 (1 - u_{mn}^2/k_0^2 a^2)^{1/2}] \tag{3.155}$$

This shows that away from cutoff the attenuation of modes near the balanced hybrid mode condition falls with frequency as f^{-2}. Assuming the wall resistance and conductance to vary as $f^{-1/2}$ leads to an overall dependence on $f^{-3/2}$. In this respect these modes are similar to the TE_{0n} modes in a smooth circular waveguide. However, exact numerical results presented by Clarricoats and Olver (1984) show that the attenuation factor of the dominant HE_{11} mode in a corrugated waveguide is lower than that of the H_{01} in a smooth wall waveguide of comparable size. We can show roughly that this is really the case by writing down the attenuation factor of the HE_{11} from (3.155):

$$\alpha|_{HE_{11}} \simeq -\mathrm{Imag}((B+X)/2)\,[2{\cdot}4048^2/a^3 k_0^2]$$

and the attenuation factor of the H_{01} from (3.146)

$$\alpha|_{H_{01}} \simeq \mathrm{Real}(Z/\eta_0)\,[3{\cdot}832^2/a^3 k_0^2]$$

A comparison of the two modes shows that for equal wall resistances the attenuation of the HE_{11} mode is lower by a factor of about 2·5. This property, together with the low crosspolar fields of the HE_{11} mode (as will be shown in chapter 5), gives great practical significance to operation near the balanced hybrid mode condition.

3.5 Cylindrical waveguides with noncircular cross sections

In an interesting paper by Wait (1967*a*), the author points out that elementary modes with a single transverse wavenumber are generally coupled by the finite wall impedance in waveguides having any cross section other than circular. As a result a natural mode which satisfies the boundary conditions in such waveguides is generally composed of a sum of elementary modes. Each elementary mode is characterised by its own transverse field distribution although all modes have the same longitudinal wavenumber. This phenomenon has been described by Wait as a fundamental difficulty in the analysis of cylindrical waveguides with impedance walls. In this section we briefly explain this difficulty with reference to elliptic and rectangular waveguides and find out special conditions which lead to simple mode solutions.

3.5.1 Elliptical waveguides

Consider a cylindrical waveguide with an elliptical cross section. In terms of elliptic coordinates (u, v, z) the waveguide surface coincides with $u = u_0$ and has a surface impedance Z and admittance Y (Fig. 3.8). Looking for the natural modes of the guide, we impose the field dependence $\exp(i\omega t - i\beta z)$. Apart from this common factor, general expressions for the longitudinal fields may take the forms:

$$E_z = \sum_m A_m R_{e,m}(q, u) C_m(q, v) \tag{3.156}$$

$$H_z = \sum_m B_m R_{0,m}(q, u) S_m(q, v) \tag{3.157}$$

where all symbols are those defined in section 1.5 and the functions $R_{e,m}$ and $R_{o,m}$ are the radial Mathieu functions of first kind which are finite on the guide axis. The transverse fields are obtainable from (1.27)—(1.28). Of interest are the angular components displayed below:

$$E_v = (-i\beta c^2/4q) h^{-1}(u, v)\ \partial E_z/\partial v + (i\omega\mu c^2/4q) h^{-1}(u, v)\ \partial H_z/\partial u$$
$$H_v = (-i\beta c^2/4q) h^{-1}(u, v)\ \partial H_z/\partial v - (i\omega\varepsilon c^2/4q) h^{-1}(u, v)\ \partial E_z/\partial u$$

Using (3.156)—(3.157) in the above two equations and substituting in the boundary conditions

$$E_v = ZH_z \text{ and } H_v = -YE_z \text{ at } u = u_0,$$

we get the following two equations for $0 \leq v \leq 2\pi$:

$$i\omega\mu c^2 \sum_m B_m R'_{0,m} S_m(v) - i\beta c^2 \sum_m A_m R_{e,m} C'_m(v) = 4q \sum_m Zh(v) B_m R_{o,m} S_m(v) \tag{3.158}$$

$$i\omega\varepsilon c^2 \sum_m A_m R'_{e,m} C_m(v) + i\beta c^2 \sum_m B_m R_{o,m} S'_m(v) = 4q \sum_m Yh(v) A_m R_{e,m} C_m(v) \tag{3.159}$$

where the prime denotes differentiation with respect to argument. The radial functions as well as the metric coefficient are to be evaluated at $u = u_0$. The key observation in the above two equations is that it is not permissible to equate individual terms of the summations on both sides of the equations. This is so

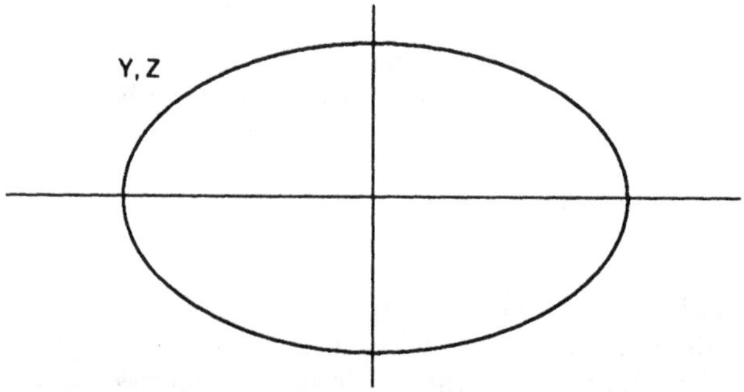

Fig. 3.8 An elliptical cylindrical waveguide with constant impedance wall

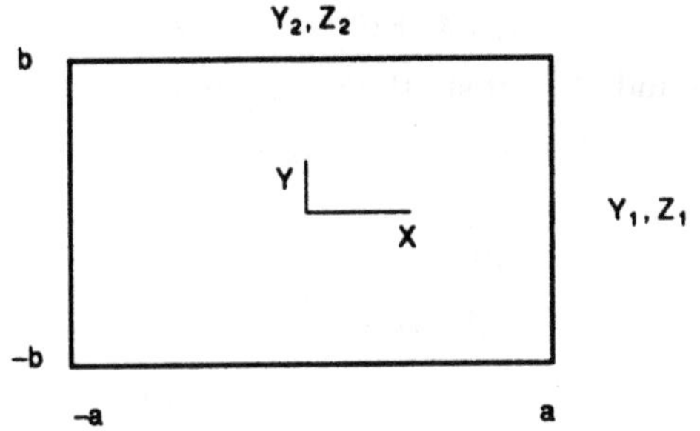

Fig. 3.9 A rectangular waveguide with general constant impedance walls

since h is not independent of v and $S_m(v)$ is not the same function as $C'_m(v)$. Likewise $C_m(v)$ and $S'_m(v)$ are not the same functions of v. Therefore, the summation signs cannot be removed, signifying that the natural mode is generally a sum of elementary modes. Obviously in the circular cylindrical waveguide, the angular wave functions reduce to the *sine* and *cosine* functions and h is independent of ϕ leading to a decoupling between the elementary modes. On the other hand, in noncircular guides the intrinsic mode coupling can only be removed if $\beta = 0$ and Z and Y are chosen to be functions of v such that $Z(v)\,h(v)$ and $Y(v)\,h(v)$ are constants. The modes will then reduce to elementary TE_m or TM_m modes.

3.5.2 Rectangular waveguides

The general remarks made on the cylindrical elliptical guide apply quite well to the cylindrical rectangular guide. Thus, in general, there is no elementary mode solution in a rectangular waveguide with finite wall impedances. However, the analysis made by Dybdal *et al.* (1971) reveals that elementary mode solutions are allowed if a special relationship holds between the wall impedances of a rectangular guide. In the following we present a straightforward proof of this statement in our own notation. Next we consider several special cases of wall impedance combinations.

So, consider a cylindircal rectangular waveguide with side walls at $x = \pm a$ and upper and lower walls at $y = \pm b$, as illustrated in Fig. 3.9. The four walls are characterised by wall impedances and admittances in the fashion:

$$Z_1 = \pm E_y/H_z \text{ and } Y_1 = \mp H_y/E_z \tag{3.160}$$

for the side walls $x = \pm a$ respectively, and

$$Z_2 = \mp E_x/H_z \text{ and } Y_2 = \pm H_x/E_z \tag{3.161}$$

for the upper and lower walls $y = \pm b$ respectively.

Now we derive the condition to be satisfied by the wall parameters in order to have elementary modal solutions. An elementary mode solution for any of the field components, say E_z, is obtained by separation of variables in the form:

$$E_z = f(k_x x)g(k_y y)\exp(-i\beta z) \tag{3.162}$$

where

$$k_x^2+k_y^2+\beta^2=k^2\equiv\omega^2\mu\varepsilon \tag{3.163}$$

The functions $f(.)$ and $g(.)$ satisfy the differential equations

$$f''(k_x x)+f(k_x x)=g''(k_y y)+g(k_y y)=0 \tag{3.164}$$

where a prime denotes differentiation with respect to the argument. A compatible H_z, to go along with E_z of (3.162), must have the form

$$H_z=\eta^{-1}\Lambda f'(k_x x)g'(k_y y)\ \exp(-i\beta z) \tag{3.165}$$

where Λ is the mode hybrid factor and η is the unbounded wave impedance of the medium. The transverse field components are obtainable from (1.27)—(1.28); therefore:

$$E_x=Cf'(k_x x)g(k_y y)\,(k_x-\Lambda k_y k/\beta)\ \exp(-i\beta z) \tag{3.166}$$

$$E_y=Cf(k_x x)g'(k_y y)\,(k_y+\Lambda k_x k/\beta)\ \exp(-i\beta z) \tag{3.167}$$

$$H_x=-C\eta^{-1}f(k_x x)g'(k_y y)\,(k_x\Lambda+k_y k/\beta)\ \exp(-i\beta z) \tag{3.168}$$

$$H_y=-C\eta^{-1}f'(k_x x)g(k_y y)\,(k_y\Lambda-k_x k/\beta)\ \exp(-i\beta z) \tag{3.169}$$

where

$$C=-i\beta/(k_x^2+k_y^2) \tag{3.170}$$

and relation (3.164) has been used to substitute for f'' and g'' wherever they occurred. Using (3.162)—(3.169) in the boundary conditions (3.160) and (3.161), we establish the following relations:

$$Z_1/\eta=C\Lambda^{-1}[f(k_x a)/f'(k_x a)]\,(k_y+\Lambda k_x k/\beta) \tag{3.171}$$

$$Y_1\eta=C[f'(k_x a)/f(k_x a)]\,(k_y\Lambda-k_x k/\beta) \tag{3.172}$$

$$Z_2/\eta=-C\Lambda^{-1}[g(k_y b)/g'(k_y b)\,(k_x-\Lambda k_y k/\beta) \tag{3.173}$$

$$Y_2\eta=-C[g'(k_y b)/g(k_y b)]\,(k_x\Lambda+k_y k/\beta) \tag{3.174}$$

Each of the functions f and g must be either even or odd in order to simultaneously satisfy the boundary conditions on every two opposite walls.

Now, forming the products Z_1Y_1 and Z_2Y_2 and using (3.163) and (3.170), we easily establish the relationship

$$Z_1Y_1+Z_2Y_2=1 \tag{3.175}$$

which is equivalent to that derived by Dybdal *et al.* (1971) and referred to as the impedance compatibility relation. When this relation is satisfied, elementary mode solutions having fixed transverse wavenumbers are guaranteed. Otherwise, a natural mode will be composed generally of an infinite weighted sum of elementary modes. Examples of waveguides whose wall parameters do not meet (3.175) are shown in Fig. 3.10 On the other hand examples which do satisfy (3.175), and hence exhibit elementary mode solutions, are shown in Fig. 3.11.

In Fig. 3.10*a*, the four walls are highly, but imperfectly, conducting. Therefore $Z_1\simeq Y_1^{-1}$ and $Z_2\simeq Y_2^{-1}$. Substitution in the LHS of (3.175) shows that the equation is not satisfied. The four walls in Fig. 3.10*b* are transversely corrugated. It follows that $Z_1=Z_2=0$, while Y_1 and Y_2 can be finite. It is thus clear that (3.175) is not satisfied either in this case. Therefore, both guides of Fig. 3.10 do not possess elementary mode solutions.

Now we show that when two opposite walls of the rectangular guide are perfect electric (or magnetic) conductors (Fig. 11*a*), it is guaranteed that elementary mode solutions exist irrespective of the impedances of the other two walls. This cannot be seen directly from the impedance compatibility relation (3.175); hence we resort to the boundary conditions given in (3.171)—(3.174). For example if $Z_1 = 1/Y_1 = 0$ (electric side walls), then (3.171) and (3.172) are simultaneously satisfied by having:

$$f(k_x a) = 0 \tag{3.176}$$

which determines the set of solutions for k_x independently of k_y. The other two boundary conditions (3.173)—(3.174) determine k_y, and the hybrid factor Λ. So, we first eliminate Λ between these two equations to get, after some algebra,

$$[ug(u)/g'(u) + X_2 t^2/v]\,[ug'(u)/g(u) - B_2 t^2/v] = -(k_x \beta b^2/v)^2 \tag{3.177}$$

where

$$u = k_y b,\ t = (k_x^2 + k_y^2)^{1/2} b,\ v = kb,\ Z_2/\eta_0 = iX_2 \text{ and } Y_2\eta_0 = iB_2$$

Solution of this equation determines all the discrete eigenvalues of u or k_y. Together with the eigenvalues of k_x given by (3.176), they form the complete set of modal propagation constants of the guiding structure. The modal hybrid factors are obtained from either (3.173) or (3.174).

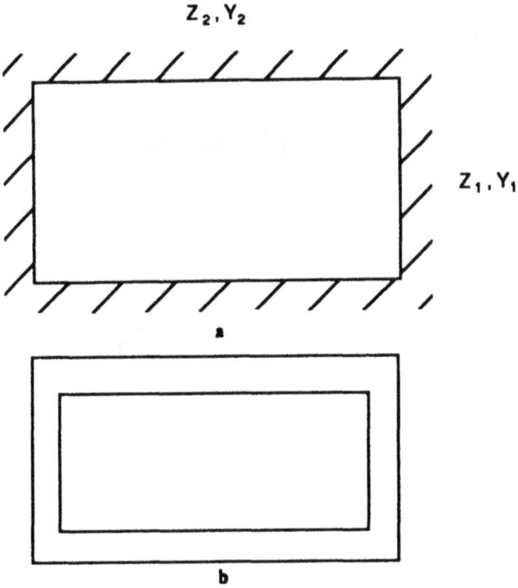

Fig. 3.10 Examples of rectangular waveguides which do not satisfy the mode compatibility relationship (3.175)

a Imperfectly conducting walls
b Transversely corrugated walls

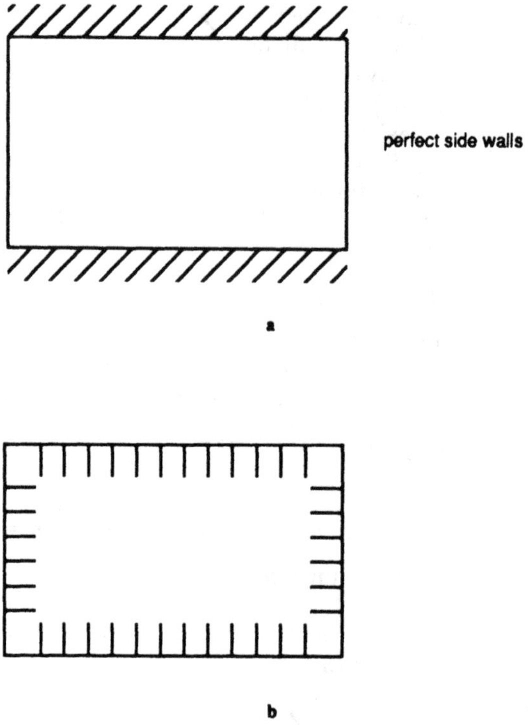

Fig. 3.11 Examples of rectangular waveguides which do satisfy the mode compatibility relationship (3.175)

a Two opposite walls are perfect electric (magnetic) conductors
b Longitudinally slotted walls

As an application of (3.177), consider the case of lossy upper and lower walls such that $Z_2 = 1/Y_2$, or $X_2 = -1/B_2$. (3.177) then reduces to

$$\left[\frac{X_2 g'(u)}{g(u)}\right] + \left[\frac{g(u)}{X_2 g'(u)}\right] = -\left[\frac{v}{ut^2}\right]\left[\left(\frac{k_x \beta b^2}{v}\right)^2 + \frac{t^4}{v^2} + u^2\right] = -(u/v + v/u)$$

In this form, the solutions are obviously given by

$$X_2 g'(u)/g(u) = -u/v \tag{3.178}$$

or

$$X_2 g'(u)/g(u) = -v/u \tag{3.179}$$

The corresponding hybrid factors are obtained from (3.173):

$$\Lambda_1 = vk_x/u\beta \tag{3.180}$$

and

$$\Lambda_2 = -u\beta/vk_x \tag{3.181}$$

A close look at the field components in (3.167)—(3.169) reveals that Λ_1 implies that $H_y=0$ and Λ_2 implies that $E_y=0$. Therefore, the solution given by (3.178) and (3.180) is *TM* to y and the solution given by (3.179) and (3.181) is *TE* to y.

Another example of a rectangular guide which supports elementary mode solutions is the longitudinally slotted guide shown in Fig. 3.11*b*. A large number of longitudinal ridges per wavelength imposes a zero longitudinal electric field at the four walls of the guide $x=\pm a$ and $y=\pm b$. It follows that this waveguide supports E modes (i.e. *TM* to z) exactly as a waveguide with smooth perfectly conducting walls. This explains the name E guide given to this waveguide by Dybdal *et. al.* (1971). However, the longitudinally slotted waveguide supports also H (or *TE* to z) modes as demonstrated by McIsaac (1974) who also shows that the dominant mode in the E guide can actually be an H mode! To show this, note that the relevant boundary conditions for the H modes are (3.171) and (3.173) with Λ tending to ∞; hence they reduce to the forms

$$X_1 = -[f(k_xa)/f'(k_xa)]kk_x/(k_x^2+k_y^2) \tag{3.182}$$

$$X_2 = -[g(k_yb)/g'(k_yb)]kk_y/(k_x^2+k_y^2) \tag{3.183}$$

where $X_{1,2} = -iZ_{1,2}/\eta_0$. These two equations are to be solved simultaneously for k_x and k_y.

To shed some light on the H modes, consider the special case of a square guide having $X_1=X_2=X$. The lowest order H mode will then have $k_x=k_y$, and we can choose for H_z an odd or even distribution as follows:

$$H_z = \begin{matrix} \sin(k_xx)\,\sin(k_yy) \\ \cos(k_xx)\,\cos(k_yy) \end{matrix}$$

for which (3.182) reduces to

$$\begin{matrix} \cot(k_xa) \\ \tan(k_xa) \end{matrix} = \pm 2k_xX/k$$

which may be readily solved for a given X. It is interesting to find the cutoff frequency of the H modes. So, we write $k=k_c=\sqrt{2}k_x$, whence this equation is easily solved for k_c:

$$k_ca = \sqrt{2}\,\cot^{-1}(\sqrt{2}X) \text{ for odd } H \text{ modes} \tag{3.184a}$$

and

$$k_ca = \sqrt{2}\,\tan^{-1}(-\sqrt{2}X) \text{ for even } H \text{ modes} \tag{3.184b}$$

It follows that when X is positive, corresponding to inductive walls, the dominant H mode is odd. Conversely, when X is negative for capactive walls, the dominant H mode is even. In either cae, the cutoff of the dominant H mode satisfies the inequality $k_ca \leqslant \pi/\sqrt{2}$. Since the cutoff wavenumber of the dominant E mode, the E_{11} mode, is given by $k_ca=\pi/\sqrt{2}$, it is clear that the dominant mode in a square longitudinally corrugated waveguide is an H mode; a result derived by McIssac (1974). An interesting special case is when $X=\infty$. In this case the first of equations (3.184) gives $k_ca=0$ which means that the dominant H mode is a *TEM* mode! Actually this result is independent of the cross sectional shape; that is, for any longitudinally slotted guide, the dominant mode is *TEM* if the wall impedance equals ∞. This fact finds application in the design of high gain radiators as will be further discussed in chapter 5.

3.6 Mode orthogonality in waveguides with constant impedance walls

Orthogonality of modes in closed waveguides with perfectly conducting walls has been demonstrated in section 2.9. The mode orthogonality property extends also to waveguides with constant surface impedance or admittance walls. It is our purpose in this section to prove this statement. To this end, consider a uniform cylindrical waveguide with an arbitrary cross section S and a side wall which is characterised by a constant (or mode independent) surface impedance Z and a constant surface admittance Y, defined precisely by:

$$Z = E_\tau / H_z, \text{ and } Y = -H_\tau / E_z \tag{3.185}$$

where τ is a transverse coordinate which is locally tangential to the side walls. With ρ a unit vector normal to the side wall and $\hat{\tau}$ a unit tangential vector, the set $(\hat{\rho}, \hat{\tau}, \hat{z})$ form a right handed set of mutually orthogonal unit vectors. We start by rewriting the Lorentz reciprocity relationship (2.74) applied to the fields of two modes:

$$\nabla.(\mathrm{E}_n \times \mathrm{H}'_m - \mathrm{E}'_m \times \mathrm{H}_n) = 0 \tag{3.186}$$

where $(\mathrm{E}_n, \mathrm{H}_n)$ are the vector fields of the n th mode in the waveguide with side wall parameters Z and Y, and $(\mathrm{E}'_m, \mathrm{H}'_m)$ are the vector fields of the m th mode in an identical waveguide except for wall parameters Z' and Y'. Each of these fields is composed of a transverse and a longitudinal component denoted here by lower case letters. So, for example, $\mathrm{E}_n = \mathrm{e}_n + e_{nz}\hat{z}$ and so forth. Now integrate (3.186) over a volume bounded by two cross sections at z and $z + dz$ and the side wall. Using the divergence theorem (A1.18), we get for $dz \to 0$

$$0 = \frac{\partial}{\partial z} \int_S (\mathrm{E}_n \times \mathrm{H}'_m - \mathrm{E}'_m \times \mathrm{H}_n).\hat{z}\, \mathrm{d}S + \oint_C (\mathrm{E}_n \times \mathrm{H}'_m - \mathrm{E}'_m \times \mathrm{H}_n).\hat{\rho}\, \mathrm{d}C \tag{3.187}$$

where the first integral is taken over the cross section S and the second is a closed integral over the contour C bounding S. Noting that the only modal fields involved in the first integral are the transverse fields, one can use these in place of the total fields. We also note that the derivative $(\partial/\partial z)$ is equivalent to the factor $-i(\beta_n + \beta'_m)$. Evaluating the contour integral on C by expanding the integrand in terms of the field components tangential to the side wall, we get

$$\oint_C (\mathrm{E}_n \times \mathrm{H}'_m - \mathrm{E}'_m \times \mathrm{H}_n).\hat{\rho}\, \mathrm{d}C = \oint_C \{(e_{n\tau}h'_{mz} - e'_{m\tau}h_{nz}) - (e_{nz}h'_{m\tau} - e'_{mz}h_{n\tau})\}\, \mathrm{d}C$$

$$= \delta Y \oint_C e_{nz}e'_{mz}\, \mathrm{d}C - \delta Z \oint_C h_{nz}h'_{mz}\, \mathrm{d}C \tag{3.188}$$

where $\delta Y = Y' - Y$ and $\delta Z = Z' - Z$, and the boundary conditions (3.185) has been used to obtain the last equality. Using (3.188) in (3.187), the latter becomes

$$\int_S (\mathbf{e}_n \times \mathbf{h}'_m - \mathbf{e}'_m \times \mathbf{h}_n).\hat{z}\, dS = \frac{-i}{\beta_n + \beta'_m}\left(\delta Y \oint_C e_{nz} e'_{mz}\, dC - \delta Z \oint_C h_{nz} h'_{mz}\, dC\right) \tag{3.189}$$

By reversing the direction of propagation of the n th mode the following changes are incurred: $\beta_n \rightarrow -\beta_n$, $\mathbf{h}_n \rightarrow -\mathbf{h}_n$, $e_{nz} \rightarrow -e_{nz}$, whence (3.189) becomes

$$\int_S (\mathbf{e}_n \times \mathbf{h}'_m + \mathbf{e}'_m \times \mathbf{h}_n).\hat{z}\, dS = \frac{-i}{\beta_n - \beta'_m}\left(\delta Y \oint_C e_{nz} e'_{mz}\, dC + \delta Z \oint_C h_{nz} h'_{mz}\, dC\right) \tag{3.190}$$

These two equations are generalisation of those obtained by Dragone (1977) in the context of mode orthogonality and coupling in transversely corrugated waveguides. Adding the two equations gives:

$$\int_S (\mathbf{e}_n \times \mathbf{h}'_m).\hat{z}\, dS = \frac{-i}{\beta_n^2 - \beta_m'^2}\left(\beta_n \delta Y \oint_C e_{nz} e'_{mz}\, dC + \beta'_m \delta Z \oint_C h_{nz} h'_{mz}\, dC\right) \tag{3.191}$$

In the case when the two modes belong to the same waveguide, i.e. when $Y = Y'$ and $Z = Z'$, the RHS is equal to zero provided that $\beta_n \neq \pm\beta_m$. Therefore, for modes in the same guide:

$$\int_S (\mathbf{e}_n \times \mathbf{h}_m).\hat{z}\, dS = N_n \delta_{nm} \tag{3.192}$$

where

$$N_n = \int_S (\mathbf{e}_n \times \mathbf{h}_n).\hat{z}\, dS \tag{3.193}$$

and δ_{nm} is equal to zero except when $n = m$, whence it is equal to unity. The mode normalization factor N_n is evaluated by taking the limit of (3.191) as δY tends to zero, δZ tends to zero, $m = n$, and therefore β'_m tends to β_n. This results in

$$N_n = \frac{i}{2}\,\frac{\oint_C e_{nz}^2\, dC + (dZ/dY)\oint_C h_{nz}^2\, dC}{[\partial\beta_n/\partial Y + (\partial\beta_n/\partial Z)(dZ/dY)]} \tag{3.194}$$

Equation (3.192) is a statement of mode orthogonality in a waveguide with mode independent impedance walls. On the other hand, coupling between modes in different waveguides with impedance walls is given by (3.191). It is important to note that the property of mode orthogonality facilitates the solution of the problems related to mode excitation and scattering as discussed in chapter 2; namely, all that has been said in sections 2.10—2.12 about modal expansion of fields, mode excitation and mode scattering at a discontinuity is perfectly applicable to waveguides with constant impedance walls.

3.7 Mode characteristics in conical horns with impedance walls

A conical horn with an impedance wall is depicted in Fig. 3.12. In a spherical coordinate frame, the conical wall coincides with the surface $\theta = \theta_1$ and is characterised by a surface impedance Z and admittance Y. It is our task to find the allowable modes of propagation inside the horn. It turns out that, in order to obtain valid modal solutions by using separation of variables, it is necessary that both Z and Y vary linearly with the radial distance r. As in cylindrical waveguides, the modes are generally hybrid for finite Z and Y unless the fields have a zero azimuthal dependence.

With the aid of equations (1.136)—(1.141), we can write down explicitly the field components of a hybrid mode as:

$$E_r = \frac{\nu(\nu+1)}{ikr^2} \hat{H}^{2}_{\nu}(kr) P^m_\nu(\cos\theta) \cos m\phi \tag{3.195}$$

$$\eta H_r = \frac{\nu(\nu+1)}{ikr^2} \Lambda \hat{H}^{(2)}_{\nu}(kr) P^m_\nu(\cos\theta) \sin m\phi \tag{3.196}$$

$$rE_\theta = -\hat{H}^{(2)}_\nu(kr) P^m_\nu(\cos\theta)(\sin\theta)^{-1}[\hat{\beta} F_m(\nu;\theta) + m\Lambda] \cos m\phi \tag{3.197}$$

$$rE_\phi = \hat{H}^{(2)}_\nu(kr) P^m_\nu(\cos\theta)(\sin\theta)^{-1}[\Lambda F_m(\nu;\theta) + m\hat{\beta}) \sin m\phi \tag{3.198}$$

$$\eta r H_\theta = -\hat{H}^{(2)}_\nu(kr) P^m_\nu(\cos\theta)(\sin\theta)^{-1}[\Lambda\hat{\beta} F_m(\nu;\theta) + m] \sin m\phi \tag{3.199}$$

$$\eta r H_\phi = -\hat{H}^{(2)}_\nu(kr) P^m_\nu(\cos\theta)(\sin\theta)^{-1}[F_m(\nu;\theta) + m\Lambda\hat{\beta}] \cos m\phi \tag{3.200}$$

where:

$$F_m(\nu;\theta) = \sin\theta[\partial P^m_\nu(\cos\theta)/\partial\theta]/P^m_\nu(\cos\theta) \tag{3.201}$$

$\eta = (\mu/\varepsilon)^{1/2}$ is the unbounded wave impedance, Λ is the mode hybrid factor and $\hat{\beta}$ is a normalised, radially dependent, phase parameter and is defined by:

$$\hat{\beta} = iH'^{(2)}_\nu(kr)/H^{(2)}_\nu(kr) \tag{3.202}$$

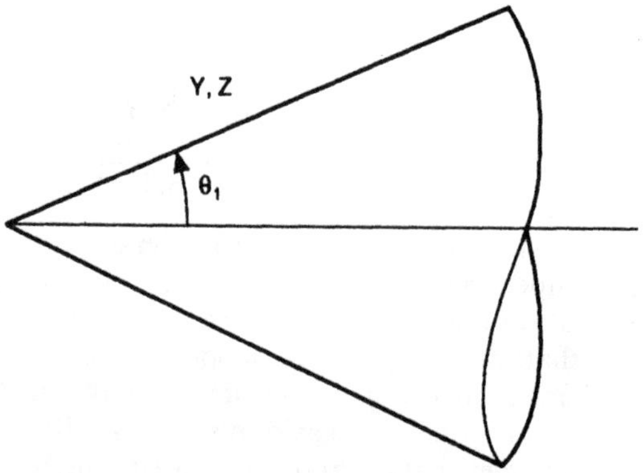

Fig. 3.12 A conical horn with constant impedance wall

The boundary conditions at the wall $\theta=\theta_1$ require that:

$$E_\phi/H_r=Z \tag{3.203}$$

$$-H_\phi/E_r=Y \tag{3.204}$$

where Z and Y are assumed constants in the sense that they are independent of the separation variables ν and m. Application of the above two boundary conditions, using (3.195)–(3.202), leads to:

$$F_m(\nu;\ \theta_1)-X\nu(\nu+1)\ \sin\ \theta_1=-m\beta\Lambda \tag{3.205}$$

$$F_m(\nu;\ \theta_1)-B\nu(\nu+1)\ \sin\ \theta_1=-m\beta\Lambda \tag{3.206}$$

where X and B are normalised reactance and susceptance defined by:

$$X=-iZ/\eta kr \tag{3.207}$$

$$B=-iY\eta/kr \tag{3.208}$$

Multiplying equations (3.205) and (3.206) together yields:

$$[F_m(\nu;\ \theta_1)-X\nu(\nu+1)\ \sin\ \theta_1][F_m(\nu;\ \theta_1)-B\nu(\nu+1)\ \sin\ \theta_1]=(m\beta)^2 \tag{3.209}$$

This is the modal equation for determining the separation variables ν and is a generalisation of that derived by Clarricoats and Saha (1971*b*) for corrugated conical horns. In order to validate the principle of separation of variables, this equation must be independent of the radial coordinate r, and hence the parameters X, B and β should also be independent of this coordiante. It is then concluded from (3.207)—(3.208) that both Z and Y must vary linearly with r, a condition which was reported early by Felsen (1959) in a treatment of a similar situation. The propagation parameter β is clearly a function of r as seen from equation (3.202). It takes mainly imaginary values for $kr<\nu$, mainly real values for $kr\gg\nu^2$ and complex values for intermediate values of kr. This behaviour reflects the well known phenomenon of gradual cutoff of radial and spherical modes. If, however, kr is sufficiently larger than ν^2, then β becomes weakly dependent on r as can be verified from the asymptotic evaluation of (3.202) which leads to:

$$\beta\simeq 1-\nu(\nu+1)/2(kr)^2 \tag{3.210}$$

So, under the condition $kr>\nu^2$, the modal eigenvalues ν are almost independent on r. By referring to the modal field expressions (3.195)—(03.200), it is clear that, unless the mode is pure TE or TM to r, i.e. $\Lambda=\infty$ or 0, the variation of β with r results in a varying transverse field distribution. As a consequence, the orthogonality between modes at any given r is not exactly satisfied (Mahmoud and Clarricoats, 1982), or in other words, the modes are intrinsically coupled. Of course, if kr is sufficinetly large, the conical structure approaches a cylinder for which the modes are readily orthogonal.

Returning back to (3.205) and (3.206) and subtracting one from the other, we get:

$$m\beta(\Lambda-\Lambda^{-1})=\nu(\nu+1)\ \sin\ \theta_1(B-X) \tag{3.211}$$

which can be solved as a quadratic algebraic equation, yielding Λ in terms of the normalised surface parameters; namely:

$$\Lambda = w(B-X) \pm [w^2(B-X)^2+1]^{1/2} \tag{3.212}$$

where $w = \nu(\nu+1)\sin\theta_1/2m\beta$.

Now we are in a position to consider some of the important special cases which should give more insight into mode characteristics in conical horns.

Smooth perfectly conducting wall

For a perfect electic wall, $X=0$ and $B=\infty$. Therefore $\Lambda=0$ or $\Lambda=\infty$ as can be deduced from (3.212), signifying that the modes fall into pure *TM* and *TE* (to *r*) categories. Equation (3.206) applies to *TM* modes (with $\Lambda=0$); hence this reduces to $F_m(\nu;\theta_1)=0$, or:

$$P_\nu^m(\cos\theta_1)=0 \tag{3.213}$$

Similarly, equation (3.205) applies to *TE* modes (with $\Lambda^{-1}=0$) and reduces to:

$$(\partial/\partial\theta)P_\nu^m(\cos\theta)|_{\theta=\theta_1}=0 \tag{3.214}$$

For a given θ_1 and m, (3.213) and (3.214) define the modal eigenvalues ν_n, $n=1, 2 \ldots$ for *TM* and *TE* modes respectively. Modes are thus labelled by TM_{mn} and TE_{mn}. In general (3.213) and (3.214) are solved numerically for the eigenvalues ν_n.

Balanced hybrid mode condition

From (3.212) it is clear that, when $X=B$, the mode hybrid factor Λ assumes either one of the values ± 1; i.e. the *TE* and *TM* parts of the mode have equal weighting. The condition $X=B$ may then be referred to as the balanced hybrid mode condition. Some of the main mode characteristics under this condition have been already mentioned in section 3.4 in connection with cylindrical guides. Similar characteristics in conical horns will now be shown.

With $X=B$, the modal equation (3.209) reduces to:

$$F_m(\nu;\theta_1) - B\nu(\nu+1)\sin\theta_1 \pm m\beta = 0 \tag{3.215}$$

where, from (3.205)–(3.206), the upper and lower signs refer respectively to $\Lambda=\pm 1$. With kr sufficiently larger than ν^2, β is well approximated by unity and the ν values obtained from solving (3.215) will then be independent of r, in conformity with the concept of separation of variables. If in addition $B=Z=0$, as in the case of a corrugated wall at balanced hybrid mode condition, (3.215) reduces to the simpler form:

$$F_m(\nu;\theta_1) \pm m = 0;\ \Lambda=\pm 1 \tag{3.216}$$

A look at the transverse field distributions given by (3.197)–(3.200) reveals that, under the above condition, these fields vanish at the cone wall $\theta=\theta_1$. This explains the low attenuation of modes operating near the balanced hybrid mode condition. The field tapering of the lowest order mode on the horn's aperture makes this mode a good radiator, with lower side lobe radiation pattern and negligible edge diffraction. The aperture fields will also exhibit a high polarisation purity as will be explained in chapter 5.

Equation (3.216) has been solved numerically for the three lower order modes of the conical guide by Clarricoats and Saha (1971*b*). Approximate, but closed form, solutions for the modal equations will now be obtained by expressing the Legendre function in terms of a cylindrical Bessel function.

Approximate solutions of the modal equation
For sufficiently small values of θ, it can be shown that the Legendre function is well approximated by the cylindrical Bessel function [Narasimhan *et al.*, 1970 and Knop *et. al.*, 1986], so that:

$$P_\nu^m(\cos\theta) \simeq C_{m\nu} J_m(q\theta) \tag{3.217}$$

where $q=[\nu(\nu+1)]^{1/2}$ and $C_{m\nu}$ is a constant which may be obtained by comparing the asymptotic expressions of both sides for large values of ν. So, as ν tends to ∞ and $\theta \simeq \sin\theta$,

$$P_\nu^m(\cos\theta) \rightarrow (\nu+\tfrac{1}{2})^{1/2}(2/\pi\theta)^{1/2}\cos\chi$$

$$J_m(q\theta) \rightarrow (-)^m(2/\pi q\theta)^{1/2}\cos\chi$$

where $\chi=(\nu+\frac{1}{2})\theta-\pi/4+m\pi/2$, and q is approximated by $q \simeq \nu+\frac{1}{2}$. We therefore deduce that $C_{m\nu}=(-)^m q$.

Using the approximate relation (3.217) in the modal equations (3.213), (3.214) and (3.216), these equations reduce to:

$$J_m(q\theta_1)=0 \tag{3.218}$$

$$(\partial/\partial\theta)J_m(q\theta)|_{\theta=\theta_1}=0 \tag{3.219}$$

for *TM* and *TE* modes, respectively, in a smooth perfectly conducting wall cone, and

$$q\theta_1 J_m'(q\theta_1)/J_m(q\theta_1) \pm m = 0 \tag{3.220}$$

for balanced hybrid modes. Solutions of (3.218)—(3.220) are respectively given by:

$$q = x_{mn}/\theta_1 \simeq \nu+\tfrac{1}{2} \tag{3.221}$$

$$q = x'_{mn}/\theta_1 \simeq \nu+\tfrac{1}{2} \tag{3.222}$$

$$\left.\begin{aligned} q &= x_{m-1,n}/\theta_1 \simeq \nu+\tfrac{1}{2} \text{ for } HE_{mn} \text{ modes, and} \\ &= x_{m+1,n}/\theta_1 \simeq \nu+\tfrac{1}{2} \text{ for } EH_{mn} \text{ modes} \end{aligned}\right. \tag{3.223}$$

where x_{mn} and x'_{mn} are the n th zeros of $J_m(x)$ and $J_m'(x)$ respectively. Equation (3.223) has been tested against exact solutions for the lowest three order modes and is found to deviate by no more than 1% for flare angles θ_1 up to 60° (Mahmoud and Clarricoats, 1982).

3.8 Problems

3.1 A constant impedance surface can be described in terms of an impedance matrix relating the tangential E components to the tangential H components on the surface; e.g. for a circular cylindrical surface:

$$E_\phi = Z_{\phi\phi}H_\phi + Z_{\phi z}H_z$$

$$E_z = Z_{z\phi}H_\phi + Z_{zz}H_z$$

Find the relation between the elements of this Z matrix and the Z, Y, p descriptors given in (3.30)—(3.31). In the special case $p=0$, show that $Z_{\phi\phi}=Z_{zz}=0$, $Z_{\phi z}=Z$ and $Z_{z\phi}=-Y^{-1}$.

3.2 Sketch Z and Y as functions of β/k_0 for a planar interface between air and a dielectric medium with $\varepsilon_r = 6$ in the range $0 \leqslant \beta/k_0 \leqslant 1$, β being the phase constant along the interface; hence the indicated range of β/k_0 corresponds to that between normal and grazing incidence to the interface. Repeat for $\varepsilon_r = 1{\cdot}4$ and comment on the results. (Use equations (3.8) and (3.9) with $\lambda = 0$.)

3.3 Derive equations (3.20)—(3.25) giving the reflection parameters of a plane wave incident on a planar interface characterised by impedance Z and admittance Y.

3.4 Find the surface impedance and admittance of earth at a certain location having $\varepsilon_r = 6$ and $\sigma = 0{\cdot}01$ mho/m, at the frequencies 1,10,100 and 1000 MHz.

3.5 A cylindrical surface with transverse inner corrugations has a radius equal to 1·5 of a free space wavelength. Find the slot depth that will yield zero surface admittance. Assume many corrugations per wavelength.

3.6 Find the surface impedance of a longitudinally corrugated cylindrical surface having a radius of 1·5 free space wavelength and a slot depth of a quarter wavelength, when the longitudinal wavenumber $\beta/k_0 = 0{\cdot}7$ or $0{\cdot}95$. Consider the two cases when the slots are air filled and when filled with a dielectric with $\varepsilon_r = 2{\cdot}5$ and comment.

3.7 A plane wave is incident on a planar corrugated surface with the plane of incidence parallel to the slot ridges and the angle of incidence is near grazing. If the slots are filled with a dielectric with $\varepsilon_r = 2{\cdot}5$, find the slot depth which will make the surface behave as a magnetic wall for *TE* waves.

Consider a rectangular waveguide with two opposite walls behaving as magnetic walls and the other two as electric walls. Show that a *TEM* mode can be supported in such a guide.

3.8 Prove the approximate relations (3.64) and (3.65) by using the asymptotic expressions of the Bessel functions.

3.9 Consider a dielectric coated circular cylinder of radius b and dielectric thickness t. If the surface $\rho = b$ is characterised by an impedance $Z(b)$ and admittance $Y(b)$, find expressions for $Z(a)$ and $Y(a)$ at the dielectric air interface. Show that in the limit $a \gg 1$ and $t \ll a$, $Z(a)$ and $Y(a)$ may be approximated by:

$$\hat{Z}(a) \simeq \frac{\hat{Z}(b)\,(1 + \tan vt/2va) + \mathrm{i}\tan vt}{\mathrm{i}Z(b)\tan vt + (1 - \tan vt/2va)}$$

with a similar expression obtained by replacing Z by Y. Here $\hat{Z}(a) = Z(a)v/\omega\mu$, $\hat{Y}(a) = Y(b)v/\omega\varepsilon$, and similar definitions apply for $\hat{Z}(b)$ and $\hat{Y}(b)$.

Consider the limiting case of a planar coated surface ($a \to \infty$) and show that the above relations can be derived from an equivalent transmission line representation.

3.10 Show that the phase velocity of the Zenneck wave on a lossy homogeneous earth is always greater than c. Sketch the phase velocity and attenuation of the Zenneck wave between 1 Mhz and 1 GHz, for an earth with $\varepsilon_r = 6$ and $\sigma = 0{\cdot}01$ mho/m.

3.11 Derive the approximate relations (3.83) and (3.85). Write a computer program to solve the modal equation (3.82) for u of the surface wave modes. Use

(3.83) or (3.85) as initial guessses for the roots. Take Y_1 and Y_2 to represent pure inductive surfaces.

3.12 Consider *TE* modes in a constant impedance parallel plane waveguide. For *TE* modes, *E* is totally polarised parallel to the impedance planes. Find the modal equation and solve it approximately under the condition $Z\eta_0^{-1}/k_0 d \ll 1$, where d is the guide width. Show that the attenuation factor α_n (neper/m) of mode n is given by

$$\alpha_n/k_0 = \frac{\text{Real}(Z\eta_0^{-1})(n\pi)^2}{(k_0 d)^3(\beta_{n0}/k_0)}$$

where $\beta_{n0}/k_0 = (1-(n\pi/k_0 d)^2)^{1/2}$

Compare with the attenuation of *TM* modes; namely the attenuation of the lowest order *TE* and *TM* modes when $k_0 d = 4\pi$ and $Z = Y^{-1} = \eta_0/3$.

3.13 Derive equations (3.94) and (3.95). Find the attenuation of the first lower order three modes in the earth-ionosphere guide given: frequency = 15 kHz, earth's conductivity = 0·01 mho/m, ionospheric reflecting layer height = 70 km and $\sigma_{\text{ionosphere}}/\omega\varepsilon_0 = 1$.

3.14 Compare equation (3.117) with the result that you would obtain by applying equation (2.93) of section 2.11. Therefore, show that the modal normalising factor N_p is given by:

$$N_p \equiv -(\partial D/\partial\beta)|_{\beta_p} u_p R_1(u_p)/\omega\varepsilon_0.$$

In this context N_p is defined by:

$$N_p = (\beta_p/\omega\varepsilon_0)\int_0^d f_p^2(x)\,\mathrm{d}x$$

3.15 Consider a cylindrical surface $\rho = a$ having a constant outer surface reactance X. Find the surface wave supported by this surface; namely, take H to be directed totally along the axial direction z, so that

$$H_z = H_\nu^{(2)}(k\rho)\exp(-i\nu\phi)$$

where the fields are assumed to be independent of z. Show that ν is determined from:

$$X/\eta_0 = H_\nu^{(2)\prime}(ka)/H_\nu^{(2)}(ka)$$

Since ν is expected to be of the same order as ka, you may use the Airy function approximation for the Hankel function. Show then that the azimuthal phase constant ν/a is approximated for large ka by

$$\nu/a \simeq k(1 + X^2/\eta_0^2)^{1/2}$$

[*Hint:* Use equation (A2.55); see a discussion of a related problem by Wait (1960)].

3.16 Derive equations (3.143) and (3.144) for a circular waveguide with a low impedance wall. Compute the attenuation and phase velocity of the TE_{11} mode in a 6 cm diameter guide operating at 10 GHz. Take the wall material to be copper with $\sigma = 5{\cdot}7\times 10^7$ mho/m. Find the ratio of the attenuation due to the longitudinal currents and due to the circumferential currents. Find also the attenuation of the TE_{01} mode

3.17 Derive equations (3.153)—(3.155) which are approximate solutions to (3.134) for $B \ll 1$ and $X \ll 1$. Compare the attenuation of a smooth wall circular guide (operating in the TE_{11} mode) with a transversely corrugated guide (supporting the HE_{11} mode). The two guides have the same inner diameter $a=5$ cm. The corrugated guide has a corrugation depth of 0·83 cm and the wall conductivity for both guides $=5{\cdot}7\times 10^7$ mho/cm. The frequency $=9$ GHz.

3.18 Compare the attenuation of the TE_{11} mode in a smooth wall circular waveguide with a dielectric coated guide supporting the HE_{11} mode. Both guides have the same diameter d such that $k_0 d=20$, and same wall material. The dielectric material is lossless and has a relative permittivity $\varepsilon_r=2{\cdot}56$ and a thickness t such that $k_0\ (\varepsilon_r-1)^{1/2}t=\pi/4$.

3.19 Use equations (3.171)—(3.174) to derive the mode compatibility relationship (3.175) in a rectangular waveguide with impedence walls.

3.20 Derive equations (3.177) and those following it up to (3.181).

3.21 Derive equations (3.180) and (3.181) by assuming TM or TE to y modes right at the outset of the analysis of the rectangular guide of Fig. 3.10*a*.

3.22 Solve (3.182) for the propagationg H modes in a square waveguide with longitudinally slotted walls whose reactance $=0{\cdot}7\ \eta_0$ and with side equal to a when $ka=2\pi$; namely, find the phase constant β for each of these modes and sketch their field distribution.

3.23 A rectangular waveguide with longitudinally slotted walls has dimensions $a\times b$. The side walls (with dimension b) have a normalised reactance X_1 and the other two walls have a normalised reactance $=X_2$. If $X_1 a=X_2 b$, show that valid solutions of H modes are given by:

$$H_z=\begin{matrix}\sin(ux/a)\ \sin(uy/b)\\ \cos(ux/a)\ \cos(uy/b)\end{matrix}$$

with u determined from

$$\pm X_1=\begin{matrix}\cot(u)\\ \tan(u)\end{matrix}\ ka/[u(1+a^2/b^2)]$$

3.24 Consider H modes in a longitudinally corrugated circular cylindrical waveguide. Show that the modal equation for the transverse wavenumber k_ρ is given by:

$$iZ/\omega\mu a=J'_m(k_\rho a)/k_\rho a J_m(k_\rho a)$$

where Z is the wall impedance and a is the inner radius. Find the conditions under which a TEM mode can exist.

3.25 Consider a junction between a smooth, perfectly conducting wall waveguide and a waveguide with wall parameters Z' and Y'. The two guides have identical cross sectional shapes and areas S. Show that, in this case, (3.191) does not apply directly. Instead, the inner product $\langle \mathrm{e}_n, \mathrm{h}'_m\rangle$ is given by

$$\langle \mathrm{e}_n, \mathrm{h}'_m\rangle=\frac{-i}{\beta_n^2-\beta'^2_m}\left[\beta_n \oint_C h_{n\tau} e'_{nz}\,dC+\beta'_m Z' \oint_C h_{nz} h'_{mz}\,dC\right]$$

Write a similar expression for $\langle e'_m, h_n \rangle$.

Use the above expressions to find the mode scattering coefficients for a junction between two parallel plate waveguides. The guide on the right of the junction has transversely corrugated metallic walls and the guide on the left has smooth walls. Take the incident mode from the left guide to be the *TEM* mode.

3.26 Using the differential equations of the Legendre function and the Bessel function, show that the $P_v^m(\cos\theta)$ can be expressed in terms of $J_m(q\theta)$ for small θ. Hence compare between the modal equations (3.134) and (3.209) for circular cylindrical and conical waveguides; namely, show that the former equation can be obtained from the latter.

3.27 Write a computer program to compute $P_v^m(\cos\theta)$ from the following identity:

$$P_v^m(\cos\theta) = (-2)^{-m}[(v+m)(v+m-1)\ldots(v-m+1)](\sin\theta)^m$$
$$\times F(1+m+v, m-v; 1+m; \tfrac{1}{2}\sin^2\theta)$$

where $F(a, b; c, z)$ is the hypergeometric function (e.g. Abramowitz and Stegun, 1965). The above identity applies for integer m, but arbitrary v. For large values of v, you may compute $P_{v-n}^m(.)$ and $P_{v-n-1}^m(.)$ where n is a suitable integer $\langle |v|$ and then use the recurrence relation

$$(v-m+1)P_{v+1}^m(x) = (2v+1)xP_v^m(x) - (v+m)P_{v-1}^m(x)$$

3.28 Solve the modal equation (3.215) by perturbation analysis, starting from the solutions of (3.216) and the condition $Bv(v+1)\sin\theta_1 \ll 1$.

3.29 Consider a conical section bounded by the conical surface $\theta = \theta_1$ and the spherical sections r and $r+dr$. Integrate the reciprocity relationship (3.186) over the volume contained inside these surfaces by taking:

$$(\mathrm{E}, \mathrm{H}) = (\mathrm{m}_{omv}, i\mathrm{n}_{omv}/\eta_0) \text{ and } (\mathrm{E}', \mathrm{H}') = (\mathrm{m}_{omw}, i\mathrm{n}_{omw}/\eta_0)$$

and using the divergence theorem to prove the identity:

$$\int_0^{\theta_1} [P_v^m(\cos\theta)P'^m_w(\cos\theta) + m^2 P_v^m(\cos\theta)P_w^m(\cos\theta)/\sin^2\theta]\sin\theta\, d\theta$$

$$= \frac{\sin\theta_1}{\nu(\nu+1) - w(w+1)}[\nu(\nu+1)P_v^m(\cos\theta_1)P'^m_w(\cos\theta_1)$$

$$-w(w+1)P_w^m(\cos\theta)P'^m_v(\cos\theta)]$$

where $P'^m_v(\cos\theta) \equiv (\partial/\partial\theta)P_v^m(\cos\theta)$

Application of this identity appears in chapter 5.

3.9 References

ABRAMOWITZ, M., and STEGUN, I.A. (1965): 'Handbook of mathemtaical functions' (Dover Publications, New York)

BARLOW, H.M., and BROWN, J. (1962): 'Radio surface waves' (Clarendon Press, Oxford)

CLARRICOATS, P.J.B., and SAHA, P.K. (1971*a*): 'Propagation and radiation behaviour of corrugated feeds. Pt. 1: Corrugated waveguide feeds', *Proc. IEE*, **118**, pp. 1167–1176

CLARRICOATS, P.J.B., and SAHA, P.K. (1971*b*): 'Progagation and radiation behaviour of corrugated feeds, Pt. 2: Corrugated conical horn feeds', *Proc. IEE*, **118**, pp. 1177–1186

CLARRICOATS, P.J.B., OLVER, A.D., and CHONG, S.L. (1975*a*): 'Attenuation in corrugated circular waveguides. Pt. 1: Theory', *Proc. IEE*, **122**, pp. 1173–1179

CLARRICOATS, P.J.B., OLVER, A.D., and CHONG, S.L. (1975*b*): 'Attenuation in corrugated circular waveguides. Pt. 2: Experiment', *Proc. IEE*, **122**, p. 1180

CLARRICOATS, P.J.B., and OLVER, A.D. (1984): 'Corrugated horns for microwave antennas', IEEE Electromagnetic waves series 18, (Peter Peregrinus Ltd.)

CLEMMOW, P.C. (1966): 'The plane wave spectrum representation of electromagnetic fields' (Pergamon Press)

DRAGONE, C. (1977): 'Reflection, transmission and mode conversion in a corrugated feed', *Bell Syst. Tech. J.*, **56**, pp. 835–867

DRAGONE, C. (1980): 'Attenuation and radiation characteristics of the HE_{11} mode', *IEEE Trans.* **MTT–28**, pp. 704–710

DRAGONE, C. (1981): 'High frequency behaviour of waveguides with finite surface impedance', *Bell Syst. Tech. J.*, **60**, pp. 89–115

DYBDAL, R.B., PETERS, L., and PEAKE, W.K. (1971): 'Rectangular waveguides with impedance walls' *IEEE Trans.*, **MTT–19**, pp. 2–9

FELSEN, L.B. (1969): 'Electromagnetic properties of wedge and cone surfaces with a linearly varying surface impedance', *IEEE Trans.*, **AP–7**, pp. S231–S243

HILL, D.A., and WAIT, J.R. (1978): 'Excitation of the Zenneck surface wave by a vertical aperture', *Radio Science*, **13**, pp. 969–977

HILL, D.A., and WAIT, J.R. (1980): 'On the excitation of the Zenneck surface wave over the ground at 10 MHz', *Ann. Telecomm.*, **35**, pp. 179–182

KING, R.J., and WAIT, J.R. (1976): 'Electromagnetic ground wave propagation theory and experiment' *in* 'Symposia Mathematica' (Academic Press, New York) Vol. 18, pp. 107–208

KNOP, C.M., CHENG, Y.B., and OSTERTAG, E.L. (1986): 'On the fields in a conical horn having an arbitrary wall impedance', *IEEE Trans.* **AP–34**, pp. 1092–1098

MAGNUS, W., and OBERHETTINGER, F. (1949): 'Formulas and theorems for the special functions of mathematical physics' (Chelsea Publishing Co., New York, NY) chap 4

MAHMOUD, S.F., and CLARRICOATS, P.J.B. (1982): 'Radiation from wide flare-angle corrugated conical horns', *IEE Proc.*, Pt. H, **129**, pp. 221–228

McISSAAC, P.R. (1974): 'Comments on rectangular waveguides with impedance walls', *IEEE Trans.*, **MTT–22**, pp. 972–973

NARASIMHAN, M.S., and RAO, B.V. (1970): 'Hybrid modes in corrugated conical horns', *Electron. Lett.*, **6**, pp. 32–34

WAIT, J.R. (1964): 'Electromagnetic surface waves', *in* 'Advances in radio research' (Academic Press, London) Vol. 1, pp. 157–217

WAIT, J.R. (1967*a*): 'A fundamental difficulty in the analysis of cylindrical waveguides with impedance walls', *Electron. Lett.*, **3**, pp. 87–88

WAIT, J.R. (1967*b*): 'On the theory of shielded surface waves', *IEEE Trans.*, **MTT–15**, pp. 410–414

WAIT, J.R. (1968): 'The whispering gallery nature of the earth ionosphere waveguide at VLF', *IEEE Trans.*, **AP–16**, p. 147

WAIT, J.R. (1969): 'Electromagnetic radiation from conical structures', *in* COLLIN and ZUCKER, (Ed.): 'Antenna theory' (McGraw Hill) Chap 12, pp. 483–522

WAIT, J.R. (1970): 'Electromagnetic waves in stratified media' (Pergamon Press)

WAIT, J.R., and HILL, D.A. (1979): 'Excitation of the HF surface wave by vertical and horizontal antennas', *Radio Science*, **17**, pp. 767–779

WAIT, J.R. (1986): 'Introduction to antennas and propagation' (Peregrinus Ltd, London) sec. 6.8

YEH, C. (1964): 'An application of Sommerfield's complex order wave functions to an antenna problem', *J. Math. Phys.* **5**, pp. 344–350

Chapter 4

Open waveguides

4.1 Introduction

Open waveguides are characterised by imperfectly reflecting, or open, boundaries. Thus, beside power guided along the waveguide, a certain portion can escape away from the guiding structure. The natural modes of propagation on an open waveguide fall into two main categories: guided modes whose power flow is confined to the near vicinity of the guide and radiation modes which account for radiated power.

Examples of open waveguides include the single constant impedance surface, the dielectric planar slab and the dielectric rod. A vast variety of planar waveguides particularly used in millimetric wave circuits are identified as open waveguides. Besides, most natural waveguides such as atmospheric ducts, the earth ionosphere waveguide and tunnels within the earth's medium can be considered as open waveguides.

The simplest open waveguide structure to start with is the single planar impedance surface, considered earlier in section 3.3. Besides the surface wave mode studied earlier, we shall derive radiation modes which are often called pseudomodes. Pseudomodes form a continuous spectrum over the transverse wavenumber and account for radiation as will be seen below. Together, the surface wave mode and the continuous spectrum of pseudomodes constitute the complete spectrum of modes in terms of which the fields of an arbitrary source can be expanded. In general, there is more than one surface wave mode on an open waveguide and they form a discrete spectrum over the transverse wavenumber.

In section 4.2 we derive the complete spectrum of modes for each of the single planar impedance surface, the planar dielectric slab and the dielectric rod. Orthogonality of modes in a given waveguide and mode coupling at a junction between two different waveguides are treated in section 4.3. In the rest of the chapter we present an analytical account of guided modes in a few examples of millimetric waveguides.

4.2 Mode spectrum on open waveguides

In this section we deal with the complete spectrum of modes on three open waveguide structures which allow rigorous analysis. These are the planar constant impedance surface, the planar dielectric slab and the dielectric rod waveguides as depicted in Fig. 4.1*a*, *b*, *c*.

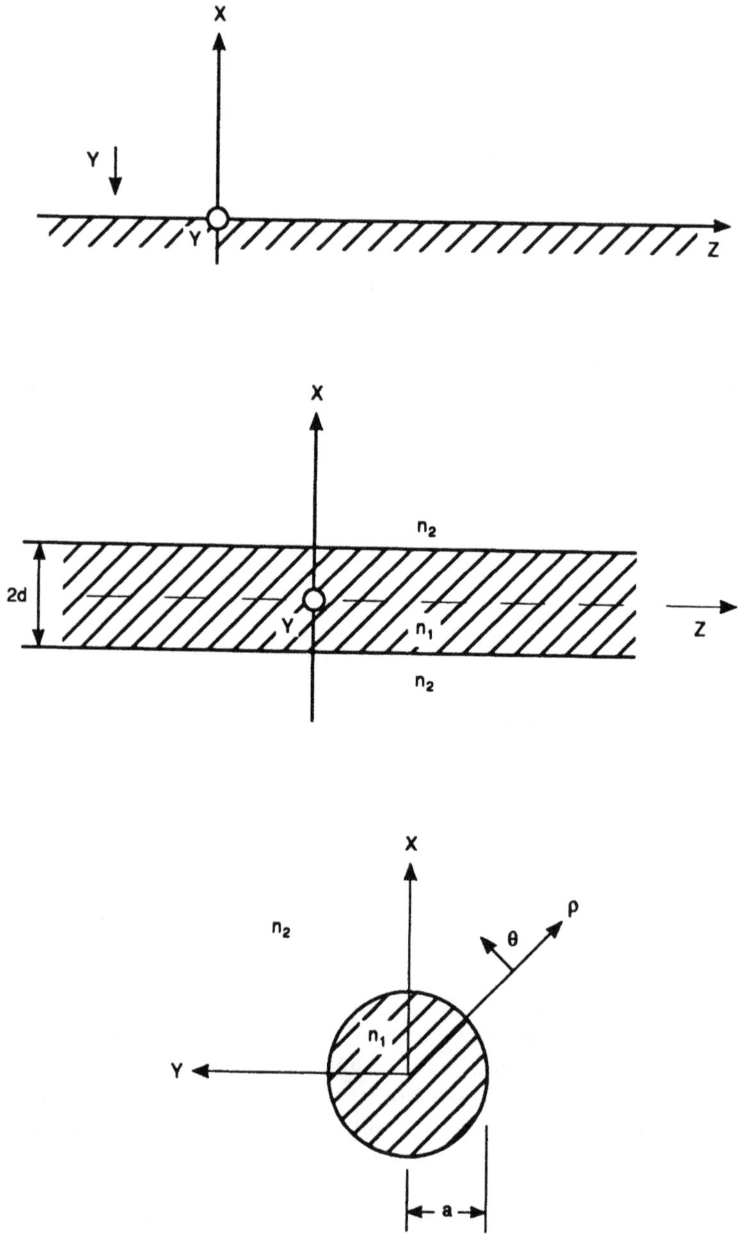

Fig. 4.1 Examples of open waveguides

a Planar constant impedance surface
b Dielectric slab
c Dielectric rod

4.2.1 The planar constant impedance surface

As demonstrated in section 3.3, a surface wave mode can be supported by a single surface with a reactive impedance; the mode being *TM* on an inductive surface and *TE* on a capactive one. Since a surface wave mode does not radiate, it cannot be the sole wave solution on an open waveguide. Therefore, one expects other types of modes that do account for radiation of power away from the guide. In the following we look for these radiation modes. To be specific, let us consider a planar surface with a constant surface admittance $Y = iB$, where B is a negative real number for an assumed lossless inductive surface. Referring to Fig. 4.1*a*, we can choose the fields to be uniform in the y direction and propagating along z. Restricting attention to *TM* modes, the only nonvanishing component of the magnetic field is h_y. We wish to find all possible modal solutions which must satisfy the wave equation (1.34) for h_y, the boundary condition at the inductive surface $x=0$, and the radiation condition at $x=\infty$. The latter states that h_y, and all other field components, should decay as fast, or faster than, a cylindrical wave of the form $\exp(-ikr)/r^{1/2}$; $r=(x^2+z^2)^{1/2}$, as r tends to ∞. For a surface wave mode, all fields decay exponentially with x and this certainly satisfies the radiation condition. So, one writes down h_y for the surface wave mode on a planar admittance surface in the form:

$$h_{ys}(x, z) \equiv \phi_s(x, z) = \alpha_s \exp(-\alpha_s x) \exp(-i\beta_s z) \tag{4.1}$$

where, in order to satisfy the radiation condition, $\mathrm{Real}(\alpha_s) > 0$, and from the wave equation:

$$\beta_s^2 = k_0^2 + \alpha_s^2 \tag{4.2}$$

Obtaining the longitudinal electric field component e_z from the Maxwellian equation (1.2) and imposing the boundary condition $Y = h_y/e_z$ at $x=0$, we determine the surface wave eigenvalues:

$$\alpha_s = -ik_0\Delta \tag{4.3}$$

$$\beta_s = k_0(1-\Delta^2)^{1/2} \tag{4.4}$$

where

$$\Delta = (\eta_0 Y)^{-1} \tag{4.5}$$

This completely describes the surface wave mode of the planar impedance surface, but now the question arises as to whether this is the only allowable mode on the structure. The answer would be 'yes', if we insist to impose the radiation condition on the modal fields. Recognising however that the radiation condition is only a sufficient, but not a necessary, condition for each individual mode, Shevchenko (1971) proposed to relax this condition for an individual mode. In other words, the radiation condition must be satisfied by the total field on the structure, but not necessarily by each individual mode. Thus, instead of imposing a cylindrical wave-like behaviour on each mode, we may only require that the modal fields be finite at infinite values of r. This allows us to find a new set of modes, other than the surface wave mode, that will account for radiation from the open structure; i.e. we can now write the following allowable modal solutions:

$$h_y(x, z; \lambda) \equiv \phi(x, z, \lambda) = [\exp(i\lambda x) + R(\lambda)\exp(-i\lambda x)]\exp(-i\beta z) \tag{4.6}$$

where the transverse wavenumber λ must be restricted to real values for ϕ to be finite as x tends to ∞. Hence λ can assume any value between 0 and ∞, with negative values being redundant. The form (4.6) suggests that $\phi(x, z; \lambda)$ is the result of a plane wave incident on the planar impedance surface from ∞ and reflected back with a reflection coefficient $R(\lambda)$. The latter is obtainable from the admittance boundary condition at $x=0$ and is given by:

$$R(\lambda) = (\lambda - k_0\Delta)/(\lambda + k_0\Delta) \tag{4.7}$$

For a fixed value of λ, the solution (4.6) is called a 'pseudomode' by Shevchenko (1971). Now the cross-sectional functions $\phi_s(x, z)$ and $\phi(x, z; \lambda)$, for a fixed z, form a complete set in terms of which an arbitrary realisable field can be expanded. In other words, the total field of an arbitrary source can be expanded as a continuous spectrum of pseudomodes in addition to the surface wave mode. Thus, the most general expression for H_y on the impedance plane is:

$$H_y(x, z) = A_s\phi_s(x, z) + \int_0^\infty A(\lambda)\phi(x, z; \lambda)\, d\lambda \tag{4.8}$$

where A_s is the amplitude of the surface wave mode and $A(\lambda)$ is the amplitude of the pseudomode of transverse wavenumber λ. ϕ_s and ϕ are defined in (4.1) and (4.6) respectively. While the surface wave mode accounts for the guided waves on the structure, the continuous spectrum of pseudomodes accounts for the radiated waves and the evanescent waves. The former are represented by the part of the spectrum given by $\lambda \leqslant k_0$, and the latter by the part $\lambda > k_0$. The range of integration in (4.8) can be changed to cover both positive and negative values of λ by using (4.6). Hence (4.8) may be rewritten as:

$$H_y(x, z) = A_s\phi_s(x, z) + \int_{-\infty}^\infty A(\lambda)R(\lambda) \exp(-i\lambda x - i\beta z)\, d\lambda \tag{4.9}$$

where $R(\lambda)$ is defined as equal to unity for negative values of λ; otherwise it is given by (4.7).

Now, it remains to show that the continuous spectrum in (4.8) or (4.9) does satisfy the radiation condition. To this end, we convert the coordinates from cartesian to cylindrical by the trnsformation:

$$x = \rho \sin\theta, \text{ and } z = \rho\cos\theta.$$

and introduce the complex angle ψ by:

$$\lambda = k_0 \sin\psi, \text{ and } \beta = k_0\cos\psi.$$

In terms of ρ, θ and ψ, the continuous spectrum term in (4.9), denoted below by H_{yc}, acquires the form:

$$H_{yc}(x, z) = \int_{-\infty}^\infty k_0 A(k_0\sin\psi)R(k_0\sin\psi) \exp[-ik_0\rho\cos(\psi-\theta)].\, d(\sin\psi) \tag{4.10}$$

The integration applies over all real values of $\sin\psi$ between $-\infty$ and $+\infty$. The part corresponding to $|\sin\psi| > 1$ represents evanescent pseudomodes for which $\cos\psi = -i|\cos\psi|$, and will die away in the far zone $k_0\rho \gg 1$. The other part of the spectrum corresponds to real values of ψ. In this part, unless ψ is close to θ, the exponential term in (4.10) is rapidly oscillating for $k_0\rho \gg 1$, and this leads to cancellation among the spectrum components. Therefore, the main contribution to the integral comes from the parts of the spectrum near to the point $\psi = \theta$. This point is called the stationary phase point, and the integration procedure based on expanding the exponent in (4.10) around this point is known as the stationary phase integration (Clemmow, 1966). Following this procedure, as demonstrated in problem 4.2, we get:

$$H_{yc}(\rho, \theta) = A(k_0 \sin\theta) R(k_0 \sin\theta) \cos\theta . (2\pi)^{1/2} . \times [\exp(-ik_0\rho + i\pi/4)/(k_0\rho)^{1/2}] \quad (4.11)$$

which is readily a diverging cylindrical wave, and, hence satisfies the radiation condition. Thus we have shown that, although the pseudomodes do not individually satisfy the radiation condition, their total field does satisfy this condition as it should.

Finally, it is worth noting that the radiation field is not always represented by a continuous spectrum. Actually, a cylindrical or spherical surface of constant impedance has a purely discrete spectrum representing both guided and radiative fields. On this point the reader is referred to the extensive work of Wait (e.g. 1960, 1964, 1970) on electromagnetic wave propagation on the earth's surface (see also probl. 4.3).

4.2.2 The planar dielectric slab

The second example of an open waveguide to be considered is a planar dielectric slab surrounded by a medium of lower dielectric constant, as depicted in Fig. 4.1*b*. In optics terminology, we use the refractive index n in place of the relative dielectric constant ε_r, where $n = \varepsilon_r^{1/2}$. So, in Fig. 4.1*b*, we denote the refractive index of the dielectric slab by n_1 and that of the surrounding medium by n_2, and we require that n_1 be greater than n_2. Waves supported by the structure can be separated into *TE* and *TM* waves with respect to x. In the following we consider only *TE* waves, but a similar treatment can be pursued for *TM* waves. For *TE* to x waves, e may be taken totally in the y direction in which all fields are assumed uniform; i.e. $\mathrm{e} = e_y\hat{y}$, and $\partial/\partial y = 0$. Owing to symmetry of the slab structure about the $x = 0$ plane, the modes fall into even and odd groups for which e_y is an even and an odd function of x respectively. Now we find the surface wave modes and the pseudomodes of the structure.

Surface wave modes

Surface waves are characterised by an exponential field decay in the transverse direction on both sides of the slab. Hence, the only component of e, denoted here by e_{ys} may be expressed by:

$$\begin{aligned} e_{ys} \equiv \psi_s &= a_s \exp(-\alpha_s(x-d)) \exp(-i\beta_s z), \; x \geqslant d \\ &= \pm a_s \exp(\alpha_s(x+d)) \exp(-i\beta_z z), \; x \leqslant -d \end{aligned} \quad (4.12)$$

The $\pm$ sign refers to even and odd modes respectively. The corresponding field dependence inside the slab is:

$$e_{ys} \equiv \psi_s = B \begin{smallmatrix}\cos\\ \sin\end{smallmatrix} (\varkappa x) \exp(-i\beta_s z), \; |x| \leqslant d \quad (4.13)$$

The wavenumbers α_s, β_s and $\varkappa$ are related, from the wave equation, by:

$$\beta_s^2 + \varkappa^2 = k_0^2 n_1^2 \tag{4.14}$$

$$\beta_s^2 - \alpha_s^2 = k_0^2 n_2^2 \tag{4.15}$$

where k_0 is the free space wavenumber. Noting that h_z is proportional to $\partial e_y/\partial x$, we deduce that the boundary conditions, requiring that e_y and h_z be continuous at $x = \pm d$, reduce to the continuity of e_y and its x derivative. Imposing these boundary conditons at $x = d$ leads to the two equations:

$$B \, {}^{\cos}_{\sin}(\varkappa d) = \alpha_s$$

$$\pm B\varkappa \, {}^{\sin}_{\cos}(\varkappa d) = (\alpha_s)^2$$

Dividing the second of these two equations by the first, we get

$$\varkappa d \tan \varkappa d = \alpha_s d \tag{4.16}$$

for even modes, and

$$\varkappa d \cot \varkappa d = -\alpha_s d \tag{4.17}$$

for odd modes. Each of (4.16) and (4.17) is a relation between $\varkappa d$ and $\alpha_s d$. Another relation between these two quantities is obtained by subtracting (4.15) from (4.14), yielding

$$(\varkappa d)^2 + (\alpha_s d)^2 = (k_0 d)^2 (n_1^2 - n_2^2) \tag{4.18}$$

The allowable even and odd *TE* modes are obtained by the simultaneous solution of (4.18) with (4.16) and (4.17) respectively. In general this is done with the aid of a computer, but fortunately in this case there is a simple graphical solution. Referring to Fig. 4.2, we plot the traces of the three equations (4.16)—(4.18) in the (κd)—$(\alpha_s d)$ plane. Only the first quadrant of the plane is of interest since, for a proper surface wave mode, α_s must be positive and the equations have even dependence on $\varkappa$. Equation (4.16) is traced as a family of branches that go to infinite values of $\alpha_s d$ at odd multiples of $(\pi/2)$ for $\varkappa d$. Similarly (4.17) is traced as a family of branches that go to infinite values of $\alpha_s d$ at even multiples of $(\pi/2)$ for $\varkappa d$. Equation (4.18) represents a circle with centre at the origin and radius V given by:

$$V = k_0 d (n_1^2 - n_2^2)^{1/2} \tag{4.19}$$

The eigenvalues α_s, $\varkappa$ and, hence, β_s of the surface wave modes are obviously given by the intersections of the V circle with the branches of (4.16) and (4.17).

Few inferences can be drawn from Fig. 4.2. First, the number of surface wave modes is finite. A surface wave mode is cutoff when $\alpha_s = 0$. As V is reduced, α_s of a given mode is reduced and the mode becomes less bounded to the slab. At the cutoff value of V, or the cutoff frequency of a given mode, α_s is zero and the mode is no longer bounded. The mode then starts to join the continuous spectrum of pseudomodes.

Pseudomodes

A pseudomode results from a plane wave incident from infinity and reflected by the guiding structure. So, appropriate expressions of e_y for TE pseudomodes outside the slab are:

$$\begin{aligned} e_y(\lambda) \equiv \psi(\lambda) &= [\exp(i\lambda(x-d)) + R(\lambda)\ \exp(-i\lambda(x-d))]\ \exp(-i\beta z)\quad x \geqslant d \\ &= \pm[\exp(-i\lambda(x+d)) + R(\lambda)\ \exp(i\lambda(x+d))]\ \exp(-i\beta z)\quad x \leqslant -d \end{aligned} \tag{4.20}$$

The $\pm$ sign stands for even and odd modes respectively. Inside the slab:

$$e_y(\lambda) \equiv \psi(\lambda) = B(\lambda)^{\cos}_{\sin}(\varkappa x),\ |x| \leqslant d \tag{4.21}$$

where,

$$\beta^2 + \lambda^2 = k_0^2 n_2^2 \text{ and } \beta^2 + \varkappa^2 = k_0^2 n_1^2.$$

Owing to symmetry about $x=0$, it is sufficient to satisfy the boundary conditions at $x=d$ only. These reduce to the continuity of e_y and its x derivative as stated earlier. The two equations thus obtained determine the reflection coefficient $R(\lambda)$. For even modes, we get:

$$R(\lambda) = (i\lambda \cot \varkappa d + \varkappa)/(i\lambda \cot \varkappa d - \varkappa), \tag{4.22}$$

and for odd modes

$$R(\lambda) = (i\lambda \tan \varkappa d - \varkappa)/(i\lambda \tan \varkappa d + \varkappa) \tag{4.23}$$

while $B(\lambda)$ is given by:

$$B(\lambda) = [1 + R(\lambda)]^{\sec}_{\operatorname{cosec}}(\varkappa d) \tag{4.24}$$

for even and odd modes respectively. Equations (4.20)—(4.24) completely characterise a pseudomode with $0 \leqslant \lambda < \infty$.

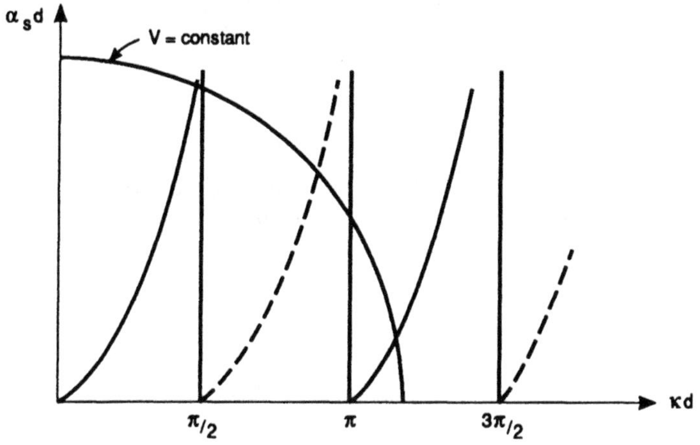

Fig. 4.2 Plot of $\alpha_s d$ versus $\varkappa d$ for a dielectric layer

Now the most general expression for E_y parallels that in (4.8) and is written as:

$$E_y = \sum_s A_s \psi_s + \int_0^\infty A(\lambda)\psi(\lambda)\,d\lambda \tag{4.25}$$

with ψ_s given by (4.12)—(4.13) and $\psi(\lambda)$ by (4.20)—(4.21). Thus, a general field is composed of a discrete sum over a finite number of surface wave modes plus a continuous sum of pseudomodes. The latter represents radiation from the structure. The modal amplitudes A_s, $s=1,2\ldots$ and $A(\lambda)$ are determined according to the source of excitation. The radiation field is given by the integral term in the above equation and may be evaluated by the method of stationary phase.

4.2.3 The dielectric circular rod

The last example of an open waveguide to be considered in this section is a dielectric circular rod embedded in a medium of lower dielectric constant. So, let the refractive index of the rod by n_1 and that of the surrounding medium be n_2, where $n_1 > n_2$; see Fig. 4.1*c*. In this case, the modes of the structure are generally hybrid, so that both e_z and h_z may exist. In the following we consider the discrete spectrum of modes (surface wave modes) and the continuous spectrum of pseudomodes.

4.2.3.1. Discrete spectrum Apart from a common factor $\exp(-i\beta z - im\phi)$, the longitudinal fields of a surface wave mode in the region outside the rod, $\rho \geqslant a$, take the form:

$$e_{zs}(m, \beta_s) = AK_m(\alpha_s \rho) \tag{4.26}$$

$$\eta_2 h_{zs}(m, \beta_s) = BK_m(\alpha_s \rho) \tag{4.27}$$

where the modified Bessel function $K_m(.)$ ensures the exponential decay of the fields for large ρ. A and B are constants, m is an integer and η_2 is the intrinsic/wave impedance in the outside medium; $\eta_2 = 120\,\pi/n_2$ ohms. Inside the rod, $\rho \leqslant a$, the fields are oscillatory and, of course, finite at $\rho = 0$; hence they are expressed by:

$$e_{zs}(m, \beta_s) = CJ_m(\varkappa\rho) \tag{4.28}$$

$$\eta_2 h_{zs}(m, \beta_s) = DJ_m(\varkappa\rho) \tag{4.29}$$

where C and D are constants. The wavenumbers β_s, α_s and $\varkappa$ have the same relations as in the case of the dielectic slab. In particular we have the following relation:

$$(\varkappa a)^2 + (\alpha_s a)^2 = (k_0 a)^2 (n_1^2 - n_2^2) = V^2 \tag{4.30}$$

where V is a normalised frequency, or merely the V number in fibre optics terminology.

The transverse field components e_ρ, h_ρ, e_ϕ and h_ϕ are obtained from the general relations (1.27)—(1.28) in terms of e_z and h_z. The boundary conditions at $\rho = a$ require the continuity of e_z, h_z, e_ϕ and h_ϕ, hence provide four homogeneous equations in the four unknown coefficients A, B, C and D. To have a nontrivial solution, the 4 × 4 determinant of the equations must vanish, thus leading to the modal equation for the wavenumber β_s. The procedure is straightforward and we trust that, by now, the reader is familiar with it. So, skipping details, we write down the resulting modal equation:

$$[F_m(u) - G_m(v)][n_1^2 F_m(u) - n_2^2 G_m(u)] = (m\beta_s/k_0)^2(v^{-2} + u^{-2})^2 \qquad (4.31)$$

where

$$F_m(u) \equiv J'_m(u)/uJ_m(u) \qquad (4.32)$$

$$G_m(v) \equiv K'_m(v)/vK_m(v) \qquad (4.33)$$

$$u = \varkappa a \text{ and } v = \alpha_s a.$$

The hybrid modes reduce to separate TE and TM modes in the case of azimuthal symmetry, $m = 0$. Equation (4.31) then breaks down to two equations:

$$F_0(u) = G_0(v) \qquad (4.34)$$

for TE_{on} modes with nonzero field components e_ϕ, h_ρ and h_z, and

$$n_1^2 F_0(u) = n_2^2 G_0(v) \qquad (4.35)$$

for TM_{on} modes with nonzero field components h_ϕ, e_ρ and e_z.

For $m > 0$, the modes are hybrid and labelled HE_{mn} or EH_{mn}. Snitzer (1961) and Clarricoats *et al.* (1971) define a modal hybrid factor by $(-iB/A)$. When this is positive the mode is considered HE and when negative it is considered EH. Although there are many other ways to classify the modes [see a good discussion by Bruno and Bridges (1988)], this definition seems to get a wide acceptance.

It is important to determine the cutoff conditon of a given mode. As surface wave modes, cutoff occurs when the modal fields cease to decay radially; i.e. when α_s or v tends to zero. Considering the $m = 1$ modes, as $v \rightarrow 0$, the RHS of (4.31) tends to n_2^2/v^4, and $G_1(v)$ tends to $[-1/v^2 + \ln(\Gamma v/2)]$ as deduced from Appendix A2. The LHS of (4.31) then becomes:

$$n_2^2/v^4 + [F_1(V)(n_1^2 + n_2^2) - 2n_2^2 \ln(\Gamma v/2)]$$

Comparing this with the RHS of the equation, it follows that the above square bracketed term must vanish as $v \rightarrow 0$. This requires that $F_1(V) = \infty$, or:

$$J_1(V) = 0 \qquad (4.36)$$

This determines the cutoff frequencies of the $m = 1$ modes. The modes are either HE_{1_n} or EH_{1_n} and have equal cutoff frequencies. However, they are not degenerate since, above cutoff, they have different propagation constants. The dominant mode is the HE_{11} and it has a zero cutoff frequency, i.e. it is never cutoff. Next higher order modes have normalised cutoff frequencies at $V_n = n$th root of $J_1(V)$.

Linearly polarised modes

An important practical example of a dielectric rod waveguide is the optical fibre line. The optical fibre, used in optical communication in the wavelength region 0·8—1·8 μm, is basically a dielectric rod, or a core, which is surrounded by a clad of slightly lower refractive index. The core is made of mainly fibre glass (silicon oxide) with n_1 near to 1·5. The clad is made of the same material, but with a suitable dopant to reduce n to a value n_2 slightly below n_1. A characteristic quantity of an optical fibre line is the fractional refractive index difference $\Delta = (n_1 - n_2)/n_1$. To reduce the number of propagating modes, and hence the signal delay dispersion on the line, the core diameter and/or Δ must be made small. In order not to have a too small core, Δ is chosen as small as 0·01—0·001. This allows the core diameter to be as large as several wavelengths, that is, in the range 6—12 µm for a monomode fibre. A good coverage of optical fibres as light transmission media can be found in several excellent books [e.g. Marcuse (1972, 1974), Midwinter (1979), Cherin (1983) and Jones (1988)]. Here, we shall only derive the main characteristics of modes on a step index fibre, that is, one in which n changes abruptly from n_1 within the core to n_2 within the clad. Now, under the condition $\Delta \ll 1$, the significant modes are almost *TEM*. This can be clearly seen by referring to the relations between β_s, $\varkappa$ and α_s. Namely, we establish the inequality

$$k_0 n_1 > \beta_s > k_0 n_2$$

and since n_1 is only slightly different from n_2, β_s is very close to both $k_0 n_1$ or $k_0 n_2$ and the mode structure is close to a *TEM* wave everywhere. Nevertheless, the fields are still oscillatory in the core and evanescent in the clad. One can easily show that the rate of decay α_s in the clad has an upper bound given by $k_0\sqrt{n_1^2 - n_2^2} \simeq k_0 n_1 \sqrt{2\Delta} \ll k_0 n_1$. This means that the modes are weakly guided modes. Now we show that under the condition of small Δ, or for weckly guided modes, the modal fields can be derived, with an excellent approximation, from a single scalar function. The modes are then approximately linearly polarised. To this end, we adopt a cartesian frame of coordinates (x, y, z) with z coincident with the fibre axis. Since the modes resemble the *TEM* mode approximately, the dominant modal fields can be assumed to be e_x and h_y. The other field components are of first or second order smallness. Apart from a common factor $\exp(-i\beta_s z)$, e_x can be written as:

$$e_x(\rho, \phi) = \psi(\rho) \cos m\phi \qquad (4.37)$$

where (ρ, ϕ) are the usual polar coordinates. Now, let us postulate that $e_y(\rho, \phi) = 0$; that is the e field is x polarised in the cross sectional planes. Using the divergence equation of the e field, we get:

$$(\partial/\partial x)e_x(\rho, \phi) - i\beta_s e_z(\rho, \phi) = 0$$

and since $\partial/\partial x = \cos\phi\,\partial/\partial\rho + (\sin\phi/\rho)\partial/\partial\phi$, we get for e_z:

$$e_z(\rho, \phi) = [(\partial\psi/\partial\rho) \cos m\phi \cos\phi - m\psi \sin m\phi \sin\phi/\rho]/i\beta_s \qquad (4.38)$$

Noting that the fields are slowly varying in directions transverse to z, e_z is deemed to be of first order smallness relative to e_x. This is true for modes having small integer values of m. Such modes are the important modes of the structure since they are appreciably excited by most means of excitation. Now that all the electric field components are known, one obtains the magnetic field by using (1.1).

Thus:

$$h_x = (i/\omega\mu)\,(\partial e_z/\partial y) \tag{4.39}$$

$$h_y = (\beta_s/\omega\mu)e_x = (\beta_s/\omega\mu)\psi \cos m\phi \tag{4.40}$$

$$h_z = (-i/\omega\mu)\,(\partial e_x/\partial y) \tag{4.41}$$

Investigation of these equations reveals that h_z is a first order small and h_x is a second order small quanitity. So if we neglect h_x compared to other field components and write down the field components tangential to the core–clad interface $\rho = a$, we find that $e_\phi = -e_x \cos\phi$ and $h_\phi \simeq h_y \sin\phi$. Since h_y is proportional to e_x, the simultaneous continuity of h_ϕ and e_ϕ across the interface is satisfied by the continuity of e_x or ψ alone. The other two field components that are tangential to the core–clad interface are e_z and h_z. Investigation of (4.38) and a similar expression for h_z reveals that they are continuous across the interface if ψ and $\partial\psi/\partial\rho$ are continuous. We thus conclude that for a small differential refractive index ($\Delta \ll 1$), the modal fields are derivable from a single scalar field (ψ), and the core–clad boundary conditions reduce to only two conditions, which are the continuity of ψ and $\partial\psi/\partial\rho$.

Applying the above argument to a step index fibre with a single clad, we write for ψ:

$$\psi(\rho) = AJ_m(\varkappa\rho) \qquad \rho \leqslant a$$
$$= BK_m(\alpha_s\rho) \quad \rho \geqslant a$$

where A and B are constants. Imposing the continuity of both ψ and $\partial\psi/\partial\rho$ at $\rho = a$, we easily find that:

$$\varkappa J_m'(\varkappa a)/J_m(\varkappa a) = \alpha_s K_m'(\alpha_s a)/K_m(\alpha_s a) \tag{4.42}$$

The primes on the Bessel functions signify differentiation with respect to the arguments. Another form of (4.42) is obtained by expressing the derivatives of the Bessel functions in terms of the m and $(m-1)$ order functions to get:

$$\kappa a \frac{J_{m-1}(\kappa a)}{J_m(\kappa a)} + \alpha_s a \frac{K_{m-1}(\alpha_s a)}{K_m(\alpha_s a)} - = 0 \tag{4.43}$$

This is the modal equation whose solution along with (4.30) determines the set of eigenvalues $\varkappa$, or β_s. The cutoff frequencies are obtained by letting α_s tend to zero, whence the second term in (4.43) also tends to zero. The cutoff equation then becomes:

$$J_{m-1}(V) = 0 \tag{4.44}$$

The linearly polarised modes are labelled LP_{mn} where n is the order of the root in (4.44). The LP_{01} mode is the dominant mode and is the same as the HE_{11} mode with zero cutoff frequency. The next higher order mode is the LP_{11} with a cutoff $V = 2{\cdot}4048$ (the first zero of J_0). This is followed by the LP_{02} and LP_{21} modes which have the same cutoff $V = 3{\cdot}84$ (first zero of J_1). Sketches of the field distributions of these modes are shown in Fig. 4.3.

4.2.3.2. Continuous spectrum of modes To derive the radiation modes of the dielectric rod, let a cylindrical wave of a given transverse wavenumber be incident on the rod from the outside medium. Since the modes are generally

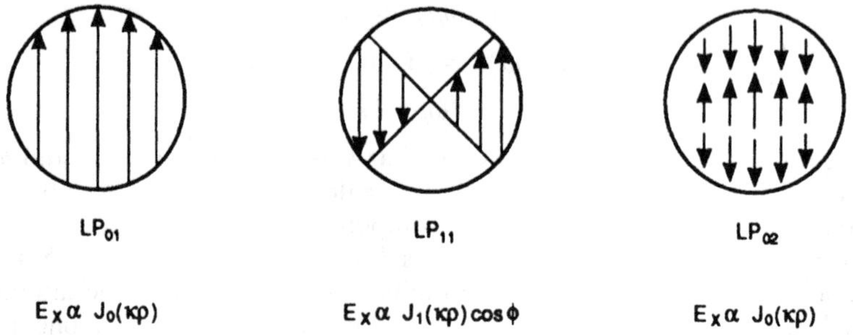

Fig. 4.3 Sketch of E field lines for *LP* modes

hybrid, we generally have both e_z and h_z. Apart from a common factor $\exp(-i\beta z - im\phi)$, a spectral component of these fields may be expressed in the outside region $\rho \geqslant a$ by:

$$e_z(\lambda, m) = A(\lambda)J_m(\lambda\rho) + C(\lambda)H_m^{(2)}(\lambda\rho) \tag{4.45}$$

$$h_z(\lambda, m) = B(\lambda)J_m(\lambda\rho) + D(\lambda)H_m^{(2)}(\lambda\rho) \tag{4.46}$$

The first term in each of the above equations represent a cylindrical wave that would exist everywhere in the absence of the rod. The second term stands for an outgoing scattered wave resulting from the presence of the rod. Because the fields of a single spectral component, or a pseudomode, are required to be finite at infinite values of ρ, λ should be restricted to real values, that is $0 \leqslant \lambda < \infty$. The corresponding fields inside the rod, $\rho \leqslant a$, are:

$$e_z(\lambda, m) = F(\lambda)J_m(\varkappa\rho) \tag{4.47}$$

$$h_z(\lambda, m) = G(\lambda)J_m(\varkappa\rho) \tag{4.48}$$

The following relations hold among the wavenumbers $\varkappa$, λ, β and k_0:

$$\lambda^2 + \beta^2 = k_0^2 n_2^2 \tag{4.49}$$

$$\varkappa^2 + \beta^2 = k_0^2 n_1^2 \tag{4.50}$$

The boundary conditions at $\rho = a$ provide four equations in the six unknown coefficients A, B, C, D, F and G. Thus we are free to choose any two of them and classify the pseudomodes accordingly. We arbitrarily classify pseudomodes into two classes:

(i) $A(\lambda) = 1$ and $B(\lambda) = 0$
(ii) $B(\lambda) = 1$ and $A(\lambda) = 0$

Referring to (4.45)—(4.46), we immediately identify class (i) pseudomode as one that results from an incident *TM* wave, and likewise class (ii) results from an incident *TE* wave. This is the same classification as has been suggested by Snyder (1971). Note that, although the incident wave in any of these classes is a pure *TE* or *TM*, the scattered wave is generally hybrid in nature, except of course when $m=0$. Finally, we mention that radiation fields of a dielectric rod are expressed by integrals over λ of (4.45)—(4.46).

4.3 Mode orthogonality and mode coupling in open waveguides

As in closed waveguides, an orthogonality relationship can be established for different modes on open waveguides. Specifically, orthogonality can exist between surface wave modes, between surface wave and pseudomodes and among pseudomodes. In the course of proving mode orthogonality, we also find the coupling between modes on different waveguides. Such coupling terms arise in problems involving longitudinal discontinuities between dissimilar guides.

A cylindrical open waveguide may be generally modelled as a structure which is uniform along a straight axis z, but is generally inhomogeneous in the plane transverse to z. The boundary of the cross section is considered to lie at ∞. This is depicted in Fig. 4.4 in which ε and μ are functions of the transverse coordinates (in the plane of the paper) and the cross section S is bounded by a contour C of infinite extent: i.e. $R \to \infty$. In a more general situation, part of C can be a constant impedance surface such as in the case of the single planar impedance surface treated in the last section.

Now consider two cylindrical open waveguides of common axis z characterized by (ε, μ) for the first one (ε', μ') for the second. Let (E, H) be the total fields of a given mode on the first waveguide. The mode can be either a surface wave mode or a pseudomode. Similarly we take (E′, H′) to be the total fields of any mode on the second waveguide. The corresponding phase constants of these modes are β and β', so that the modal fields vary as

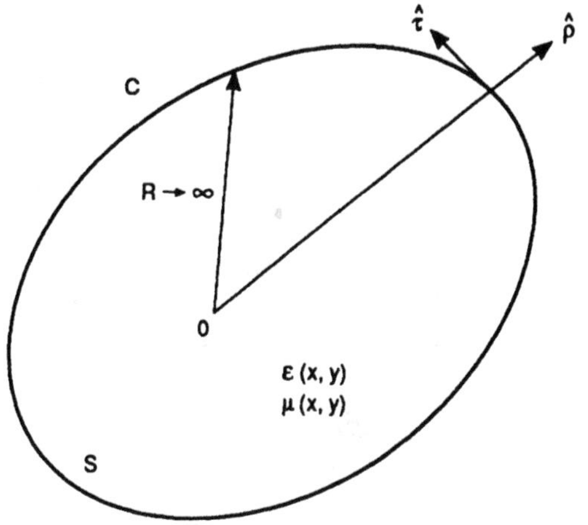

Fig. 4.4 A general representation of an open waveguide

Contour C lies totally or partially at infinity

$\exp(-i\beta z)$ and $\exp(-i\beta' z)$ respectively. Next we apply Maxwell's equations (1.1) and (1.2) to get the identity

$$\nabla.(\mathrm{E}\times\mathrm{H}'-\mathrm{E}'\times\mathrm{H})=i\omega\Delta\varepsilon\mathrm{E}.\mathrm{E}'-i\omega\Delta\mu\mathrm{H}.\mathrm{H}' \tag{4.51}$$

in a source free region. In the above $\Delta\varepsilon=\varepsilon-\varepsilon'$ and $\Delta\mu=\mu-\mu'$.

A general cylindrical coordinate system (ρ, τ, z) as shown in Fig. 4.4 can be adopted. The total modal field E is decomposed as $\mathrm{E}=\mathrm{e}+e_z\hat{z}$, where e is the transverse vector field given, in turn, by $\mathrm{e}=e_\rho\hat{\rho}+e_\tau\hat{\tau}$. Similar definitions apply to the other fields H, E′ and H′. Now integrate (4.51) over a volume bounded by two cross sectional planes at z and $z+dz$, such that $dz\rightarrow 0$, and the cylindrical side surface with perimeter C and length dz. Changing the volume integral on the LHS of the equation into a closed surface integral according to the divergence theorem, we get:

$$(\partial/\partial z)\left[\int_S(\mathrm{exh}'-\mathrm{e}'\mathrm{xh}).\hat{z}\,\mathrm{d}S\right]+\int_C(\mathrm{ExH}'-\mathrm{E}'\times\mathrm{H}).\hat{\rho}\,\mathrm{d}C$$

$$\int_S i\omega(\Delta\varepsilon\mathrm{E}.\mathrm{E}'-\Delta\mu\mathrm{H}.\mathrm{H}')\,\mathrm{d}S \tag{4.52}$$

Replacing the differential operator $(\partial/\partial z)$ by $(-\mathrm{i}\beta-\mathrm{i}\beta')$ and rewriting the contour integral in terms of the τ and z field components, the above equation becomes:

$$\begin{aligned}-i(\beta+\beta')\int_S(\mathrm{exh}'-\mathrm{e}'\mathrm{xh}).\hat{z}\,\mathrm{d}S=&\int_C(e_zh_\tau'-e_z'h_\tau)\,\mathrm{d}C\\&-\int_C(e_\tau h_z'-e_\tau'h_z)\,\mathrm{d}C\\&+i\omega\int_S\Delta\varepsilon(\mathrm{e}.\mathrm{e}'+e_ze_z')\,\mathrm{d}S\\&-i\omega\int_S\Delta\mu(\mathrm{h}.\mathrm{h}'+h_zh_z')\,\mathrm{d}S\end{aligned} \tag{4.53}$$

By reversing the direction of propagation in the first waveguide, we obtain another equation with β replaced by $-\beta$. This further implies the reversal of h, h_τ and e_z. Manipulation of the so-obtained equation together with (4.53) leads to the following result:

$$\langle\mathrm{e},\mathrm{h}'\rangle=[i\beta'I_{1C}+i\beta I_{2C}+\beta'I_{1S}-\beta I_{2S}]/(\beta^2-\beta'^2) \tag{4.54}$$

where the inner product on the LHS is defined by:

$$\langle\mathrm{a},\mathrm{b}\rangle\equiv\int_S\mathrm{a}\times\mathrm{b}.\hat{z}\,\mathrm{d}S \tag{4.55}$$

and:

$$I_{1C}=\int_C (e_\tau h'_z - e'_\tau h_z)\,\mathrm{d}C \tag{4.56}$$

$$I_{2C}=\int_C (e_z h'_\tau - e'_z h_\tau)\,\mathrm{d}C \tag{4.57}$$

$$I_{1S}=\omega\int_S (\Delta\varepsilon \mathrm{e}.\mathrm{e}' - \Delta\mu h_z h'_z) \tag{4.58}$$

$$I_{2S}=\omega\int_S (\Delta\varepsilon e_z e'_z - \Delta\mu \mathrm{h}.\mathrm{h}') \tag{4.59}$$

This gives the coupling between modes in different waveguides. Seemingly, we are not simplifying matters by expressing a single integral in (4.54) by four integrals. However, we should remember that the surface integrations on the RHS of the equation need to be evaluated only on a finite region over which $\Delta\varepsilon$ and $\Delta\mu$ are nonzero. The line integrals are easy to evaluate and in many cases equal to zero. We shall deomonstrate the use of (4.54) by an example later in this section. Next, we discuss mode orthogonality.

4.3.1 Mode orthogonality

To prove mode orthogonality in a given open waveguide, we apply (4.54) to the case $\Delta\varepsilon=0$ and $\Delta\mu=0$. Therefore, the surface integrals I_{1S} and I_{2S} vanish. Taking the contour C at ρ equals infinity, it becomes clear that, if one or both of the two modes considered is a surface wave mode, then the contour integrals are identically zero. Thus, we conclude that:

$$(\mathrm{e}_m, \mathrm{h}_n) = N_m\delta_{mn} \tag{4.60}$$

and

$$\langle \mathrm{e}_m, \mathrm{h}(\lambda)\rangle = \langle \mathrm{e}(\lambda), \mathrm{h}_m\rangle = 0,\ 0\leqslant\lambda<\infty \tag{4.61}$$

Here, $(\mathrm{e}_m, \mathrm{h}_m)$ are the transverse fields of the mth surface wave mode, whereas $(\mathrm{e}(\lambda), \mathrm{h}(\lambda))$ are those of a pseudomode with transverse wavenumber λ. N_m is a mode normalising factor and δ_{mn} is zero unless $m=n$ when it is unity. The above two equations are statements of orthogonality between different surface wave modes and between a surface wave mode and a pseudomode.

When the two modes considered in (4.54) are pseudomodes, the contour integrations do not identically vanish as C recedes to ∞, and one needs to perform these integrations carefully. We shall do that in connection with an example following this discussion. We shall find out that the inner product of two pseudomodes on the same open structure has the form:

$$\langle \mathrm{e}(\lambda_a), \mathrm{h}(\lambda_b)\rangle = N(\lambda_a)\delta(\lambda_a-\lambda_b) \tag{4.62}$$

where $\delta(.)$ is the Kronecker delta function and $N(\lambda)$ is a pseudomode normalising factor.

4.3.2 Example

We consider a dielectric slab waveguide which undergoes an abrupt change of thickness as shown in Fig. 4.5. With a surface wave mode incident normally from the left hand side, it is desired to determine the scattered surface wave and

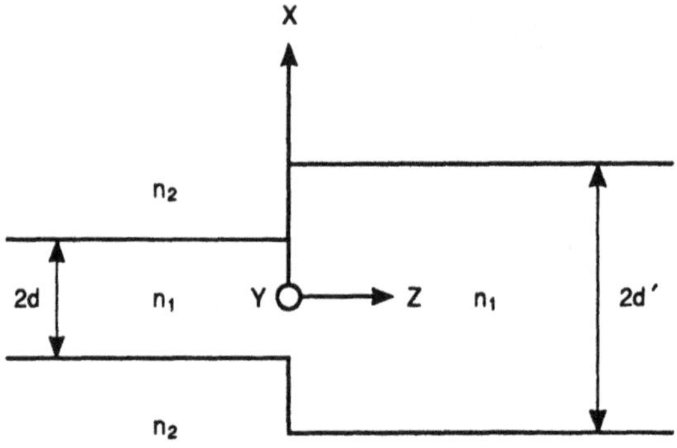

Fig. 4.5 A step discontinuity in a dielectric slab waveguide

radiation modes on both sides of the junction. This is a prototype problem arising in the design of many optical and microwave components such as couplers, filters and antennas (e.g. Bhartia and Bahl, 1984). Let M and N be the number of allowable surface wave modes on the left and right sides of the junction respectively. First, we determine the mode normalising factors N_s and $N(\lambda)$ and then the coupling terms between modes on different sides from equation (4.54). We then proceed to determine the scattered mode amplitudes by imposing the boundary conditions at the junction.

Because of two dimensional nature of the problem, the cross section integrations in (4.54) and the following equations reduce to one dimensional integration over x. The contour integrations over C reduce to end points evaluation of the integrand. In the following, we consider only TE to z excitation. Evaluating N_s for surface wave modes, $s=1, 2, \ldots M$, we have:

$$N_s = \langle \mathrm{e}_s, \mathrm{h}_s \rangle = -\int_{-\infty} e_{ys} h_{xs}\, \mathrm{d}x = (2\beta_s/\omega\mu) \int_0^{\infty} \psi_s^2\, \mathrm{d}x$$
$$= (\beta_s/\omega\mu)\,[\alpha_s + (\alpha_s^2/\varkappa^2 d)\,(V^2 + \alpha_s d)] \tag{4.63}$$

where (4.12), (4.13) and the modal equation have been invoked. It turns out that (4.63) applies to even as well as odd modes. A similar relation exists for the modes on the right side of the junction with primed parameters d', α_s' . . . etc.

Next, consider pseudomodes on the left hand side of the junction. Applying (4.54) to two pseudomodes:

$$\langle \mathrm{e}(\lambda), \mathrm{h}(\lambda') \rangle = i\beta' I_{1_C}/(\beta^2 - \beta'^2)$$
$$= \frac{-2\beta'/\omega\mu}{\beta^2 - \beta'^2} - [\psi(\lambda)\partial\psi(\lambda')/\partial x - \psi(\lambda')\partial\psi(\lambda)/\partial x]_{x\to\infty}$$

On using (4.20) for $\psi(\lambda)$ and noting that $\beta^2 - \beta'^2 = \lambda'^2 - \lambda^2$, we get:

$$\langle e(\lambda), h(\lambda') \rangle = (2\beta'/\omega\mu) P(\lambda, \lambda')$$

where

$$P(\lambda, \lambda') = \lim_{x\to\infty}\left[\frac{\sin(\lambda-\lambda')x}{\lambda-\lambda'}(R(\lambda)+R(\lambda'))+\frac{\sin(\lambda+\lambda')x}{\lambda+\lambda'}(1+R(\lambda)R(\lambda'))\right.$$
$$\left.-i\frac{\cos(\lambda+\lambda')x}{\lambda+\lambda'}(1-R(\lambda)R(\lambda'))+i\frac{\cos(\lambda-\lambda')x}{\lambda-\lambda'}(R(\lambda)-R(\lambda'))\right]$$

On taking the limit $x\to\infty$, only the first term on the RHS survives and it is proportional to $\delta(\lambda-\lambda')$; hence we get:

$$\langle \mathrm{e}(\lambda), \mathrm{h}(\lambda')\rangle = N(\lambda)\delta(\lambda-\lambda') = (2\beta'/\omega\mu)\pi[R(\lambda')]\delta(\lambda-\lambda') \tag{4.64}$$

Now we turn attention to inner products of modes on opposite sides of the junction. Consider first two surface wave modes and apply (4.54) to get:

$$\langle \mathrm{e}_m, \mathrm{h}'_n\rangle = \frac{\beta'_n}{\beta_m^2-\beta_n'^2}I_{1s} = \frac{2\beta'_n\omega\varepsilon_0(n_2^2-n_1^2)}{\beta_m^2-\beta_n^2}\int_d^{d'}\psi_m\psi'_n\,\mathrm{d}y \tag{4.65}$$

Similarly, the inner product between a surface wave mode on one side and a pseudomode on the other is:

$$\langle \mathrm{e}_m, \mathrm{h}'(\lambda)\rangle = \frac{2\beta'\omega\varepsilon_0(n_2^2-n_1^2)}{\beta_m^2-\beta'^2}\int_d^{d'}\psi_m\psi'(\lambda)\,\mathrm{d}y \tag{4.66}$$

The inner product, or coupling, between two pseudomodes on opposite sides of the junction is also obtainable from (4.54):

$$\begin{aligned}\langle \mathrm{e}(\lambda), \mathrm{h}'(\lambda')\rangle &= [i\beta' I_{1_C}+\beta' I_{1_S}]/(\beta^2-\beta'^2)\\ &= (2\beta'/\omega\mu)\pi[R(\lambda)+R'(\lambda')]\delta(\lambda-\lambda')\\ &\quad -i(2\beta'/\omega\mu)[R(\lambda)-R'(\lambda')]/(\lambda'-\lambda)\\ &\quad +\frac{2\beta'}{\beta^2-\beta'^2}\omega\varepsilon_0(n_2^2-n_1^2)\int_d^{d'}\psi(\lambda)\psi'(\lambda')\,\mathrm{d}y\end{aligned} \tag{4.67}$$

Equations (4.65)—(4.67) give inner products of all possible combination of modes. It is also useful to note the following relation which applies in the present case of dielectric discontinuity under *TE* excitation:

$$\langle \mathrm{e}', \mathrm{h}\rangle = (\beta/\beta')\langle \mathrm{e}, \mathrm{h}'\rangle \tag{4.68}$$

and is valid for surface wave modes and pseudomodes.

Now it is in order to determine the scattered mode amplitudes. To this end, we impose the boundary conditions at the junction plane $z=0$; namely, the continuity of the total transverse electric and magnetic fields. For a unit incident surface wave mode, let the amplitudes of the reflected surface waves be Γ_s, $s=1,2\ldots M$, and of the transmitted surface waves by T_s, $s=1,2\ldots N$. There will also be continuous spectra of reflected and transmitted pseudomodes whose amplitudes are denoted by $\Gamma(\lambda)$ and $T(\lambda)$, $0 \leq \lambda < \infty$. We therefore write for $z=0$:

$$\mathrm{e}_1 + \sum_1^M \Gamma_s \mathrm{e}_s + \int_0^\infty \Gamma(\lambda)\mathrm{e}(\lambda)\,\mathrm{d}\lambda = \sum_1^N T_s \mathrm{e}'_s + \int_0^\infty T(\lambda)\mathrm{e}'(\lambda)\,\mathrm{d}\lambda \ldots \tag{4.69}$$

$$\mathrm{h}_1 - \sum_1^M \Gamma_s \mathrm{h}_s - \int_0^\infty \Gamma(\lambda)\mathrm{h}(\lambda)\,\mathrm{d}\lambda = \sum_1^N T_s \mathrm{h}'_s + \int_0^\infty T(\lambda)\mathrm{h}'(\lambda)\,\mathrm{d}\lambda \tag{4.70}$$

To solve for the modal amplitudes, several approaches have been tried in the literature. Marcuse (1970) derived a closed form solution for a small discontinuity. Mahmoud and Beal (1975) converted the continuous spectrum into a discrete spectrum by expanding $\Gamma(\lambda)$ and $T(\lambda)$ as summations of Laguerre polynomials. Thus the methods developed for closed waveguides, such as those mentioned in chapter 2, can readily be applied. Rulf (1975) deduced a singular integral equation for $T(\lambda)$ whose solution leads to all modal amplitudes. Rozzi (1978) obtained an integral equation for the total magnetic field on the junction plane and solved it by means of the Rayleigh–Ritz method. Morishita *et al.* (1979) chose to minimise the mean square error in satisfying the boundary conditions at the junction plane. Gelen *et al.* (1979) devised an iterative solution for the modal amplitudes. Shigesawu and Tsuji (1986) have suggested dividing the continuous spectrum into its radiative and evanescent parts and discretising each part separately.

Here, we present solutions to the discontinuity problem in the limit of a small discontinuity. We also review the methods of Rozzi and of Gelin *et al.* for an arbitrary discontinuity. So, reverting to (4.69) and (4.70), we take the inner product of the first with h'_n, $n=1, 2 \ldots N$, and the inner product of the second with e'_n, and on using the orthogonality relations (4.60)—(4.61), we get:

$$\langle \mathrm{e}_1, \mathrm{h}'_n \rangle + \sum_s \Gamma_s \langle \mathrm{e}_s, \mathrm{h}'_n \rangle + \int_0^\infty \Gamma(\lambda) \langle \mathrm{e}(\lambda), \mathrm{h}'_n \rangle\,\mathrm{d}\lambda = T_n N'_n \tag{4.71}$$

$$\langle \mathrm{e}'_n, h_1 \rangle - \sum_s \Gamma_s \langle \mathrm{e}'_n, \mathrm{h}_s \rangle - \int_0^\infty \Gamma(\lambda) \langle \mathrm{e}'_n, \mathrm{h}(\lambda) \rangle,\,\mathrm{d}\lambda = T_n N'_n \tag{4.72}$$

Similarly, take the inner product of (4.69) with $\mathrm{h}'(\lambda')$ and (4.70) with $\mathrm{e}'(\lambda')$ and invoke (4.62) to get:

$$\langle \mathrm{e}_1, \mathrm{h}'(\lambda') \rangle + \sum_s \Gamma_s \langle \mathrm{e}_s, \mathrm{h}'(\lambda') \rangle + \int_0^\infty \Gamma(\lambda) \langle \mathrm{e}(\lambda), \mathrm{h}'(\lambda') \rangle\,\mathrm{d}\lambda = T(\lambda') N'(\lambda') \tag{4.73}$$

$$\langle \mathrm{e}'(\lambda'), \mathrm{h}_1 \rangle - \sum_s \Gamma_s \langle \mathrm{e}'(\lambda'), \mathrm{h}_s \rangle - \int_0^\infty \Gamma(\lambda) \langle \mathrm{e}'(\lambda'), \mathrm{h}(\lambda) \rangle\,\mathrm{d}\lambda = T(\lambda') N'(\lambda') \tag{4.74}$$

Now consider the small discontinuity approximation. In this case it is justifiable to assume that $\langle \mathrm{e}_s, \mathrm{h}'_n \rangle$ and the like terms are of first order smallness for $s \neq n$. The reflection parameters Γ_s and $\Gamma(\lambda)$ are also first order small quantities. So, neglecting second order quantities, (4.71) and (4.72) reduce to the approximate forms:

$$T_n N'_n \simeq \langle e_1, h'_n\rangle + \Gamma_n \langle e_n, h'_n\rangle \simeq \langle e'_n, h_1\rangle - \Gamma_n \langle e'_n, h_n\rangle$$

Solving for T_n and Γ_n:

$$T_n \simeq \frac{\langle e_1, h'_n\rangle\langle e'_n, h_n\rangle + \langle e'_n, h_1\rangle\langle e_n, h'_n\rangle}{N'_n[\langle e_n, h'_n\rangle + \langle e'_n, h_n\rangle]} \simeq \frac{\langle e_1, h'_n\rangle + \langle e'_n, h_1\rangle}{2N'_n} \tag{4.75}$$

$$\Gamma_n \simeq \frac{\langle e'_n, h_1\rangle - \langle e_1, h'_n\rangle}{\langle e_n, h'_n\rangle + \langle e'_n, h_n\rangle} \tag{4.76}$$

Using similar approximations in (4.73) and (4.74), we end up with:

$$T(\lambda) \simeq \{\langle e_1, h'(\lambda)\rangle + \langle e'(\lambda), h_1\rangle\}/2N'(\lambda) \tag{4.77}$$

$$\Gamma(\lambda) \simeq \{\langle e'(\lambda), h_1\rangle - \langle e_1, h'(\lambda)\rangle\}/\{M(\lambda) + M'(\lambda)\} \tag{4.78}$$

where $M(\lambda)$ and $M'(\lambda)$ are defned by:

$$\langle e(\lambda), h'(\lambda')\rangle \simeq M(\lambda)\delta(\lambda - \lambda'), \text{ and } \langle e'(\lambda'), h(\lambda)\rangle \simeq M'(\lambda)\delta(\lambda - \lambda')$$

The above formulae are good for small longitudinal discontinuities, such as the case of a small step of dielectric height (see Fig. 4.4) or a small change of n_1.

In the case of an arbitrarily large discontinuity, an iterative solution can be outlined as follows. Starting with the approximate evaluations of T_n and $T(\lambda)$ given by (4.75) and (4.77), the reflection coefficients are evaluated from the relations:

$$(\delta_{1_n} + \Gamma_n)N_n = \sum_s T_s\langle e'_s, h_n\rangle + \int_0^\infty T(\lambda)\langle e'(\lambda), h_n\rangle \, d\lambda \tag{4.79}$$

$$\Gamma(\lambda)N(\lambda) = \sum_s T_s\langle e'_s, h(\lambda)\rangle + \int_0^\infty T(\lambda')\langle e'(\lambda'), h(\lambda)\rangle \, d\lambda' \tag{4.80}$$

which are derived by forming the inner products of (4.69) with h_n and $h(\lambda)$ respectively. In (4.79), $\delta_{1_n} = 0$ unless $n = 1$, when it is equal to unity. Once the reflection coefficients are obtained from the above equations, more accurate approximations for the T^ns and $T(\lambda)$ are obtained from (4.71) and (4.73). These are 'plugged' in again in (4.79)—(4.80) to update the Γ's and so on. This iterative solution has been suggested by Gelin *et al.* (1979).

Another approach adopted by Rozzi (1978) is based on forming an integral equation for the total transverse magnetic field H_t at the junction plane $z = 0$. To derive this equation, use is made of (4.70) to express the various reflection and transmission coefficients in terms of H_t. Substituting the results in (4.69), the following integral equation is obtained:

$$2e_1(r) = \int_S Z(r, r') \times H_t(r').\hat{z} \, dS \tag{4.81}$$

where r is the position vector in the transverse plane S and $Z(r; r')$ is the diadic vector given by:

$$Z(r, r') = \sum_1^M (N_s)^{-1} e_s(r) e_s(r') + \sum_1^N (N_s')^{-1} e_s'(r) e_s'(r')$$
$$+ \int_0^\infty d\lambda (N(\lambda))^{-1} e(\lambda; r) e(\lambda, r')$$
$$+ \int_0^\infty d\lambda (N'(\lambda))^{-1} e'(\lambda; r) e'(\lambda, r') \tag{4.82}$$

which is a somewhat generalised form of the integral equation derived by Rozzi (1978). In the special case of the dielectric slab of Fig. 4.5, the position vector r can be replaced by the coordinate x and all the vector fields turn into scalar functions. The integral equation (4.81) can be solved by the Rayleigh–Ritz method (e.g. Harrington, 1961), involving the expansion of all field functions into a complete set of orthonormal basis functions. One choice of this set for the dielectric slab discontinuity is the set of Laguerre functions. An alternative choice, which is particularly good for an incident surface wave, is the set of surface wave functions on both sides of the junction, augmented by the set of Laguerre functions. Numerical examples are given by Rozzi (1978).

4.4 Millimetric waveguides

For several decades, the microwave band of frequencies, covering centimetric waves, has been utilised in telecommunication, radar and remote sensing systems. With the growing demand on these systems, the microwave band has become completely congested. This created a pressing need to explore the millimetric wave band for the provision of a much desired wide bandwidth. Research efforts since the early 1970s have resulted in several novel types of millimetric waveguides and their associated millimetric wave circuit components. The main requirements of a millimetric waveguide structure are low loss, reasonable size, reasonable dimensional tolerances, ease of fabrication and suitability of circuit component integration. While the oversize metallic waveguides can provide decreasing attenuation with frequency, they are all bulky and not suitable for integration. This is the reason that several types of millimetric waveguides have evolved in order to meet the integrated circuit applications. These types include a variety of planar dielectric waveguides: the fin lines, the H-guide and the groove guide. In the following, we review the basic properties of some of these waveguides, all of which come under the category of open waveguides. We focus attention on the lower order guided modes since they are the most relevant to integrated circuit design.

4.4.1 The H-guide

It has been shown in section 2.5 that attenuation caused by the longitudinal currents in metallic closed waveguides increases with frequency as $f^{1/2}$ while that caused by the transverse currents fall as $f^{-3/2}$. Since the longitudinal currents in a rectangular waveguide supporting the TE_{10} mode are carried by the broad walls, normal to the E field lines, these walls are responsible for most of the attenuation in the millimetric band. In an H-waveguide, these walls are simply removed and the side walls are extended as in Fig. 4.6*b*. In order still to confine the field within the guide, a dielectric slab is placed across the two remaining walls. The resulting configuration appears to be like the letter H. The attenuation due to conductor losses in and H-waveguide is normally small and the major contribution comes from dielectric losses. This can also be reduced by the choice of a low loss dielectric. It has also been shown that lamination of the dielectric slab can considerably reduce the dielectric losses (Tischer, 1970). In the following we analyse the dominant modes in an H-guide and simultaneously introduce the so-called transverse resonance method of mode analysis. This method looks into the modal field dependence in the

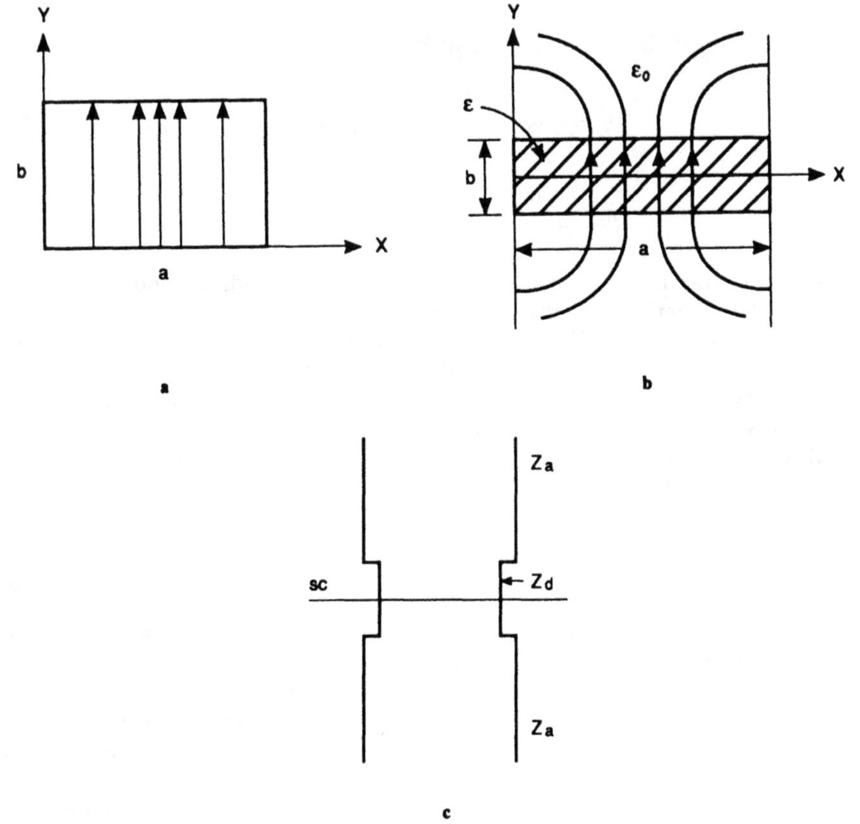

Fig. 4.6

a Rectangular waveguide
b An H waveguide
c Equivalent transverse transmission line to the H waveguide

transverse direction. It leads to the modal equation without having to write down the detailed field distribution.

Because of material inhomogeneity, the modes of an H-waveguide are hybrid; i.e. having both e_z and h_z. However, they can be classified as *TE* or *TM* to y, the direction normal to the air–dielectric interface. Considering *TM* to y modes, we write for e_y, apart from a common $\exp(i\omega t - i\beta z)$ factor:

$$e_y = \psi(y)\sin(m\pi x/a) \tag{4.83}$$

which readily satisfies the boundary conditions on the walls for integer values of m. The other field components are expressible in terms of e_y or $\psi(v)$ by analogy with (21.24) and (1.25):

$$e_x = \frac{m\pi/a}{\beta^2 + (m\pi/a)^2}\psi'(y)\cos(m\pi x/a) \tag{4.84}$$

$$e_z = \frac{-i\beta}{\beta^2 + (m\pi/a)^2}\psi'(y)\sin(m\pi x/a) \tag{4.85}$$

$$h_x = \frac{-\omega\varepsilon\beta}{\beta^2 + (m\pi/a)^2}\psi(y)\sin(m\pi x/a) \tag{4.86}$$

$$h_z = \frac{i\omega\varepsilon(m\pi/a)}{\beta^2 + (m\pi/a)^2}\psi(y)\cos(m\pi x/a) \tag{4.87}$$

where $\psi'(y)$ stands for the y derivative of ψ. The boundary conditions at the air–dielectric interface $y = \pm b/2$ require the continuity of e_x, e_z, h_x and h_z. However, in view of (4.84)—(4.87), these four conditions reduce to only two; namely the continuity of $\varepsilon\psi(y)$ and $\psi'(y)$. Now, instead of proceeding in the normal way by applying the boundary conditions and deducing the modal equation, we shall develop the transverse resonance method of mode analysis. To this end, a transmission line analogy is devised to represent the mode behaviour in the transverse direction y. First a transverse wave impedance, or a characteristic impedance of the equivalent transmission line, extending along y is defined by:

$$Z = e_z/h_x = -e_x/h_z = -\psi'(y)/i\omega\varepsilon\psi(y) \tag{4.88}$$

Applying this to a travelling wave $\exp(-ik_y y)$ in the dielectric region, the transverse impedance is $Z_d = k_y/\omega\varepsilon$. For an evanescent wave $\exp(-\alpha y)$ in the air region, the transverse impedance is $Z_a = -i\alpha/\omega\varepsilon_0$. We can therefore represent the H-guide in the transverse direction by a 3-cascaded section of transmission lines as shown in Fig. 4.6*c*. The middle section has a characteristic impedance Z_d and the other two have Z_a. Furthermore if we confine attention to even modes (in e_y), we can place a short circuit at $y = 0$ without distrubing the mode. Now the modal equation can be obtained from the transmission line representation as follows. Considering the plane $y = b/2$, let the impedance looking downwards be Z_{down} and the impedance looking up be Z_{up}. The boundary conditions at this plane are equivalent to the equation:

$$Z_{down} + Z_{up} = 0 \tag{4.89}$$

which is the transverse resonance equation. For even modes, the transmission line representation in Fig. 4.6*c* interprets (4.89) as:

$$(k_y/\omega\varepsilon)i\tan(k_y b/2)+(-i\alpha/\omega\varepsilon_0)=0$$

or

$$k_y\tan(k_y b/2)=\alpha(\varepsilon/\varepsilon_0) \quad (4.90)$$

with

$$k_y^2+\alpha^2=k_0^2(\varepsilon/\varepsilon_0-1)$$

This is the modal equation for even modes. It is seen that the transverse resonance method is a short cut procedure for obtaining the modal equation without the detailed description of the modal fields. It is particularly suitable in the case of multilayer structures. In the present problem $\psi(y)$ is obviously expressed by:

$$\begin{aligned}\psi(y)&=\cos k_y y,\ |y|\leqslant b/2\\ &=(\varepsilon/\varepsilon_0)\cos(k_y b/2)\exp[-\alpha(|y|-b/2)],\ |y|\leqslant b/2\end{aligned}$$

We have assumed that the guide is high enough in the y direction so that the fields will have died out before reaching the open end. One can now get the power flow down the guide, and the power loss in the conducting walls and in the dielectric. Doing this exercise (see probl. 4.15), we arrive at the mode attenuation factor α_z (along z) as:

$$\alpha_z=\alpha_c+\alpha_d$$

where

$$\alpha_c=\frac{2(\omega\varepsilon_0/\sigma)(m\pi/a)^2}{[\beta^2+(m\pi/a)^2]\beta ad}(p/q) \quad (4.91)$$

$$\alpha_d=\tan\delta\{[\beta^2+(m\pi/a)^2]/2\beta\}(p'/q) \quad (4.92)$$

α_c is the attenuation constant due to losses in the walls, having conductivity σ and skin depth d, and α_d is the attenuation caused by the dielectric with loss tangent $\tan\delta$. In the above p, p' and q are the dimensionless parameters

$$p=(\varepsilon/\varepsilon_0)[1+\sin(k_y b)/k_y b]+[1+\cos(k_y b)]/\alpha b$$

$$q=[1+\sin(k_y b)/k_y b]+(\varepsilon/\varepsilon_0)[1+\cos(k_y b)]/\alpha b$$

$$p'=[1+\sin(k_y b)/k_y b]+\{k_y^2/[\beta^2+(m\pi/a)^2]\}\{1-\sin(k_y b)/k_y b\}$$

Investigation of (4.91) and (4.92) reveals that, for sufficiently high frequencies, α_c decays like $f^{-3/2}$ while α_d increases linearly with f. Thus the dielectric losses dominate in the millimetric wave band. To reduce the dielectric losses, Tischer (1970) proposed laminating the dielectric slab and predicted a posible reduction of the attenuation by about one order of magnitude.

4.4.2 The groove guide

Like the H-guide, the groove guide (Fig. 4.7) is derived from conventional rectangular waveguides by removing the broad walls in order to reduce wall losses. The grooves are then introduced to trap the fields in the central region,

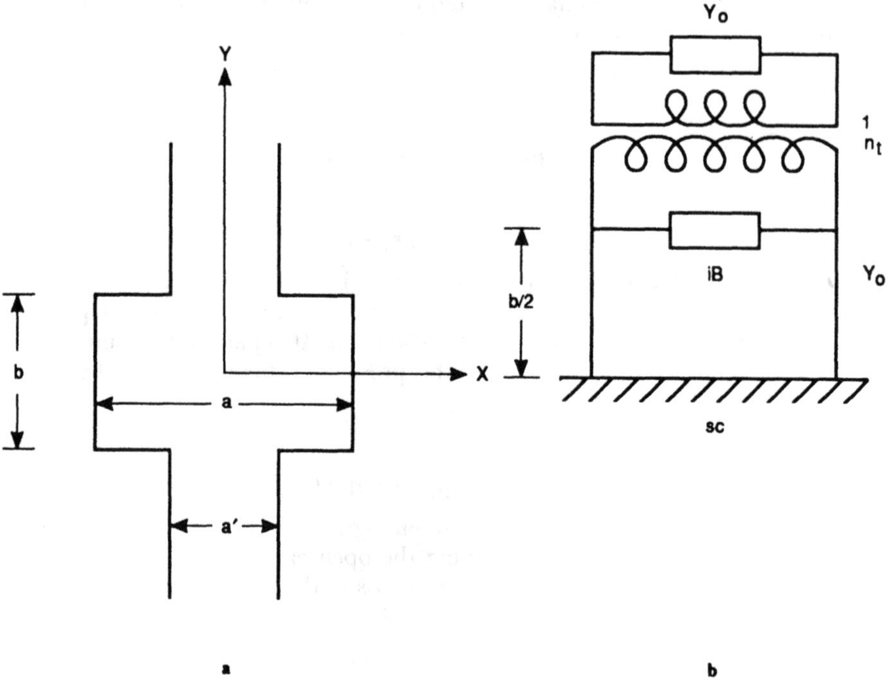

Fig. 4.7 The groove waveguide

a Geometry
b An equivalent circuit for even *TE* modes

hence preventing radiation from the open ends. This guide was first introduced by Nakahara and Kurauchi (1964). Futher studies have since followed, including the work of Oliner and Lampariello (1985) who presented a simple, yet accurate, equivalent transverse circuit for the dominant mode. The groove waveguide has the advantage over the H-guide that it has no dielectric losses, and hence lower attenuation. It is thus more suited to applications requiring long runs of waveguides. One of these applications is in the design of a travelling wave leaky antenna which needs a guide length of several tens of wavelengths. Meanwhile, the attenuation due to guide losses should be much less than losses due to radiation (Oliner and Lampariello, 1985).

Since the groove guide is homogeneously filled, the basic modes are either *TE* or *TM* to the longitudinal direction z. The dominant mode is the TE_{10} mode which resembles the same mode in the rectangular waveguide. In the following we present a rather self contained analysis for the lower order modes in the groove guide. The analysis is based on approximating the fields in the groove region, but then carrying out the rest of the analysis rigorously (Mahmoud, 1990). So, considering the TE_{1n} modes, and apart from a common factor $\exp(i\omega t - i\beta z)$, the h_z field in the groove region $|y| \geq b/2$ is aproximated by a single exponential term:

$$h_z = \sin(\pi x/a')\exp[-\alpha(|y| - b/2)] \tag{4.93}$$

In the central region $|y| \leq b/2$, a rigorous form of h_z is

$$h_z = \sum_{m=0}^{\infty} A_m \sin[(2m+1)\pi x/a] \cos k_{ym} y \tag{4.94}$$

where we consider only even modes about $y=0$. The wavenumbers involved are related by:

$$k_{ym}^2 + [(2m+1)\pi/a]^2 = (\pi/a')^2 - \alpha^2 = k_0^2 - \beta^2 \tag{4.95}$$

and the boundary conditions on the perfectly conducting side walls are readily satisfied. The other field components are derived from h_z through relations (1.20)—(1.23) in chapter 1. In particular we find that e_x and h_x are proportional to $\partial h_z/\partial y$ and $\partial h_z/\partial x$ respectively. Now the boundary conditions at the plane $y=b/2$ require the continuity of h_z, e_x and h_x. These three boundary conditions reduce to only two since the continuity of h_z at all values of x implies the continuity of h_x. Therefore, it is sufficient to apply the boundary conditions to h_z and e_x only. Continuity of h_z reads:

$$\sum_m A_m \sin[(2m+1)\pi x/a] \cos(k_{ym} b/2) = \sin(\pi x/a') \text{ for } |x| \leq a'/2 \tag{4.96}$$

and continuity of e_x reads:

$$\begin{aligned}\sum_m k_{ym} A_m \sin[(2m+1)\pi x/a] \sin(k_{ym} b/2)\\ &= \alpha \sin(\pi x/a') \qquad \text{for } |x| \leq a'/2\\ &= 0 \qquad \text{for } a/2 \geq |x| \geq a'/2\end{aligned} \tag{4.97}$$

Thus, the electric field e_x is defined over the whole interval $(-a/2, a/2)$, while the magnetic field h_z is defined only in the range $(-a'/2, a'/2)$ and is unknown outside. It is then appropriate to try to expand h_z in terms of a complete set of functions inside the region $(-a'/2, a'/2)$. We choose the set of functions $\sin[(2n+1)\pi x/a']$, $n=0, 1, \ldots$ for this purpose. On the other hand e_x is expanded in terms of the set of functions $\sin[(2n+1)\pi x/a]$ in the range $(-a/2, a/2)$. To implement these ideas, multiply (4.96) by $\sin[(2n+1)\pi x/a']$ and integrate between the limits $-a'/2$ and $a'/2$.

$$\sum_{m=0}^{\infty} A_m c_{mn} \cos(k_{ym} b/2) = (a'/2)\delta_{n0} \tag{4.98}$$

where $\delta_{n0} = 0$ for $n \neq 0$ and $= 1$ for $n=0$, and

$$c_{mn} = \int_{-a'/2}^{a'/2} \sin[(2m+1)\pi x/a] \sin[(2n+1)\pi x/a'] \, dx \tag{4.99}$$

Similarly, multiply (4.97) by $\sin[(2n+1)\pi x/a]$ and integrate over the range $(-a/2, a/2)$

$$A_n k_{yn}(a/2)\sin(k_{yn}b/2)=c_{n0}\,\alpha \tag{4.100}$$

Substituting for A_m in (4.98) from (4.100) and putting $n=0$ in the former equation, it becomes:

$$(2\alpha/a)\sum_{m=0}^{\infty}(c_{m0})^2\cot(k_{ym}b/2)/k_{ym}=a'/2 \tag{4.101}$$

This is the modal equation for the dominant mode since the only unknown is β. This can be written in a form that allows a transverse circuit representation by separating out the zeroth term in the summation and dividing both sides of the equation by $c_{00}^2(2\alpha/a)$:

$$(1/k_{y0})[-i\cot(k_{y0}b/2)]-i\sum_{m=1}^{\infty}(c_{m0}/c_{00})^2\cot(k_{ym}b/2)/k_{ym}$$

$$+(i/\alpha))/n_t^2=0 \tag{4.102}$$

The first term in this equation can be interpreted as the input admittance of a short circuited transmission line of length $b/2$ and normalised characteristic admittance $(1/k_{y0})$, representing the central region of the guide. The third term, representing the groove, is interpreted as the reflected admittance of a line of characteristic admittance (i/α) through a step up transformer of turns ratio n_t (see Fig. 4.7*b*). Finally, the second term is the admittance iB that accounts for all the higher order modes generated at the junction $y=b/2$. On comparing (4.102) with (4.101), the turns ratio n_t is deduced:

$$n_t=2c_{00}/\sqrt{aa'}=(4/\pi)(a'/a)^{3/2}\frac{\cos(\pi a'/2a)}{1-(a'/a)^2} \tag{4.103}$$

where (4.99) has been utilised to get c_{00}. The junction susceptance B given by the second term in (4.102) is dependent on the modal phase constant β through the factor k_{ym}. However, it is usually possible to approximate k_{ym} for $m\geqslant 1$ in the following fashion:

$$k_{ym}=(k_0^2-\beta^2-(2m+1)^2\pi^2/a^2)^{1/2}\simeq i(2m+1)\pi/a$$

since for the low order modes $(k_0^2-\beta^2)\ll(3\pi/a)^2$. This renders B independent of β; therefore

$$B\simeq\sum_{m=1}^{\infty}(c_{m0}/c_{00})^2\coth[(2m+1)\pi b/2a]/[(2m+1)\pi/a] \tag{4.104}$$

and is only dependent on the geometry. The coth(.) term in this equation accounts for the coupling between the evanescent fields at the two junctions $y=\pm b/2$. If b is sufficiently large, (4.104) is further simplified by setting the coth term equal to unity. Explicit evaluation of the coupling terms c_{m0} renders B as:

$$B/Y_0 \simeq k_{y0}a \sum_{m=1}^{\infty} \frac{(1-r^2)^2 \cos^2[(2m+1)\pi r/2]\,(2m+1)}{\pi[1-(2m+1)^2 r^2]^2 \cos^2(\pi r/2)} \tag{4.105}$$

where $r=a'/a$. It is easy to verify that, in the special case $a'=0$, B tends to infinity as it should. In the trivial case $a'=a$, B is equal to zero and $n_t=1$.

4.4.3 Planar dielectric waveguides

The most suitable form of millimetric and optical waveguides for circuit integration is the planar form. In millimeter wave circuit integration a ground conducting plane is usually used (see Fig. 4.8). This acts as a heat sink and, meanwhile, as a short circuit to modes with horizontal polarisation. Thus only waveguide modes having E normal to the ground plane are the dominant modes. In the case of optical integrated circuits, a ground plane is considered undesirable since it can result in excessive signal attenuation.

Several examples of planar millimeter waveguides are shown in Fig. 4.8*a*—*e*. The simplest configuration is the image line where a dielectric strip is placed on a ground plane. The fields tend to be concentrated in the strip and to decay in the surrounding air. The dominant mode is the HE_{11} mode with E polarised normally to the ground plane. Because the magnetic field is a maximum at the ground plane, the conductor ohmic losses tend to be high. In the isolated image line of Fig. 4.8*b* the dielectric slab is isolated from the ground plane by a dielectric layer of lower permittivity ε_2. The fields in this layer will tend to be evanescent towards the ground plane, hence reducing conductor losses.

In both the image and insulated image lines the fields are mainly guided by the dielectric strip. Since it is technologically difficult to manufacture a dielectric strip with sufficiently smooth side walls, additional attenuation is caused by the wall roughness. This problem is alleviated in the strip dielectric guide (SDG) shown in Fig. 4.8*c*, where a dielectric slab of permittivity $\varepsilon_1 > \varepsilon_2$ is used as the main guide for the fields, while the role of the dielectric strip (ε_2) is only to confine the fields in the horizontal direction (Itoh and Mittra, 1975). The success of this design depends on the fact that it is relatively easy to manufacture a dielectric slab wih very smooth surfaces. The inverted strip dielectric guide (ISDG), shown in Fig. 4.8*d*, has all the advantages of the SDG in addition to a lower conductor loss, since the inverted strip acts to isolate the ground plane from the main guiding dielectric slab (Itoh, 1976). Finally we note that it is easy to build coupled guide systems in any of the above configurations. An example of two coupled guides in the SDG configuration is shown in Fig. 4.8*e*. The degree of coupling is adjusted by the distance between the two strips, (Itoh and Mittra, 1975 and Itoh, 1976).

A rigorous modal analysis of any of the above structures is quite complicated. This is so since there is no simple field solution that can satisfy all boundary conditions at the different dielectric interfaces. In general, different plane wave functions are coupled at the dielectric interfaces. So, a spectral sum of plane waves is needed in each homogeneous region in order to satisfy all boundary conditions. Nevertheless, simplified and reasonably accurate solutions for the dominant modes can be found. The most popular method of approximate mode analysis is the method of effective dielectric constant (Knox and Toulios, 1970). The method provides an excellent approximation for the modal eigenvalues under the general condition that the transverse wavenumbers are small relative to the longitudinal wavenumber. In the following, we give an account of the method of effective dielectric constant by applying it to the SDG of Fig. 4.8*c* and then comparing it with a more rigorous solution.

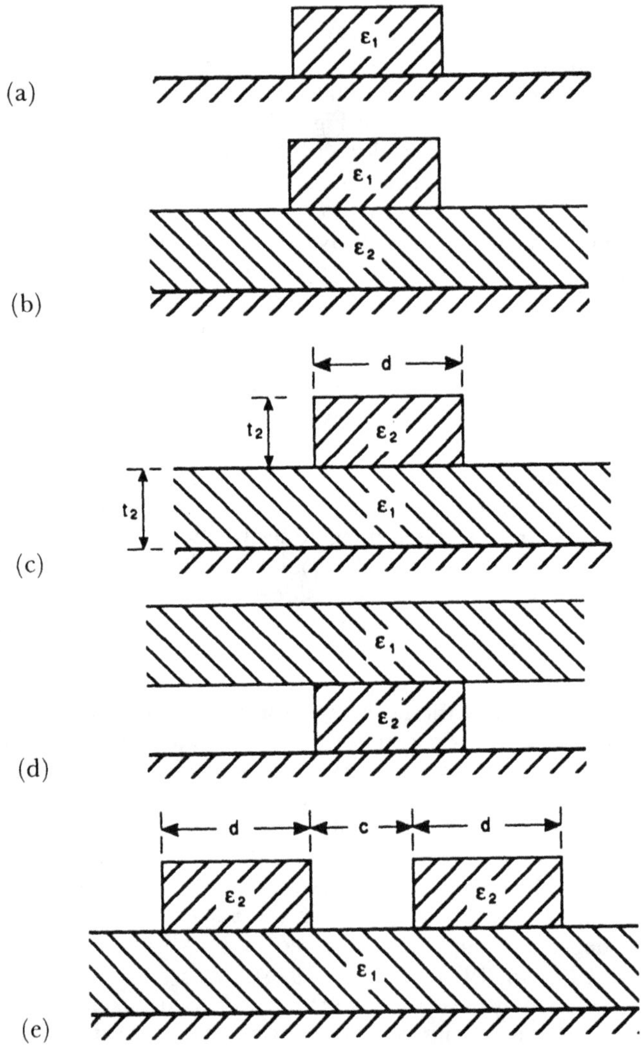

Fig. 4.8 Planar dielectric waveguides

a The image guide
b The insulated image guide
c The strip dielectric guide (SDG)
d The inverted strip dielectric guide (ISDG)
e Two coupled strip dielectric guides

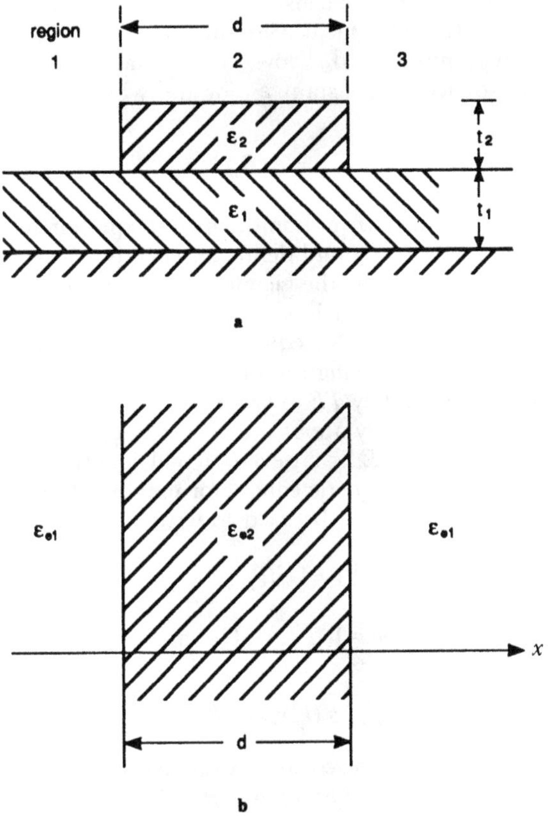

Fig. 4.9 Steps of solution by the method of effective ε

a Geometry for computing k_y
b Geometry for computing k_x

The steps of the solution by the methd of effective dielectric constant are demonstrated in Fig. 4.9*a*, *b*. for the SDG structure. The structure is divided into three regions as in Fig. 4.9*a*, each of which is assumed homogeneous and infinite in the x direction. The fields are also assumed uniform along x in the three regions. Thus we solve for the surface wave modes in each region. Considering the middle region and looking for TM to y modes, we apply the transverse resonance method to get the following modal equation:

$$(k_y/\varepsilon_{r1})\tan(k_y t_1) - (\alpha_2/\varepsilon_{r2})\frac{\alpha_0 + (\alpha_2/\varepsilon_{r2})\tanh(\alpha_2 t_2)}{(\alpha_2/\varepsilon_{r2}) + \alpha_0 \tanh(\alpha_2 t_2)} = 0 \qquad (4.106)$$

where:

$$k_y^2 + \beta^2 = k_0^2 \varepsilon_{r1}, \quad -\alpha_2^2 + \beta^2 = k_0^2 \varepsilon_{r2}, \quad -\alpha_0^2 \times \beta^2 = k_0^2$$

and $\varepsilon_{r1,2}$ are the relative dielectric constants of the two dielectric media involved. The equation can be solved for the wavenumbers k_{yn}, $n=1, 2, \ldots$ or equivalently the longitudinal wavenumbers β_n. Now, the crucial step of the method is to define an effective dielectric constant ε_{e2} for the whole of region 2 according to:

$$\varepsilon_{e2,n}=\varepsilon_0(\beta_n/k_0)^2 \tag{4.107}$$

In an exactly similar manner we obtain the modal equation for regions 1 and 3 in Fig. 9*a*, the wavenumbers β_m, and define an effective dielectric constant $\varepsilon_{e1,n}$ for any of the two regions. Now, the geometry of the problem reduces to three homogeneous regions each of which is infinite in the y direction. The dielectric constant of any of the regions is equal to the effective dielectric constant assigned to it. The resulting configuration is given in Fig. 4.9*b*. The modes in this structure are approximately *TE* to x since they have been assumed *TM* to y previously. Owing to symmetry about the $x=0$ plane, modes can be classified into even or odd modes; i.e. E_y is an even or odd function of x. For example, considering even modes, a magnetic conductor may be placed at $x=0$, and the transverse resonance method can be used to write down the modal equation:

$$k_x \tan(k_x d/2)=[k_0^2(\varepsilon_{e2,n}-\varepsilon_{e1,n})-k_x^2]^{1/2} \tag{4.108}$$

This is to be solved for k_{xm}, $m=1, 2 \ldots$. The modal phase factors β are then given by:

$$\beta_{mn}=(k_0^2\varepsilon_{e2,n}-k_{xm}^2)^{1/2} \tag{4.109}$$

The method of effective dielectric constant provides a simple and yet sufficiently accurate evaluation of β_{mn} for the first few low order modes (Knox and Toulios, 1970). It works better for large aspect ratios and low contrasts between the dielectric constants involved (Mclevige *et al.*, 1975). However, the method does not provide estimates of the modal field distributions. As pointed out by Solbach and Wolfe (1978) among others, the modal fields are generally hybrid and need to be represented by a spectral plane wave sum in every subregion of the guide in order to achieve exact field matching at the interfaces. In the following we outline a rigorous formulation for the modes based on mode matching technique. The formulation follows the work of Mittra *et al.* (1980).

First, it is convenient to close the waveguide by a suitably placed electric conductor as shown in Fig. 4.10. This will have negligible effect on the low order modes if its height h above the ground plane is chosen sufficiently large, but it will allow the use of a discrete mode sum in the different subregions of the structure. We define three subregions separated by the planes $x=\pm d/2$ as shown in the Figure. We can find the complete set of discrete modes in each of these subregions. Each set of modes is composed of *TE* and *TM* to y subsets. Let us denote by $\psi_n(y)$, $n=1, 2 \ldots$ the e_y component of TM_n modes in the region $|x|\leq d/2$, with corresponding eigenvalues k_{yn}. The latter are the y wavenumbers in the medium with $\varepsilon=\varepsilon_1$. Similarly, denote by $\phi_n(y)$ the h_y field of the TE_n modes in the same region with corresponding eigenvalues v_{yn}. Similar notations are used for the regions $x>d/2$ and $x<-d/2$ with the addition of a prime. Thus

$\psi'_n(y)$ denotes the e_y component of TM_n modes with k'_{yn} the corresponding eigenvalues and so on. Now, looking for even modes (for e_y) about the plane of symmetry $x=0$, then, apart from the phase propagation term $\exp(i\omega t - i\beta z)$, we write for e_y and h_y in the region $|x| \leqslant d/2$:

$$e_{y1}(y, x) = \sum_n A_n \psi_n(y) \cos(u_n x) \tag{4.110}$$

$$h_{y1}(y, x) = \sum_n C_n \phi_n(y) \sin(v_n x) \tag{4.111}$$

where $u_n^2 + k_{yn}^2 + \beta^2 = \omega^2 \mu \varepsilon_1 = v_n^2 + v_{yn}^2 + \beta^2$. Similarly, the corresponding fields in the region $x \geqslant d/2$ are:

$$e_{y2}(y, x) = \sum_n B_n \psi'_n(y) \exp[-\alpha_n(x - d/2)] \tag{4.112}$$

$$h_{y2}(y, x) = \sum_n D_n \phi'_n(y) \exp[-\gamma_n(x - d/2)] \tag{4.113}$$

where the fields are assumed to decay exponentially in the x direction, and $k'^2_{yn} + \beta^2 - \alpha_n^2 = \omega^2 \mu \varepsilon_1 = v'^2_{yn} + \beta^2 - \gamma_n^2$. However, if any of the α_n^s or γ_n^s turns out to be purely imaginary, this will mean that leakage occurs for this particular mode and results in a finite attenuation. The A_n, B_n . . . are modal amplitudes and the following relations hold: $u_n^2 + \alpha_n^2 = k'^2_{yn} - k_{yn}^2$ and $v_n^2 + \gamma_n^2 = v'^2_{yn} - v_{yn}^2$.

Other field components are obtainable from well known formulae [see (1.27)—(1.28)]. Now, we impose the boundary conditions at the plane $x = d/2$, which require the continuity of each of the fields e_y, h_y, e_z and h_z. This results in the four equations:

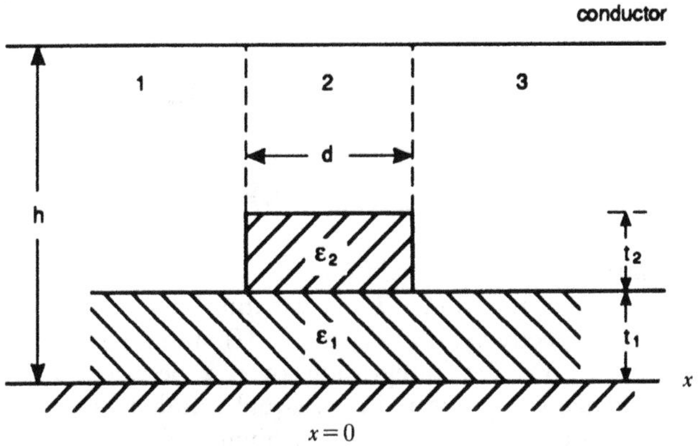

Fig. 4.10 Mode matching solution applied to a SDG

$$\sum_n A_n\psi_n(y)\cos(u_n d/2)=\sum_n B_n\psi'_n(y) \tag{4.114}$$

$$\sum_n C_n\phi_n(y)\sin(v_n d/2)=\sum_n D_n\phi'_n(y) \tag{4.115}$$

$$\beta\sum_n A_n\partial_y(\psi_n)f(u_n)+\omega\mu\sum_n C_n\phi_n(y)v_n f(v_n)$$

$$=\beta\sum_n B_n\partial_y(\psi'_n)/(\beta^2-\alpha_n^2)-\omega\mu\sum_n D_n\gamma_n\phi'_n(y)/(\beta^2-\gamma_n^2) \tag{4.116}$$

$$\omega\sum_n A_n\varepsilon(y)\psi_n(y)u_n g(u_n)+\beta\sum_n C_n\partial_y(\phi_n)g(v_n)$$

$$=\omega\sum_n B_n\varepsilon'(y)\psi'_n(y)\alpha_n/(\beta^2-\alpha_n^2)-\beta\sum_n D_n\partial_y(\phi'_n)/(\beta^2-\gamma^2) \tag{4.117}$$

where the following definitions are used:

$$f(u)=\cos(ud/2)/(\beta^2+u^2),\quad g(u)=\sin(ud/2)/(\beta^2+u^2)$$

$$\partial_y(\psi)\equiv\partial\psi(y)/\partial y$$

The functions $\varepsilon(y)$ and $\varepsilon'(y)$ describe the y dependence of ε in the regions $|x|\leqslant d/2$ respectively.

So far, the formulation is exact. However, to obtain useful numerical results, the sum terms in (4.114)—(4.117) must be truncated to finite sums. So, let us retain only M modes of the TM type and N modes of the TE type in each of the subregions $|x|\leqslant d/2$ and $|x|\geqslant d/2$. The y dependence of the above equations can be eliminated by taking their projections on a suitable complete set of basis functions. Noting the following orthogonality relations among the TE or TM eigenfunctions:

$$\int_0^h \psi_n(y)\varepsilon(y)\psi_m(y)\,\mathrm{d}y=\delta_{mn} \tag{4.118}$$

$$\int_0^h \phi_n(y)\phi_m(y)\,\mathrm{d}y=\delta_{mn} \tag{4.119}$$

with similar relations for the primed functions ψ'_n and ϕ'_n, we choose either ψ'_n or ϕ'_n as our set of basis functions. Thus we multiply (4.114) and (4.117) by $\varepsilon'(y)\psi'_m$ and ψ'_m respectively, $(m=1, 2 \ldots M)$ and (4.115) and (4.116) by $\phi'_m(y)$ $(m=1, 2, \ldots N)$. On integrating over y between 0 and h, we get $2(M+N)$ homogeneous equations in an equal number of modal amplitudes. For a nontrivial solution, we set the determinant of coefficients equal to zero and solve the resulting equation for the modal phase constants β.

Table 4.1 Propagation constants β (mm^{-1}) of guided modes in ISDG (Fig. 4.8*d*) at $f=79{\cdot}4$ GHz (after Mittra *et al.*, 1980)

Mode	1 TE 1 TM	3 TE 3 TM	5 TE 5 TM	7 TE 7 TM	Effective ε method
E_{11}	2·9718	2·9892	2·9872	2·9873	2·9906
H_{11}	2·7341	2·6210	2·7295	2·7203	2·7595
E_{21}	2·3646	2·3871	2·3910	2·3908	2·4070

The method of mode matching just outlined has been applied by Mittra *et al.* (1980) to the ISDG of Fig. 4.8*d*. Table 4.1 gives some of their numerical results for $\beta\,mm^{-1}$ of the first three lowest order modes for the following guide parameters: $\varepsilon_1=\varepsilon_2=4\varepsilon_0$, $t_1=0{\cdot}795$ mm, $t_2=0{\cdot}735$ mm, $h=3(t_1+t_2)$ and $d=2{\cdot}05$ mm. The frequency $=79{\cdot}4$ GHz. The Table shows the convergence of the mode matching solution as M and N increase. Results obtained from the method of effective ε are also included for comparison. It can be concluded that, as far as the modal phase constants are concerned, the method of effective ε is sufficiently accurate in most practical cases. However, the modal field distribution can only be obtained from a rigorous field matching solution. A rigorous solution is also needed to predict the attenuation of leaky modes as stressed by Peng and Oliner (1981) and Oliner *et al.* (1981) who show that leakage occurs for some modes on structures that contain a planar slab of dielectric (such as structures *b*, *c*, *d* and *e* in Fig. 4.8). Leakage occurs in the form of a surface wave carrying power sideways in the dielectric slab.

4.5 Problems

4.1 Prove equation (4.7) of $R(\lambda)$. Show that the surface wave mode corresponds to a pole of $R(\lambda)$.

4.2 Deduce (4.11) from (4.10) under the condition $k\rho \gg 1$.

Hint: Near the stationary point $\psi=\theta$, the following approximation is valid: $\cos(\psi-\theta)\simeq 1-x^2/2$; $x=\psi-\theta$. Then, use the identity:

$$\int_{-\infty}^{\infty} \exp(-iax^2)\,\mathrm{d}x=(i\pi/a)^{1/2}$$

4.3 Consider a cylindrically curved constant admittance surface, described by the equation $\rho=a$ and characterised by $Y=H_\phi/E_z$. Considering TM modes with

$$E_z(\rho,\phi)=AH_v^{(2)}(k\rho)\,\mathrm{e}^{-iv\phi}$$

Show that the complete spectrum of modes has only a discrete spectrum with eigenvalues v satisfying the modal equation:

$$i(\mu/\varepsilon)^{1/2}Y=H_v'^{(2)}(ka)/H_v^{(2)}(ka)$$

4.4 Consider *TM* modes on the dielectric slab of Fig. 4.1*b*. Show that the modal equation for even and odd modes in h_y are respectively:

$$\kappa d \tan(\kappa d) = u_s\, d(n_1/n_2)^2$$

$$\varkappa d \cot(\varkappa d) = -u_s(n_1/n_2)^2$$

4.5 Consider the asymmetric dielectric layer of thickness d shown in Fig. pr. 4.5, in which $n_1 > n_2 > n_3$. Show that the *TE* surface wave modes satisfy the modal equation:

$$\tan(2\varkappa d) = \frac{(\alpha_2 + \alpha_3)/\varkappa}{1 - \alpha_2\alpha_3/\varkappa^2}$$

where $\varkappa$ is the x wavenumber in the dielectric layer and α_2 and α_3 are the attenuation rates in the upper and lower media. Show that cutoff occurs when:

$$V = \tan^{-1}[(n_2^2 - n_3^2)/n_1^2 - n_3^2)]^{1/2} + m\pi$$

with $V = 2kd(n_1^2 - n_3^2)^{1/2}$. Deduce that the dominant mode ($m = 0$) propagates at all frequencies only in the symmetric case; i.e. when $n_2 = n_3$.

4.6 Deduce equations (4.22)—(4.24) for pseudomodes on a dielectric slab.

4.7 Derive the modal equation (4.31) for the surface wave modes on a dielectric rod.

4.8 Find the largest diameter of a step index monomode optical fibre to operate at the wavelength 1·3 μm and to have a differential refractive index $\Delta = 0{\cdot}003$.

4.9 Find an expression for the ratio of the power guided by the core and that guided by the clad for the LP_{01} mode on an optical fibre.

[Use the identities:

$$\int_0^z J_0^2(x)x\,dx = \tfrac{1}{2}z^2(J_0^2(z) + J_1^2(z))$$

and

$$\int_z^\infty K_0^2(x)x\,dx = \tfrac{1}{2}z^2(K_1^2(z) - K_0^2(z))]$$

4.10 Find the scattered field coefficients $C(\lambda)$ and $D(\lambda)$ appearing in (4.45)—(4.46) for an obliquely incident *TE* or *TM* wave on a dielectric rod.

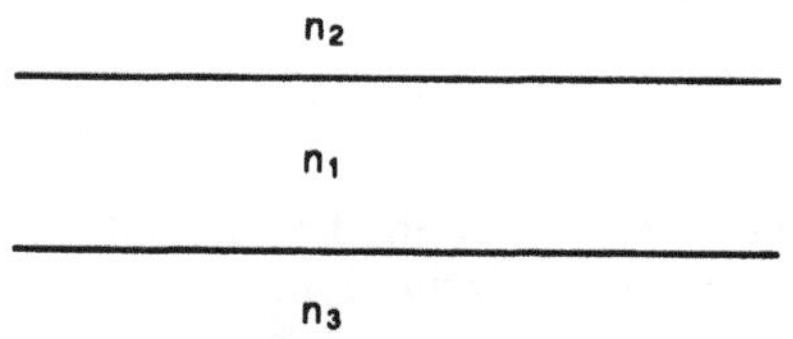

Fig. pr. 4.5 Asymmetric dielectric layer

Show that the coupling between these two waves vanishes under normal incidence (i.e. when $\beta = 0$).

4.11 Derive formula (4.54) for the coupling between modes on nonuniform open structures.

4.12 Apply (4.54) to find the coupling among *TM* modes on a planar impedance surface due to a change in its surface impedance.

4.13 Prove equations (4.63) and (4.64).

4.14 Find the coupling terms between modes on a dielectric slab due to a change in the slab refractive index n_1. The slab thickness $2d$ is assumed to remain constant.

4.15 Consider a dielectric slab waveguide, as shown in Fig. 4.1*b*, which is excited by an electric line source J_y given by:

$$J_y = I\delta(z)\delta(x - x_0) \text{ amperes/m}^2$$

Show that the excited field E_y is given by (4.25) with:

$$A_s = \tfrac{1}{2}\omega\mu I\psi_s(x_0)/\left[\beta_s \int_{-\infty}^{\infty} \psi_s^2 \, dx\right]$$

$A(\lambda) = \frac{1}{2}\omega\mu I\psi(\lambda;\, x_0)/[\beta(\lambda)N(\lambda)]$, and $N(\lambda)$ is defined by

$$\int_{-\infty}^{\infty} \psi(\lambda;\, x)\psi(\lambda';\, x)\, dx = N(\lambda)\delta(\lambda - \lambda')$$

4.16 Prove relations (4.75) and (4.76) for a weak discontinuity. Use (4.68) for further reduction of these relations.

4.17 Redo problems 4.4 and 4.5 using the transverse resonance method to obtain the modal equations.

4.18 Consider a longitudinal discontinuity in the dielectric planar slab guide caused by a slight change in thickness $2d$. Find the forward and backward radiation pattern caused by a surface wave scattering at the discontinuity. *Hint:* Use formulae (4.77) and (4.78) for $T(\lambda)$ and $\Gamma(\lambda)$, then use (4.25) and apply (4.11).

4.19 Prove the formulae for conductor and dielectric attenuation (4.91) and (4.92) in an H-guide. Comment on their behaviour versus frequency.

4.20 Plot the turns ratio n_t and the junction susceptance B given by (4.103) and (4.105) respectively versus (a'/a) between 0·2 and 1·0.

4.21 Prove the orthogonality relationships (4.118) and (4.119).

4.22 By truncating equations (4.114)—(4.117) to M *TM* modes and N *TE* (to y) modes, and using the orthogonality relations (4.118)—(4.119), show how to obtain $2(M+N)$ homogeneous equations in the unknown coefficients A_n, B_n, $n = 1, M$ and C_n, D_n, $n = 1, N$.

4.23 Apply the results of problem (4.22) to $M=1$ and $N=0$. Hence, show that the modal equation reduces to:

$$u_1 \tan(u_1 d/2) = \alpha_1 I, \text{ where}$$

$$I = \frac{\beta^2 + u_1^2}{\beta^2 - \alpha_1^2} \frac{\langle \psi_1(y), \varepsilon'(y)\psi_1'(y)\rangle}{\langle \varepsilon(y)\psi_1(y), \psi_1'(y)\rangle}$$

Verify that if $I=1$ (which is almost the case), then the above modal equation is the same as that obtained by the method of effective ε.

4.6 References

BHARTIA, P., and BAHL, I.J. (1984): 'Millimeter wave engineering and applications', (Wiley-Interscience Publications, NY)

BRUNO, W.M., and BRIDGES, W.B. (1988): 'Flexible dielectric waveguides with power cores', *IEEE Trans.* **MTT–36**, pp. 882–890

CHERIN, A.H. (1983): 'An introduction to optical fibers' (McGraw Hill, London)

CLARRICOATS, P.J.B., and SAHA, P.K. (1971): 'Propagation and radiation behaviour of corrugated feeds. Pt. 1: Corrugated waveguide feeds', *Proc. IEE*, **118,** pp. 1167–1176

CLEMMOW, P.C. (1966): 'The plane wave spectrum representation of electromagnetic fields' (Pergamon Press).

GELIN, P.H., PETENZI, M., and CITERNE, J. (1979): 'New rigorous analysis of the step discontinuity in a slab dielectric waveguide', *Electron. Lett.*, **15**, pp. 355–365

HARRINGTON, R.F. (1961): 'Time harmonic electromagnetic fields' (McGraw Hill, New York, Toronto, London)

ITOH, T., and MITTRA, R. (1975): 'New waveguide structures for millimeter wave integrated circuits'. Intern. Microwave Symp. Digest, Palo Alto, CA, May 12–14, pp. 277–279

ITOH, T. (1976): 'Inverted strip dielectric waveguide for millimeter-wave integrated circuits', *IEEE Trans.* **MITT–24,** pp. 821–827

JONES, Jr, W. (1988): 'Introduction to optical fiber communication systems' (Holt Rinehart and Winston, NY)

KING, R.J., and WAIT, J.R. (1976): 'Electromagnetic ground wave propagation theory and experiment' *in* 'Symposia Mathematica' (Academic Press, New York) Vol. 18, pp. 107–208

KNOX, R.M., and TOULIOS, P.P. (1970): 'Integrated circuits for the millimeter through optical frequency range', Proc. Symp. Submillimeter Waves, New York, March 31st–April 2nd.

MAHMOUD, S.F., and BEAL, J.C. (1975): 'Scattering of surface waves at a dielectric discontinuity on a planar waveguide', IEEE Trans., **MTT–23,** pp. 193–198

MAHMOUD, S.F. (1990): 'Modal analysis of open groove waveguide', IEEE Trans., **MTT–38,** no. 4, April.

MARCUSE, D. (1970): 'Radiation losses of tapered dielectric slab waveguides', *Bell Syst. Tech. J.*, **49**, p. 273

MARCUSE, D. (1972): 'Light transmission optics' (Van Nostrand Reinhold Co., New York)

MARCUSE, D. (1974): 'Theory of dielectric optical waveguides' (Academic Press, New York)

McLEVIGE, W.V., ITOH, T., and MITTRA, R. (1975): 'New waveguide structures for millimeter-wave and optical integrated circuits', *IEEE Trans.* **MTT–23,** pp. 788–794

MIDWINTER, J.E. (1979): 'Optical fibers for transmission' (John Wiley, New York)

MITTRA, R., HOU YUN-LI, and JAMNEJAD, V. (1980): 'Analysis of open dielectric waveguides using mode matching technique and variational methods', *IEE Trans.* **MTT–28,** pp. 36–43

MORISHITA, K., INAGAKI, S.I., and KUMAGAI, N. (1979): 'Analysis of discontinuities in dielectric waveguides by means of the least squares boundary residual method', *IEEE Trans.* **MTT–27,** pp. 410–315

NAKAHARA, T., and KURAUCHI, N. (1964): 'Transmission modes in the grooved guide', *J. Inst. Electron. Comm. Eng. Jap.*, **47**, pp. 43–51

OLINER, A.A. and LAMPARIELLO, P. (1985): 'The dominant mode properties of open groove guide: An improved solution', *IEEE Trans.* **MTT–33**, pp. 755–764
RULF, B. (1975): 'Discontinuity relation in surface waveguide', *J. Opt. Soc. Amer.*, **65,** p. 1248
ROZZI, T.E. (1978): 'Rigorous analysis of the step discontinuity in a planar dielectric waveguide', *IEEE Trans.* **MTT–26,** pp. 738–746
SHEVCHENKO, V.V. (1971): 'Continuous transitions in open waveguides' (translated from Russian by Peter Beckmann) (The Golem Press, Boulder, Col.)
SHIGESAWA, H., and TSUJI (1986): 'Mode propagation through a step discontinuity in dielectric planar waveguide', *IEEE Trans.* **MTT–34,** pp. 205–212
SNITZER, E. (1961): 'Cylindrical dielectric waveguide modes', *J. Opt. Soc. Amer.*, **51,** pp. 491–498
SNYDER, A.W. (1971): 'Continuous mode spectrum of a circular dielectric rod', *IEEE Trans.* **MTT–19,** pp. 720–727
SOLBACH, K., and WOLF, I. (1978): 'The electromagnetic fields and the phase constants of dielectric image lines', *IEEE Trans.* **MTT–26,** pp. 266–274
TISCHER, F.J. (1970): 'H-guide with laminated dielectric slab', *IEEE Trans.* **MTT–18,** pp. 9–15
WAIT, J.R. (1960): 'On the excitation of electromagnetic waves on a curved surface', *IRE Trans. Antenn. Propag., July*, pp. 445–449
WAIT, J.R. (1964):'Electromagnetic surface waves' *in* 'Advances in radio research' (Academic Press, London) Vol. 1, pp. 157–217
WAIT, J.R. (1970): Electromagnetic waves in stratified media' (Pergamon Press, New York, London)

Additional references

CHANG, Y.W., and PAUL, J.A. (1978): 'Millimeter wave image guide integrated passive devices', *IEEE Trans.* **MTT–24,** pp. 806–814
KNOX, R.M. (1976): 'dielectric waveguide microwave integrated circuits–An overview', *IEEE Trans.* **MTT–24,** pp. 806–814
MORISHITA, K., KONDOH, Y., and KUMAGAI, N. (1980): 'On the accuracy of scalar approximation technique in optical fiber analysis', *IEEE Trans.* **MTT–28,** pp. 33–36
SHINDO, S., and ITANAMI, T. (1978): 'Low loss rectangular dielectric image line for millimeter wave integrated circuits', *IEEE Trans.* **MTT–26,** p. 747–751
SODHA, M.S., and CHATAK, A.K., (1977): 'Inhomogeneous optical waveguides' (Plenum Press, NY, London)
YONEYAMA, T., and NISHIDA, S. (1981): 'Nonradiative dielectric waveguide for millimeter wave integrated circuites', *IEEE Trans.* **MTT–29,** pp. 1188–1192
YOUNG, B., and ITOH, T. (1987): 'Analysis and design of microslab waveguide', *IEEE Trans.* **MTT–35,** pp. 850–857
WAIT, J.R. (1955): 'Scattering of a plane wave from a circular dielectric cylinder at oblique incidence', *Canadian J. Phys.*, **33,** pp. 189–195

Chapter 5
Low crosspolar waveguides

5.1 Introduction

In the previous three chapters we have discussed the general properties of three classes of waveguides: closed waveguides, constant impedance wall waveguides and open waveguides. In this chapter we apply the analysis developed there to a type of waveguides used in a specific application; namely, a class of waveguides which are capable of maintaining highly polarised fields. Such waveguides with low crosspolar fields find applications in many communication systems where efficient use of the microwave spectrum of frequencies is of primary importance.

To use the microwave spectrum efficiently, it has become a common practice for satellite and terrestrial communication systems to reuse a carrier frequency to transmit two baseband signals on the same bandwidth, but with two orthogonal polarisations. In order to avoid any interference, stringent specifications must be imposed on the allowable crosspolar radiation from the antenna system. Typically, polarisation discrimination of the order of 35 dB or better must be achieved. This stringent specification also reflects on the waveguide feeding the antenna; hence the need for low crosspolar waveguides.

We start by studying the limitations of the TE_{11} mode of a circular waveguide as a low crosspolar radiator (section 5.2). We conclude that hybrid mode supporting structures can make low crosspolar radiators. The most popular of these structures is the transversely corrugated guide which is considered in section 5.3. Other low crosspolar guides are studied in sections 5.4 and 5.5.

5.2 Field polarisation of the TE_{11} and TM_{11} modes

A horn radiator can be realised by flaring out a standard circular wavegide to a suitably large aperture. For a small flare angle the aperture fields are almost the same as those of the dominant TE_{11} mode in a uniform cylindrical waveguide. In the following we study the nature of the transverse fields of the TE_{11} mode since these determine the radiated fields from the aperture. Writing the transverse distribution of h_z in the form

$$h_z = iy_0 J_1(u\rho/a)\sin\phi$$

the transverse electric field is

$$\mathrm{e}_t = (k_0 a/u)\left[\frac{J_1(u\rho/a)}{u\rho/a}\cos\phi\hat{\rho} + J_1'(u\rho/a)\sin\hat{\phi}\right] \tag{5.1}$$

where a is the aperture radius, $u = 1{\cdot}841$, $y_0 = (\varepsilon/\mu)^{1/2}$ and $k_0 = \omega\mu y_0$.

In order to study the polarisation properties of the TE_{11} mode, it is essential to express the e_t field in terms of its cartesian; instead of polar, components. Thus (5.1) takes the alternative form

$$e_t = (k_0 a/u)\{\hat{x}[J_0(u\rho/a) + J_2(u\rho/a)\cos 2\phi] + \hat{y}[J_2(u\rho/a)\sin 2\phi]\} \qquad (5.2)$$

where identities (A2·18) and (A2·19) have been invoked. (5.2) reveals that the electric field is mainly x-polarised. Actually the crosspolar component e_y is zero in the E and H planes ($\phi = 0, \pi/2$). It is maximum in the planes $\phi = \pm\pi/4$ and $\pm 3\pi/4$. The level of the copolar field e_x in the E, H and the $\phi = \pi/4$ planes and the crosspolar field e_y in the $\phi = \pi/4$ plane are plotted in Fig. 5.1*a*. A cross sectional view of the lines is shown in Fig. 5.1*b*. The main features of the TE_{11} mode as a radiator can be deduced from Fig. 5.1*a*. Firstly, since the field distributions in the E and H planes are not the same, the radiation beam is not axially symmetric. The weak field tapering towards the edges in the E plane results in high relative side lobe levels in this plane; the first side lobe level is about -18 dB relative to the on axis radiation while the corresponding level in the H plane is about -26 dB. The high edge fields in the E plane cause high edge diffraction which leads to further increase in the side lobe levels of radiation. Secondly, the crosspolar field e_y is relatively high and causes a peak crosspolar radiation of about -18 dB relative to the peak copolar radiation, as will be shown analytically in the next section. This level of crosspolar radiation is considered overwhelmingly high in applications requiring high polarisation purity. An acceptable level is well below -30 dB.

Next consider the TM_{11} mode in a circular guide. Writing e_z in the form:

$$e_z = iJ_1(u'\rho/a)\cos\phi\,\exp(-i\beta z) \qquad (5.4)$$

one can find the transverse field in cartesian form as:

$$e_t = (\beta a/u')\{\hat{x}[J_0(u'\rho/a) - J_2(u'\rho/a)\cos 2\phi] - \hat{y}[J_2(u'\rho/a)\sin 2\phi] \qquad (5.5)$$

where $u' = 3{\cdot}83172$ and $\beta = (k_0^2 - u'^2/a^2)^{1/2}$.

This again shows that the crosspolar field e_y is zero in the E and H planes ($\phi = 0$ and $\pi/2$) and is maximum in the $\pm\pi/4$ and $\pm 3\pi/4$ planes. Note, however, the minus sign in front of the y component of e_t. The copolar field is plotted in the planes $\phi = 0$, $\pi/2$ and $\pi/4$ and the crosspolar field in the plane $\phi = \pi/4$ in Fig. 5.2*a*. The transverse e lines in the cross section are shown in Fig. 5.2*b*. A comparison of Fig. 5.1*a* with 5.2*a* (or 5.1*b* with 5.2*b*) reveals that, if a weighted sum of TE_{11} and TM_{11} modes is formed, one can considerably reduce the crosspolar field e_y. Meanwhile, the copolar field at the edges of the E plane will be reduced, leading to less diffraction and lower side lobe level of radiation. This observation has been exploited by each of Potter (1963) and Satoh (1971) to design dual mode horns with low crosspolar aperture fields. The idea is to introduce a discontinuity along the horn which converts some of the TE_{11} mode power into the right amount of TM_{11} power. Since the two modes have different phase velocities, their relative phase has to be carefully adjusted so that the crosspolar fields add destructively at the horn's mouth. The discontinuity in Potter's horn (Fig. 5.3*a*) is realised by an axially symmetric step in radius at the horn's throat. In Satoh's horn a symmetric dielectric lining is introduced along the horn's flare (Fig. 5.3*b*). Because of the different phase velocities of the two

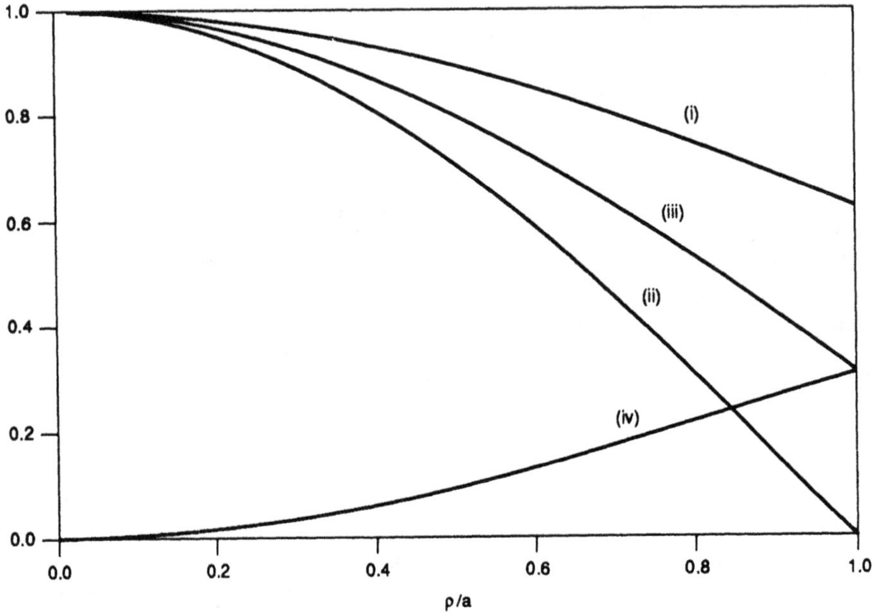

Fig. 5.1*a* Cartesian field components of the TE_{11} mode

(i) e_x in the e plane
(ii) e_x in the h plane
(iii) e_x in the plane $\phi = \pi/4$
(*iv*) e_y in the plane $\phi = \pi/4$

Fig. 5.1*b* *e* field lines of the TE_{11} mode

modes concerned, these two horns are inherently narrow band in frequency although Satoh's horn is relatively broader. The reason is that the difference in the phase velocities of the two modes is greater near the horn's throat than near the mouth. Typically Satoh's horn maintains low crosspolar levels over about 20% of the centre frequency, corresponding to 5% for Potter's horn.

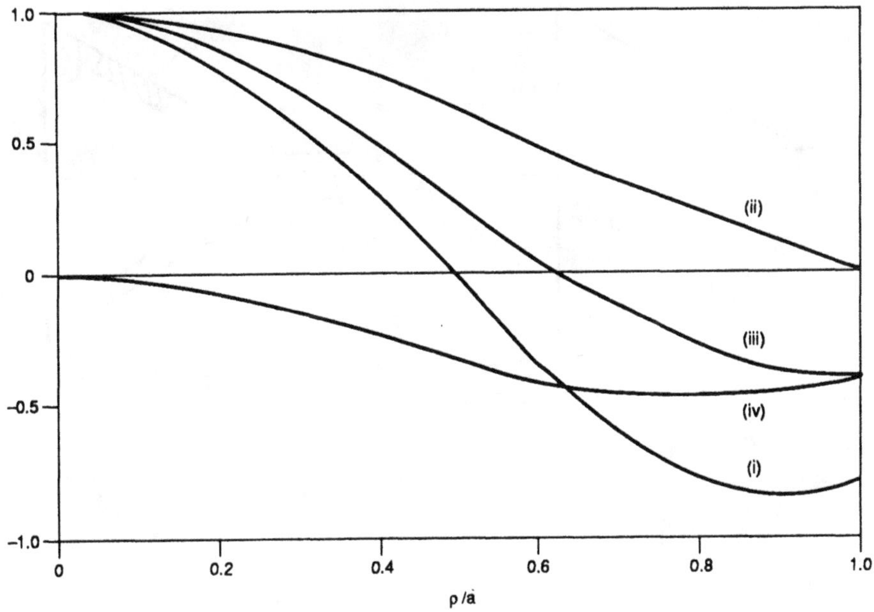

Fig. 5.2*a* Cartesian components of the TM_{11} mode

(i) e_x in the e plane
(ii) e_x in the h plane
(iii) e_x in the plane $\phi = \pi/4$
(iv) e_y in the plane $\phi = \pi/4$

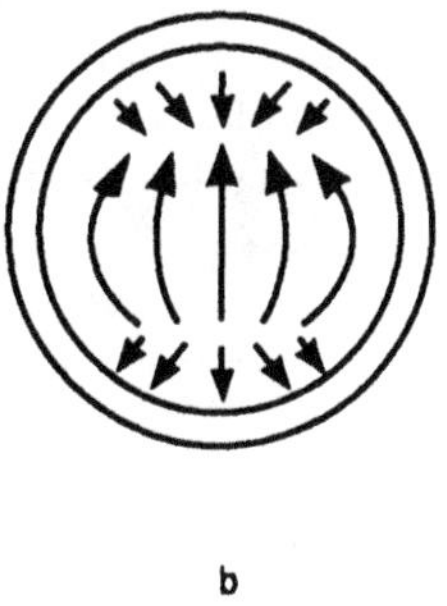

Fig. 5.2*b* *e* field lines of the TM_{11} mode

To conclude the above discussion, we note that, if a horn is designed to support a hybrid mode as its dominant mode, it can provide high polarisation purity over a broad band of frequencies. In the rest of the chapter we introduce few examples of hybrid mode supporting waveguides.

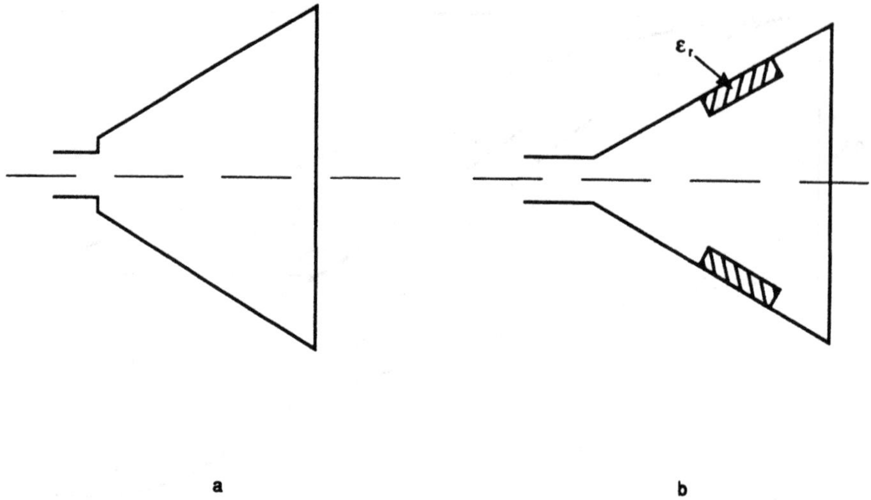

Fig. 5.3 *a* Potter's horn

b Satoh's horn, or dielectric loaded horn

5.3 The transversely corrugated horn

Perhaps the most popular microwave low crosspolar radiator is the transversely corrugated horn (Fig. 5.4). Kay (1962) in the USA was seemingly the first to introduce corrugations in the wall of a conical horn. He introduced a few quarter wavelength deep corrugations near the horn's mouth in an attempt to impede longitudinal currents which are responsible for edge diffraction. The result was an improvement in the radiation pattern symmetry about the horn's axis. Minnet and Thomas (1966) in Australia observed that wall corrugations resulted also in an improvement of polarisation purity. These discoveries have since prompted great interest in corrugated guides. Notable contributions have

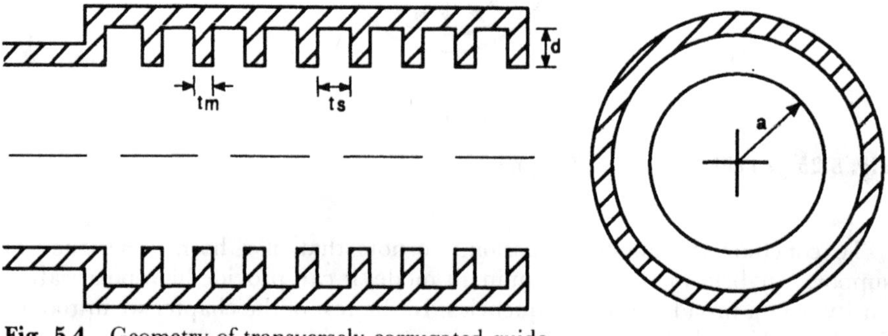

Fig. 5.4 Geometry of transversely corrugated guide

resulted from the work of Clarricoats and his colleagues at QMC, University of London, and are summarised in the excellent book by Clarricoats and Olver (1984) which also contains an extensive literature survey. Other important contributions include the work of Drogone (1977*a*, *b*) on propagation and radiation from corrugated waveguides and his later work on waveguides with general surface impedance (Dragone, 1980, 1981, 1985). Other notable workers include Narasimhan (e.g. 1970, 1971, 1974), Rudge and Adatia (e.g. 1975, 1976), Thomas (e.g. 1969, 1978) among others. (see the references). In the following we give an account of mode propagation and radiation in corrugated cylindrical and conical horn.

5.3.1 Transversely corrugated circular guide

A rigorous treatment of the transversely corrugated guide (Fig. 5.4) should account for the periodicity of the structure, and therfore call for the inclusion of space harmonics inside and outside the slots. However, under the condition of a sufficiently large number of slots per wavelength, a much simpler approach based on an average surface impedance can be pursued. This approach neglects the space harmonics which are effectively small under the aforementioned condition. An assessment of the accuracy of the surface impedance approach is given by Clarricoats and Olver (1984).

As discussed in section 3.2.4, the corrugated wall has an average surface impedance and admittance given by:

$$Z = e_\phi / h_z = 0 \tag{5.6}$$

$$Y = iy_0 B = -h_\phi / e_z = -iy_0(\cot k_0 d + 1/2k_0 a)(1 + t_m/t_s) \tag{5.7}$$

where the last bracketed term comes about owing to an averaging of the reciprocal of Y over the slotted and ridged parts. Now, the general analysis presented in section 3.4 for circular guides with impedance walls can be applied to the transversely corrugated guide. The modal equation (3.134) reduces in our case to

$$(F_1(u) - Bu^2/v)F_1(u) = \bar{\beta}^2 \tag{5.8}$$

where we have set $m=1$ for the dominant modes. As before $u = k_\rho a$, $v = k_0 a$, $\bar{\beta} = (1-(u/k_0 a)^2)^{1/2}$, and $F_1(u)$ is defined by (3.130). The mode hybrid factor in (3.135) reduces to:

$$\Lambda = Bu^2/2\bar{\beta}v \pm (1 + (Bu^2/2\bar{\beta}v^2))^{1/2} \tag{5.9}$$

for *HE* and *EH* modes respectively.

In order to study the crosspolar behaviour of the modes with $m=1$, it is appropriate to express the transverse e field in cartesian coordinates. Starting with (3.124)—(3.125), we get the forms

$$e_x = (\Lambda + \bar{\beta})J_0(u\rho/a) + (\Lambda - \bar{\beta})J_2(u\rho/a) \cos 2\phi \tag{5.10}$$

$$e_y = (\Lambda - \bar{\beta}(J_2(u\rho/a) \sin 2\phi \tag{5.11}$$

This shows that the maximum value of the crosspolar field e_y in the $\phi = \pm\pi/4$ plane is proportional $(\Lambda - \bar{\beta})$. Therefore a zero crosspolar field is attained by HE_{1n} modes if $\Lambda = \bar{\beta}$. For large aperture, $k_0 a \gg 1$, this condition becomes approximately $\Lambda = 1$, or from (5.9) $B=0$. Therefore, when the wall surface admittance vanishes, the modes become balanced hybrid modes. Actually, when $B=0$, $\Lambda=\pm 1$, corresponding to the HE_{1n} and EH_{1n} modes respectively. While the HE_{1n} will have almost zero crosspolar field, the EH_{1n} modes will have equal copolar and crosspolar fields. Under the balanced hybrid mode condition, the modal equation (5.8) reduces to

$$F_1(u) = \pm\bar{\beta} \simeq \pm 1$$

or

$$J_1'(u) = \pm J_1(u)/u$$

whose solutions are

$$u = \text{zeros of } J_0(.) = 2{\cdot}405,\ 5{\cdot}5201, \ldots \text{ for } HE_{1n} \text{ modes, and}$$

$$u = \text{zeros of } J_2(.) = 5{\cdot}1356, \ldots \text{ for } EH_{1n} \text{ modes}$$

According to Clarricoats and Saha (1971*a*) the root $u=5{\cdot}1356$ corresponds to the EH_{12} mode, the name EH_{11} is reserved for the surface wave mode that would exist if the wall surface admittance is inductive. For the latter u is purely imaginary.

Near the balanced hybrid mode condition when $B/v \ll 1$, the eigenvalues u and hybrid factors Λ will deviate slightly from their values at the balanced hybrid condition. To illustrate the behaviour of u and Λ with frequency, we plot them for the HE_{11} mode in Figs. 5.5 and 5.6. The corrugation depth d is chosen such that $B=0$ at $k_0 a = 15$. The useful bandwidth for low crosspolar radiation is determined by the range over which Λ does not deviate appreciably from unity. Values of u for the higher order modes HE_{12} and EH_{12} in the same guide are plotted in Fig. 5.7. Next, we obtain the copolar and crosspolar radiation patterns of the lower order modes of the guide.

Radiation pattern

The radiation fields are simply given by the Fourier transform of the aperture fields, namely, one can show that the cartesian components of the electric radiated field is given by (see probl. 5.2):

$$e^r_{x,y}(R, \theta, \phi) = (i \exp(-ik_0 R)/2\pi R) k_0 \cos\theta \; FT(e_{x,y}(\rho, \phi')) \qquad (5.12)$$

where $FT(.)$ stands for the Fourier transform, which takes the form

$$FT(e(\rho, \phi')) = \int_0^{2\pi}\int_0^{2\pi} e(\rho, \phi) \exp(ik_0\rho \sin\theta \cos(\phi-\phi'))\rho \, d\rho \, d\phi \qquad (5.13)$$

Substituting the cartesian components of e from (5.10)—(5.11) into (5.13), and evaluating the integral (see probl. 5.3), we get the cartesian components of the radiation fields:

$$e^r_x = A[(\Lambda+\bar{\beta})N_0(u, v\sin\theta) + (\Lambda-\bar{\beta})N_2(u, v\sin\theta)\cos 2\phi] \qquad (5.14)$$

$$e^r_y = A(\Lambda-\bar{\beta})N_2(u, v\sin\theta)\sin 2\phi \qquad (5.15)$$

where

$$N_l(u, v\sin\theta) = [a^2/(v^2\sin^2\theta - u^2)].(v\sin\theta J_l(u) J_1(v\sin\theta) - u J_l(v\sin\theta) J_1(u)] \qquad (5.16)$$

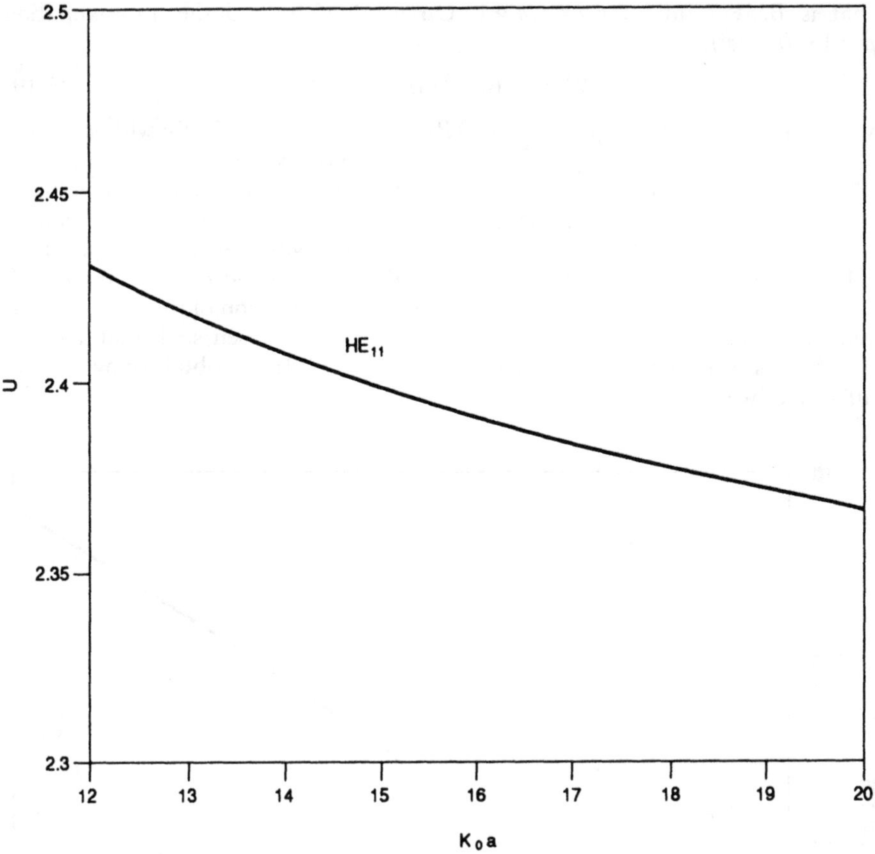

Fig. 5.5 Eigenvalue u of the HE_{11} mode in a corrugated guide

The corrugation depths are adjusted or zero Y at $k_0a=15$.

$l=0$, 2 and $A=ik_0 \cos\theta \exp(-ik_0R)/R$.

Since $N_2=0$ at $\theta=0$, the crosspolar radiation is characterised by a null on the axis $\theta=0$. It vanishes everywhere in the E and H planes and attains its maximum value in the $\phi=\pi/4$ plane. Referred to the on axis copolar radiation field, the peak crosspolar radiation is

$$XP=[(\Lambda-\hat{\beta})/(\Lambda+\hat{\beta})]\{N_2(u,\, v\sin\theta)/N_0(u,\, 0)\}_{max} \tag{5.17}$$

As noted by Dragone (1977*a*), the maximum value of XP for the HE_{11} mode under balanced hybrid condition ($u=2{\cdot}405$) occurs at $v\sin\theta=3{\cdot}67$, whence the second bracketed term in the above equation is equal to 0·26; therefore:

$$XP=0{\cdot}26|(\Lambda-\hat{\beta})/(\Lambda+\hat{\beta})| \tag{5.18}$$

To investigate the crosspolar behaviour of the HE_{11} mode with frequency, we study the above expression near the balanced hybrid mode condition, that is $B/v \gg 1$, and for $v = k_0 a \gg 1$. On using (5.9) and the approximation $\hat{\beta} \simeq 1 - (u^2/2v^2)$:

$$XP = 0{\cdot}26[Bu^2/4v + + u^2/4v^2] \quad (5.19)$$

which shows that, for a given B, the XP decreases monotonically with v: i.e. the larger the horn's aperture in wavelengths, the lower is the crosspolar radiation. The bandwidth over which this radiation is less than a specified value is determined mainly by the behaviour of the slot admittance near its resonant frequency (at which $B = 0$). For large aperture corrugated guides, say $k_0 a > 12$, a relative crosspolar level less than -40 dB is obtainable over a 30 to 40% bandwidth. To illustrate this, the peak crosspolar radiation of the HE_{11} mode is plotted versus $v = k_0 a$ in Fig. 5.8. The slot depth is chosen such that $Y = 0$ at $v = 15$. It is seen that crosspolar levels less than -42 dB are obtained over about 40% bandwidth.

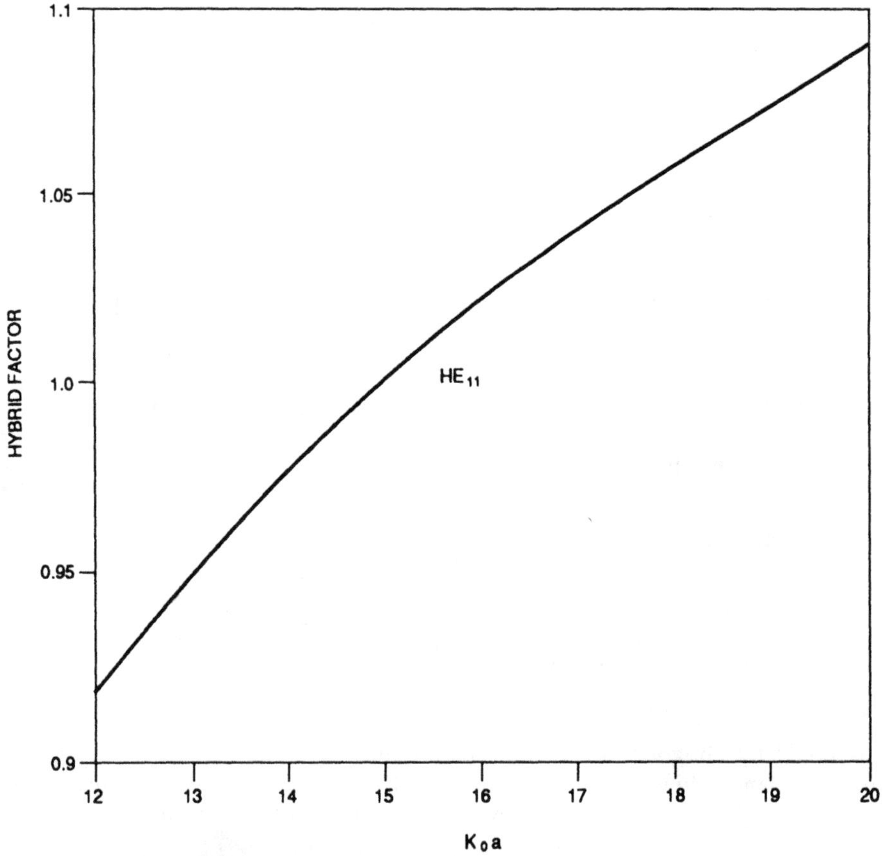

Fig. 5.6 Hybrid factor Λ of the HE_{11} mode in a corrugated guide.

Same corrugation depths as in Fig. 5.5

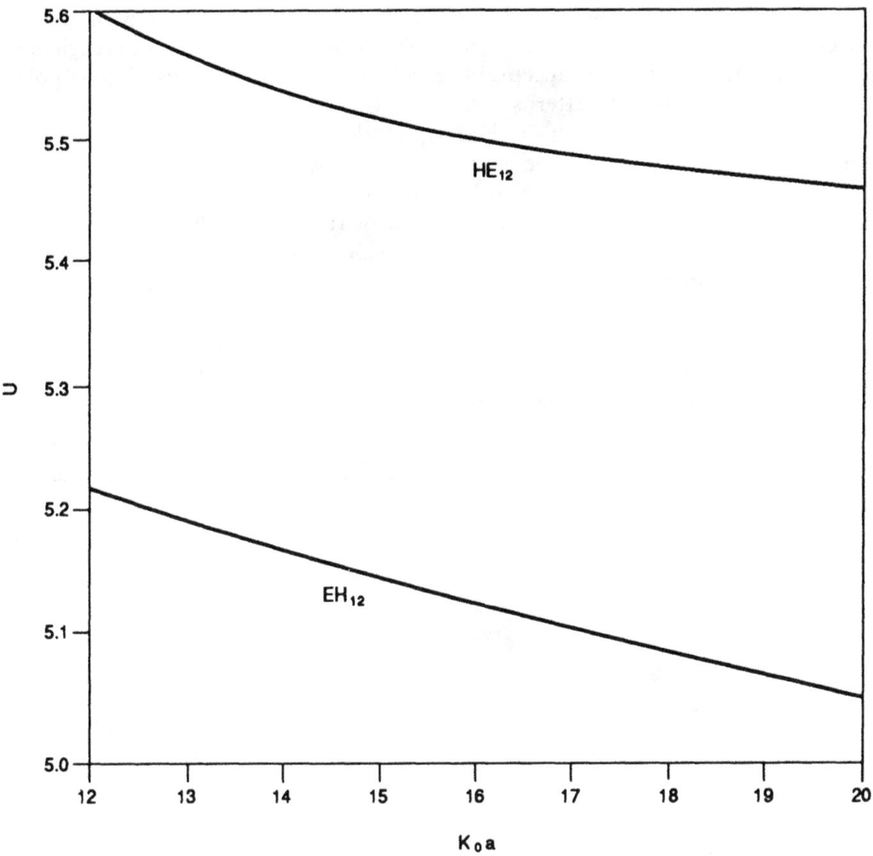

Fig. 5.7 Eigenvalue u of the HE_{12} and the EH_{12} mode in a corrugated guide.

Same corrugation depths as in Fig. 5.5

Even higher bandwidths have been achieved using ring loaded slots shown in Fig. 5.9*b* (Takeichi *et al.*, 1971, and Takeda *et al.*, 1976). Dual band corrugated guides (Fig. 5.9*c*) have been proposed and tested by Ghosh *et al.* (1982) to operate in two bands centered at two widely spaced frequencies. The corrugation depth is made to alternate between d_1 and d_2, which gives resonance at the two centre frequencies. This dual band corrugated guide has been used in a satellite receiver to receive the communication signal at 12 GHz and a tracking signal at 18 GHz. Referring to Fig. 5.9*c*, the slots act as series connected admittances Y_1 and Y_2 with an overall admittance $Y_1Y_2/(Y_1+Y_2)$. Thus, near the resonant frequency of Y_1, when this is close to zero, the other admittance does not influence the guide performance and vice versa. This explains the dual band feature of the guide.

To study the radiation pattern of the corrugated guide, the copolar field is plotted versus $v \sin \theta$ in the E plane; $\phi = 0$ in Fig. 5.10. The slot depth is adjusted for zero Y at $v=15$. Along with this, the crosspolar radiation field is plotted in the $\phi=45°$ plane. It is seen that the relative side lobe level of the copolar field is about -26 dB which is an acceptable level in most applications. The copolar and crosspolar radiation patterns of the next two higher order modes are shown in Figs. 5.11 and 5.12. Obviously the HE_{12} mode exhibits very high relative side lobe levels while the EH_{12} mode is very rich in crosspolar radiation. These two modes, if appreciably excited, can cause severe deterioration of the overall relative side lobe level and crosspolar level of radiation. They are excited at junction discontinuities or by mode conversion of the dominant HE_{11} mode. More will be said in this section about higher order mode excitation. Now it is interesting to evaluate the gain of the radiating circular aperture of the corrugated horn. Let us do this for the dominant mode at the balanced hybrid condition and for large k_0a. Under these conditions, $\Lambda \simeq \hat{\beta} \simeq 1$ and $u=2{\cdot}405$. The aperture field distribution is simplified to (see (5.10)):

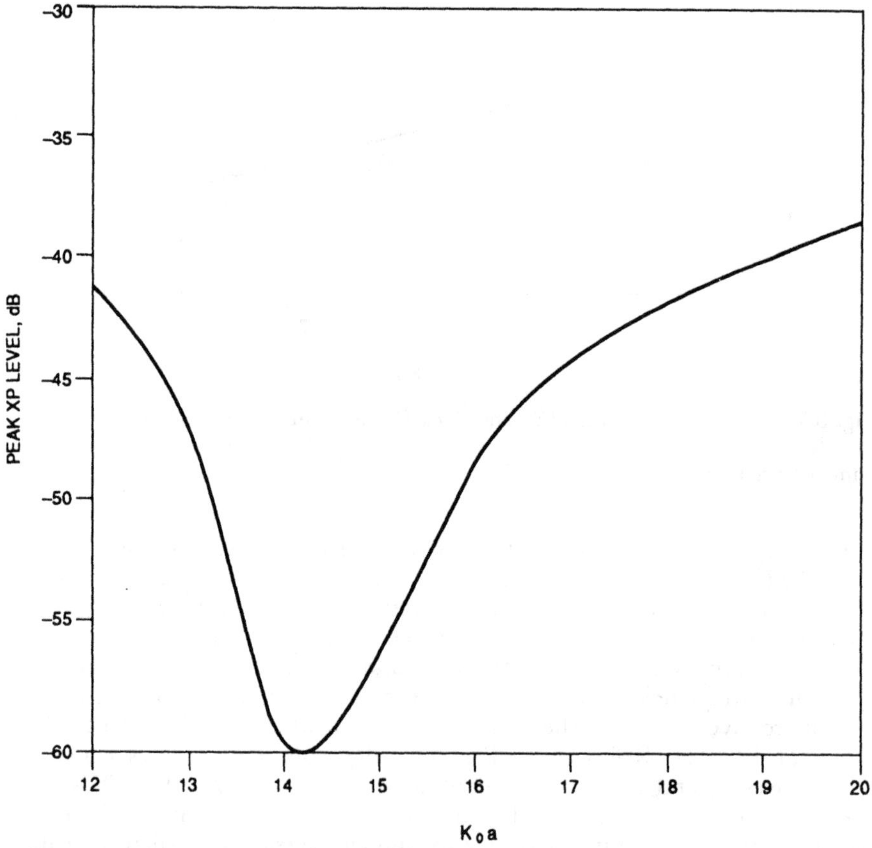

Fig. 5.8 Peak crosspolar radiation level XP due to the HE_{11} mode in a corrugated guide The wall admittance Y is adjusted to zero at $k_0a=15$. (after Mahmoud and Clarricoats, 1982)

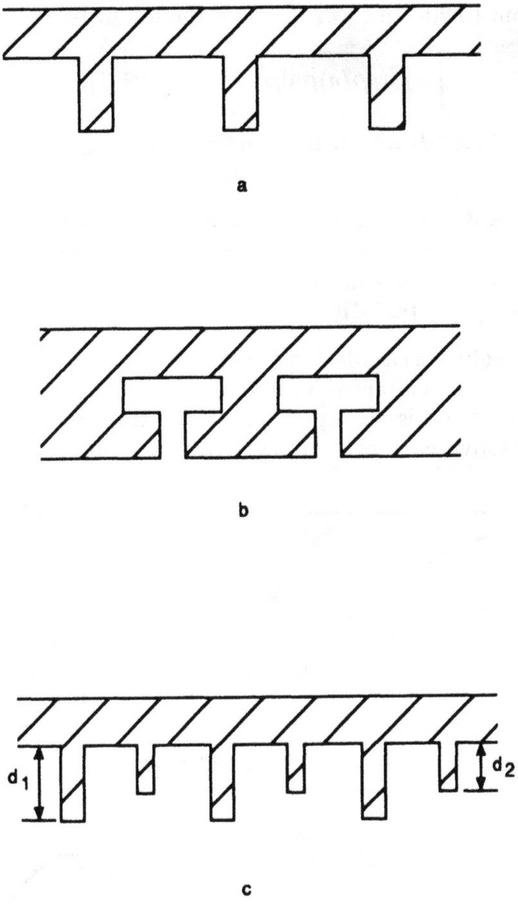

Fig. 5.9 Slot shapes

a Rectangular slot
b Ring loaded slot
c Dual depth slots

$$e_x = 2J_0(u\rho/a) \tag{5.20}$$

while the radiation field from (5.14) and (5.16) is:

$$e_x^r = \frac{2a^2uJ_1(u)}{(v\sin\theta)^2 - u^2} \cdot J_0(v\sin\theta)[-i\exp(-ik_0R)/R]k_0\cos\theta \tag{5.21}$$

The on-axis gain is defined by

$$G = 4\pi R^2|e_x^r(\theta=0)^2|\Big/\int_0^a\int_0^{2\pi} e_x^2\,\rho\,\mathrm{d}\phi\,\mathrm{d}\rho \tag{5.22}$$

Substituting from (5.20) and (5.21), using the following identity:

$$\int_0^a J_0^2(u\rho/a)\rho \, d\rho = \tfrac{1}{2}a^2(J_0^2(u) + J_1^2(u)),$$

and noting that $J_0(u) = 0$ we get the simple result:

$$G = (4/u^2)(k_0 a)^2 \qquad (5.23)$$

To assess this result, we note that the maximum possible gain which occurs under uniform illumination of an aperture of radius a is equal to $(k_0 a)^2$. The gain of the balanced hybrid mode of the corrugated guide is less than this maximum by $10 \log_{10} (2{\cdot}405^2/4) \simeq 1{\cdot}52$ dB.

5.3.2 Transversely corrugated rectangular guide

As long as an axially symmetric radiation pattern is required, the corrugated guide of circular shape is the optimum choice from the view points of good performance, relative ease of manufacture and ease of analysis and design.

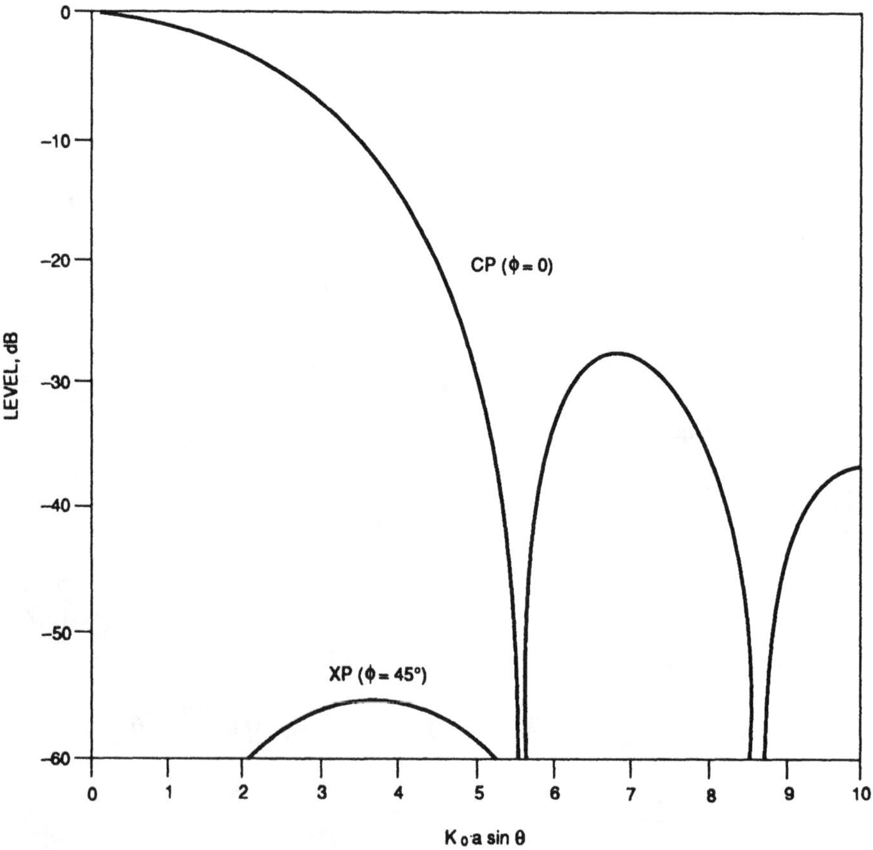

Fig. 5.10 Copolar and crosspolar radiation patterns of the HE_{11} mode: (i) e_{xr} in the e plane and (ii) e_{yr} in the plane $\phi = \pi/4$

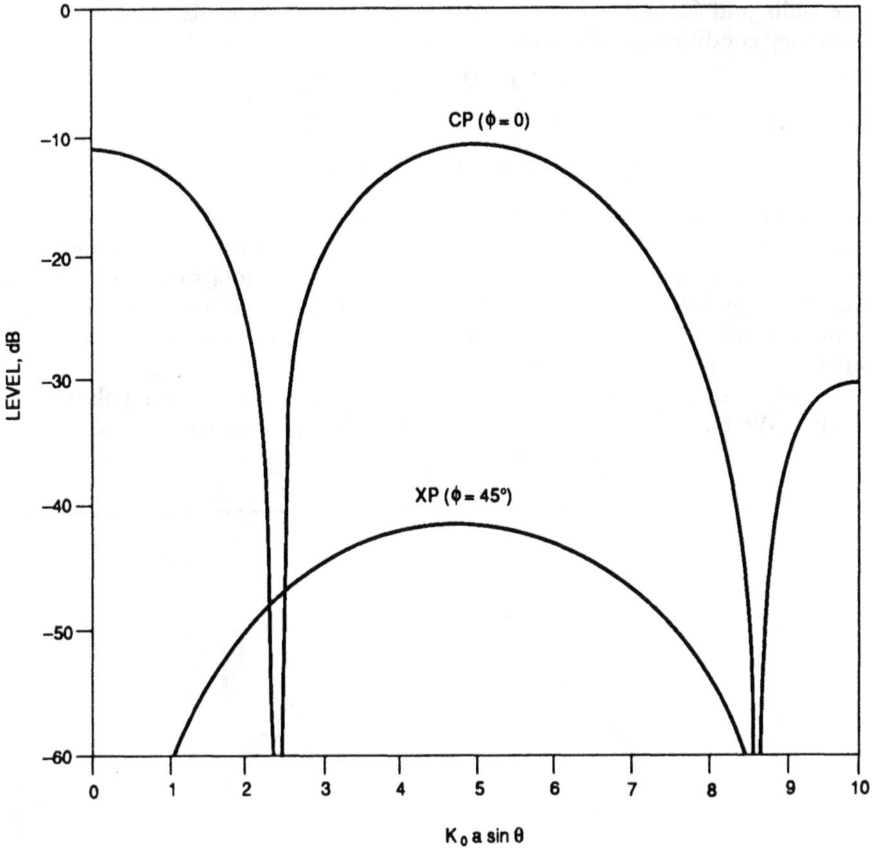

Fig. 5.11 Copolar and crosspolar radiation patterns of the HE_{12} mode: (i) e_{xr} in the e plane and (ii) e_{yr} in the plane $\phi = \pi/4$

However, when the required radiation pattern is not axially symmetric, such as in certain applications of satellite antennas requiring unequal beamwidths in the two principal planes, corrugated guides of noncircular shapes become attractive. Of these shapes, the elliptical and rectangular have received attention in the literature (e.g. Guy *et al.*, 1979, Baldwin *et al.*, 1975, and Dragone, 1985). Here, we shall consider only the rectangular shape. As discussed in section 3.5.2, the rectangular waveguide with transverse corrugations in the four walls is not amenable to simple modal analysis. However, as we shall see below, an approximate but sufficiently accurate mode description can be obtained if the inside dimensions of the guide are large relative to a wavelength. This condition is always valid near the radiating aperture of horn feeds.

With reference to Fig. 3.9, the rectangular guide is modelled as having constant impedance walls with Z_1 and Y_1 the impedance and admittance of the side walls and Z_2 and Y_2 those of the upper and lower walls. Thus, the four boundary conditions at the four walls are:

$$\text{(i) } Z_1 = \pm E_y/H_z, \quad \text{(ii) } Y_1 = \mp H_y/E_z$$

at the side walls $x = \pm a$, and

$$\text{(iii) } Z_2 = \mp E_x/H_z, \quad \text{(iv) } Y_2 = \pm H_x/E_z$$

at the upper and lower walls $y = \pm b$. In the present case, $Z_1 = Z_2 = 0$ while Y_1 and Y_2 can take on any value. Substitution in (3.175) shows that the impedance compatibility relationship is not satisfied; hence one cannot satisfy all the four boundary conditions simultaneously by elementary wave functions. To overcome this difficulty, we shall relax the least important boundary condition in order to get simple, yet sufficiently accurate, modal solutions.

To this end, let us try to find the dominant mode with major polarisation along y. We try a *TM* to y mode for which E_y is expressed by:

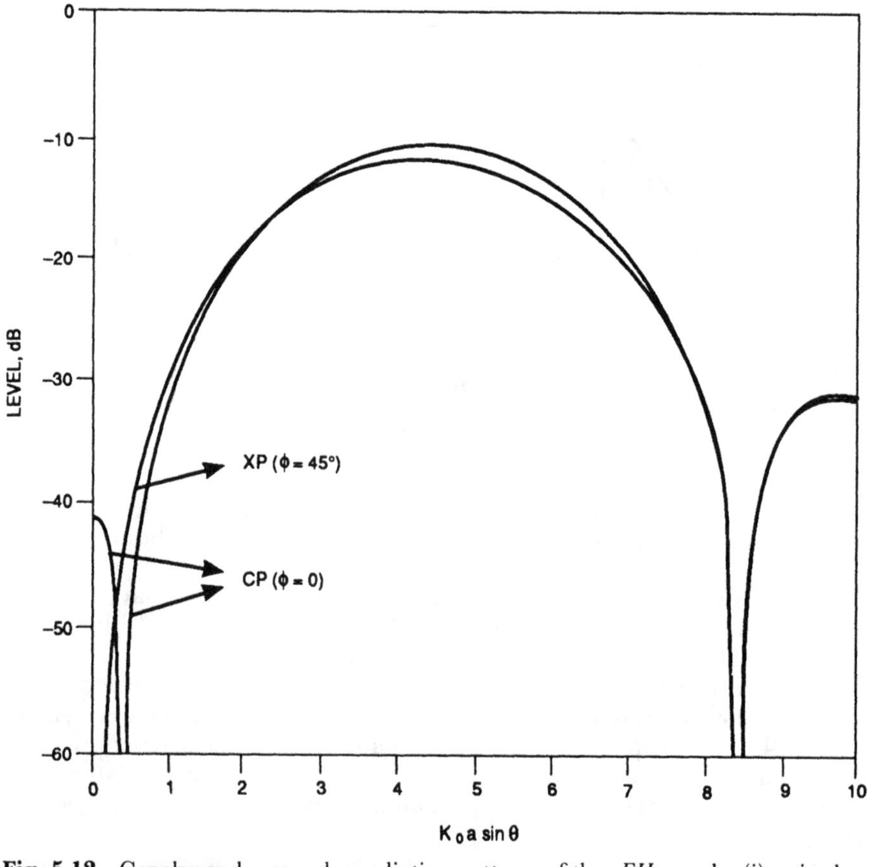

Fig. 5.12 Copolar and crosspolar radiation patterns of ther EH_{12} mode: (i) e_{xr} in the e plane and (ii) e_{yr} in the plane $\phi = \pi/4$.

$$E_y = \cos(k_x x)\ \cos(k_y y)\ \exp(i\omega t - i\beta z) \tag{5.24}$$

The other field components are listed below:

$$E_z = -i\beta(k_0^2 - k_y^2)^{-1}\partial E_y/\partial y \tag{5.25}$$

$$E_x = (k_0^2 - k_y^2)^{-1}\partial^2 E_y/\partial x\partial y \tag{5.26}$$

$$H_x = -\beta\omega\varepsilon_0(k_0^2 - k_y^2)^{-1}E_y \tag{5.27}$$

$$H_z = i\omega\varepsilon_0(k_0^2 - k_y^2)^{-1}\partial E_y/\partial x \tag{5.28}$$

where $k_x^2 + k_y^2 + \beta^2 = k_0^2$.

Assuming electrically large dimensions, i.e. $k_0 a$ and $k_0 b \gg \pi$, the transverse wavenumbers k_x and k_y are deemed to be of first order smallness relative to k_0 and β. Thus we find that the major modal fields in (5.24)—(5.28) are E_y and H_x. The longitudinal fields E_z and H_z are relatively of first order smallness while E_x is of second order smallness. We bear that in mind while applying the boundary conditions (i)—(iv) above. The first of these conditions implies that E_y vanishes at the side walls $x = \pm a$. From (5.24) this determines k_x for the dominant mode:

$$k_x = \pi/2a \tag{5.29}$$

which is readily $\ll k_0$ when $k_0 a \gg \pi$. Using (5.25), we deduce that E_z will also vanish at the side walls. Therefore, the second boundary condition, $Y_1 = H_y/E_z$ at $x = \pm a$, is immaterial since both fields are readily zero. The fourth boundary condition is:

$$Y_2 = -i\omega\varepsilon_0[E_y/(\partial E_y/\partial y)]_{y=b}$$

Writing $Y_2 = i(\varepsilon_0/\mu_0)^{1/2}B_2$, and using (5.24), this becomes:

$$B_2 = k_0\ \cot(k_y b)/k_y \tag{5.30}$$

Normally B_2 is kept close to zero in the frequency band of interest by the proper choice of the corrugation depth. Therefore, the first root of (5.30) is close to $k_y = \pi/2b$. Using a perturbation analysis about this value, we get:

$$k_y = (\pi/2b)(1 - B_2/k_0 b) \tag{5.31}$$

Now, since both k_x and k_y are determined, all the mode properties are completely specified. In this derivation, we have clearly ignored the boundary condition (iii) which requires that $E_x = 0$ at $y = \pm b$. Instead, (5.26) reveals that E_x is of the order of $(\pi/2k_0 a)$ $(\pi/2k_0 b)$, which is a second order of smallness. The relaxation of this boundary condition is therefore of little concern. One should expect the present analysis to determine the copolar field quite accurately, but perhaps not the crosspolar field. However, the order of the latter is believed to be correctly predicted. That is to say, the level of the crosspolar field E_x relative to the copolar E_y is of the order of $(\pi/2k_0 a)$ $(\pi/2k_0 b)$.

The design of a rectangular horn will require careful design of a matching section between the smooth guide feeder (having TE_{11} mode) and the horn's throat. The horn itself can be made by assembling four identical corrugated plates. For more details on the design and for useful details on the horn's manufacture, the reader is referred to Dragone (1985).

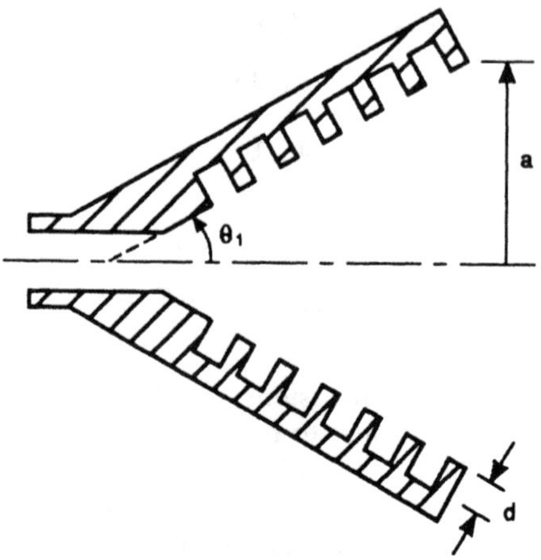

Fig. 5.13 Geometry of the corrugated conical horn

5.3.3 Corrugated conical horn

A horn radiator necessarily encompasses a flared out or a conical section which joins a narrow waveguide feed to a wider radiating aperture. We consider here modal propagation on a corrugated conical horn with a linear profile. This is characterised by a constant semiflare angle θ_1 (see Fig. 5.13). Later in this section, we discuss the properties of profiled conical horns, in which θ_1 varies along the horn's length.

With many corrugations per wavelength (greater than about six), the wall at $\theta = \theta_1$ is adequately described by average Z and Y whose values, by analogy with the corrugated cylinder, are given by

$$Z = e_\phi / h_z = 0 \tag{5.32}$$

$$Y = -h_\phi / e_z \simeq -i\eta_0^{-1}[\cot k_0 d + 1/2k_0 r \sin \theta_1] \tag{5.33}$$

Now, the structure belongs to the category of constant wall impedance guides to which the analysis of section 3.7 directly applies. In particular, the modal equation (3.209) with $X=0$ and $m=1$ reduces to

$$F_1(\nu;\, \theta_1)\, [F_1(\nu;\, \theta_1) - B\nu(\nu+1) \sin \theta_1] = \hat{\beta}^2 \tag{5.34}$$

where $F_1(\nu;\, \theta_1)$ is defined by (3.201). B is the normalised wall susceptance and $\hat{\beta}$ the normalised radial phase constant, which are defined by (3.208) and (3.202) respectively. When $k_0 r \gg \nu$, $\hat{\beta}$ approaches unity and becomes independent of r. Under this condition, and constant B, the modal equation becomes independent of r and validates the process of separation of variables.

The modal equation (5.34) was solved numerically under the balanced hybrid mode condition, $B=0$, and $\hat{\beta}=1$, by Clarricoats and Saha (1971*b*). Highly accurate solutions for the lowest three order modes have been derived in closed forms by Mahmoud and Clarricoats (1982). Thus, for the HE_{11} mode:

$$\nu + 1/2 \simeq 2{\cdot}4048/(\theta_1 \sin \theta_1)^{1/2} \tag{5.35}$$

For the HE_{12} and the EH_{12} modes:

$$\nu+1/2 \simeq 5{\cdot}5201/\theta_1 \tag{5.36}$$

$$\nu+1/2 \simeq 5{\cdot}1356/\theta_1 \tag{5.37}$$

respectively. Although, they are derived under the condition of a small flare angle, these formulae deviate from the exact solution obtained numerically by no more than 1% for θ_1 up to 60°. As we move away from hybrid balance, the eigenvalues deviate slowly from those in (5.35)—(5.37). As an illustration, the quantities $(\nu+1/2)\ \theta_1$ obtained numerically for the lowest three modes are plotted versus k_0a and for two values of θ_1 in Figs. 5.14 and 5.15. Here $B=0$ at $k_0a=15$.

The modal hybrid factor Λ is governed by (3.211) which gives for $X=0$ and $m=1$

$$\hat{\beta}(\Lambda-\Lambda^{-1})=\nu(\nu+1)\sin\theta_1 B$$

When $B=0$, $\Lambda=\pm 1$, indicating the balanced hybrid mode condition. The plus sign corresponds to the *HE* modes and the minus sign to the *EH* modes. As a result, the *HE* modes are characterised by maximum radiation on the axis $\theta=0$, while the *EH* modes have nulls on this axis. Away from this condition, Λ deviates from ± 1 and is given by (3.212).

To study the crosspolar behaviour of the corrugated conical horn, it is convenient to define two orthogonal unit vectors in the copolar and crosspolar directions by

$$\hat{\imath}_{co}=\cos\phi\hat{\Theta}-\sin\phi\hat{\Phi} \tag{5.38}$$

$$\hat{\imath}_{xp}=\sin\phi\hat{\Theta}+\cos\phi\hat{\Phi} \tag{5.39}$$

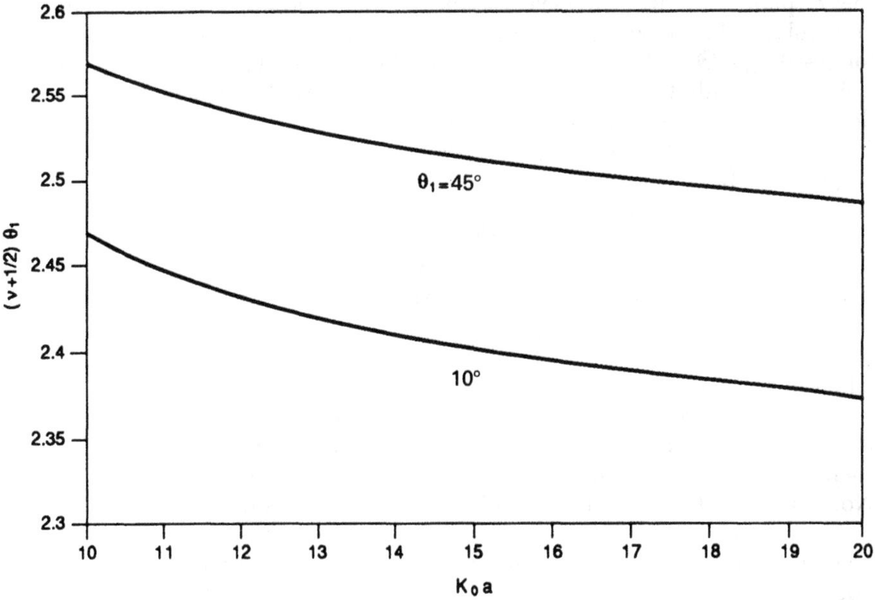

Fig. 5.14 Eigenvalues $(\nu+1/2)\ \theta_1$ for the HE_{11} mode versus k_0a

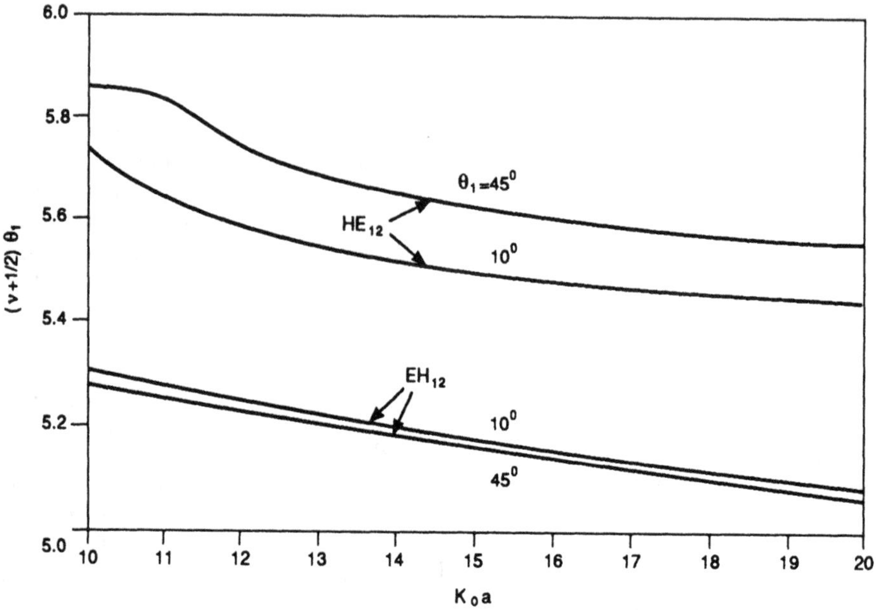

Fig. 5.15 Eigenvalues $(\nu+1/2)\ \theta_1$ for the HE_{12} and EH_{12} modes versus k_0a

These are the directions in which we would orient an e probe to measure the copolar and cropsspolar radiation from an x polarised apertue. For example, in the E plane $(\phi=0)$, we measure the copolar radiation along $\hat{\Theta}$ and the crosspolar radiation along $\hat{\Phi}$. This agrees with the above equations which give, for $\phi=0$, $\hat{\imath}_{co}=\hat{\Theta}$ and $\hat{\imath}_{xp}=\hat{\Phi}$. Likewise, in the H plane $(\phi=\pi/2)$ one measures the copolar radiation field along $\hat{\Phi}$ and the crosspolar field along $\hat{\Theta}$. More detailed discussions on these definitions are found in Ludwig (1973) and Silver (1984). Now, in terms of these newly defined vectors the modal aperture e field in (3.197)—(3.200) takes the form

$$re_t=\tfrac{1}{2}(\hat{\beta}+\Lambda)f_{\nu+}(\theta)\hat{\imath}_{co}+\tfrac{1}{2}(\hat{\beta}-\Lambda)f_{\nu-}(\theta)[\cos 2\phi\hat{\imath}_{co}+\sin 2\phi\hat{\imath}_{xp}) \qquad (5.40)$$

where $f_{\nu\pm}(\theta)$ represent the functional dependence of the copolar and crosspolar fields over the aperture and are given explicitly by:

$$f_{\nu\pm}(\theta)=\partial P^1_\nu(\cos\theta)/\partial\theta \pm P^1_\nu(\cos\theta)/\sin\theta \qquad (5.41)$$

These functions play the same roles as the J_0 and J_2 Bessel functions in the corrugated cylindrical guide. One can easily verify that $f_{\nu\pm}(\theta_1)=0$ for the HE and EH modes respectively at the balanced hybrid mode condition. To indicate their behaviour, $f_{\nu\pm}(\theta)$ are plotted for the HE_{11} mode at balance in Fig. 5.16. Note the resemblance to the J_0 and J_2 Bessel functions of the cylindrical case.

Now, with known e field on the horn's aperture $r=r_a$, the radiation field can be determined by the spherical wave expansion (SWEX) method in which the aperture field is expanded in terms of spherical wave vector harmonics. The far field is then obtained as the special case of large k_0r. The method was first introduced by Clarricoats and Saha (1971*b*) to predict the copolar radiation pattern under hybrid balance and later generalised by Mahmoud and Clarricoats (1982). In the following we give a brief account of SWEX.

Following ideas from section 1.6.2, the aperture field e_t can be expanded in terms of a complete set of the spherical wave vector functions m and n defined by (1.139)—(1.140). Hence, by analogy with (1.148) we set

$$e_t(\theta, \phi) = \sum_l (a_l m_{oml} + b_l n_{eml}) \tag{5.42}$$

where, in view of (5.40), we choose odd m and even n functions. In addition, the order m is restricted to $m=1$. By using the orthogonality relationship (1.145)—(1.146), the coefficients a_l and b_l are obtained in the form (1.149). Substituting for e_t from (5.40), we get

$$\begin{matrix} a_n \\ b_n \end{matrix} = i\pi k_0 r_a \left\{ \begin{matrix} \hat{H}_n^{(2)}(k_0 r_a) \\ -\hat{H}_n'^{(2)}(k_0 r_a) \end{matrix} \right\} [\tfrac{1}{2}(\hat{\beta} + \Lambda) T_{n+}(\theta_1) \mp \tfrac{1}{2}(\hat{\beta} - \Lambda) T_{n-}(\theta_1)] / (\gamma_n \hat{H}_n^{(2)}(k_0 r_a) \hat{H}_n'^{(2)}(k_0 r_a)) \tag{5.43}$$

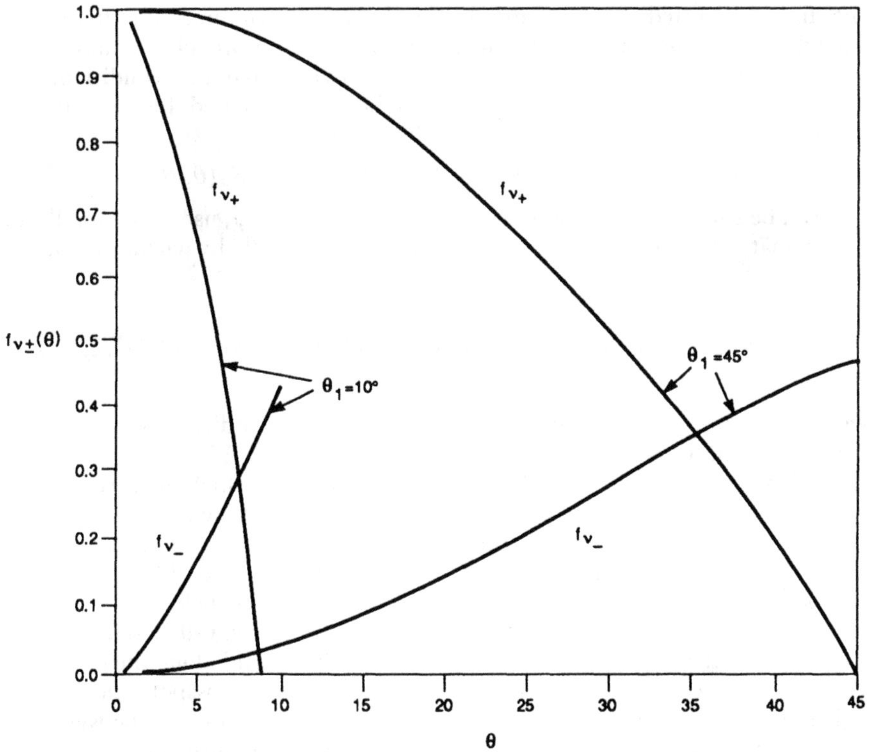

Fig. 5.16 The functions $f_{v\pm}$ (θ) for the HE_{11} mode for two values of θ_1

where

$$\gamma_n = 4\pi n^2(n+1)^2/(2n+1),$$

and

$$T_{n\pm}(\theta_1) = \int_0^{\theta_1} f_{\nu\pm}(\theta) f_{n\pm}(\theta) \sin\theta d\theta \quad (5.44)$$

The radiation field is obtained by using (5.43) in (5.42) and using the far zone approximation of the m and n functions. We then get

$$e(r, \theta, \phi) = \{\tfrac{1}{2}(\hat{\beta}+\Lambda)\hat{\imath}_{co} S_+(\theta, \theta_1) + \tfrac{1}{2}(\hat{\beta}-\Lambda)(\hat{\imath}_{co} \cos 2\phi + \hat{\imath}_{xp} \sin 2\phi).$$
$$S_-(\theta, \theta_1)\} \exp(-\hat{\imath}k_0 r)/(r/r_a) \quad (5.45)$$

where

$$S_\pm(\theta, \theta_1) = \pi \sum_n (T_{n\pm}(\theta_1)/\gamma_n) f_{n\pm}(\theta) \hat{\imath}^{n+1} / \hat{h}_n^{(2)}(k_0 r_a) \quad (5.46)$$

Clearly the crosspolar radiation attains its maximum in the plane $\phi = \pi/4$ and vanishes in the E and H planes. Relative to the boresight copolar radiation level, the crosspolar field in the $\phi = \pi/4$ plane is given by:

$$XP(\theta, \theta_1) = [(\hat{\beta}-\Lambda)/(\hat{\beta}+\Lambda)]\ [S_-(\theta, \theta_1)/S_+(0, \theta_1)] \quad (5.47)$$

The first bracketed term is the key term in determining the frequency dependence of the crosspolar radiation. With $\hat{\beta}$ approaching unity (for electrically large aperture), and near to the hybrid balance condition, the hybrid factor Λ for the HE_{1n} modes can be approximated by $1+\frac{1}{2}B\nu(\nu+1) \sin\theta_1$. Under these conditions (5.47) becomes:

$$XP(\theta, \theta_1) \simeq -(B/4)\nu(\nu+1) \sin\theta_1 [S_-(\theta, \theta_1)/S_+(\theta, \theta_1)]$$

This can be simplified more if we notice that $B = -iY\eta_0/k_0 r$, use (5.33) for Y and approximate $\nu(v+1)$ for the HE_{11} by $(2{\cdot}405)^2/\theta_1 \sin\theta_1$ by virtue of (5.35). Thus,
one gets:

$$XP(\theta, \theta_1) = (2{\cdot}405)^2 (\sin\theta_1/\theta_1)\ [S_-(\theta,\theta_1)]/S_+(\theta, \theta_1)]\ .\ [\cot k_0 d + 1/k_0 a_m]/4k_0 a_m \quad (5.48)$$

where a_m is the aperture radius; i.e. $a_m = r_a \sin\theta_1$. To indicate the frequency response of the crosspolar radiation, the peak value of XP over θ is plotted versus $\delta = f/f_0 - 1$ in Fig. 5.17. Here f_0 is the frequency at which the corrugations are a quarter of a wavelength deep. It is clear that the relative crosspolar level is a monotonically decreasing function of the aperture radius. This also determines the bandwidth of the horn which increases with increased horn aperture. It should be recognised, however, that the crosspolar fields do not only comprise that component due to the dominant HE_{11} mode. Other sources stem from higher order modes generated at the horn's throat and along the flared out section. Of these modes, the EH_{12} mode is the richest in crosspolar fields. The higher order modes are generated by scattering at the junction between the horn's throat and the smooth waveguide feeder. They are also generated along the flared section owing to partial conversion of the dominant mode. This is so

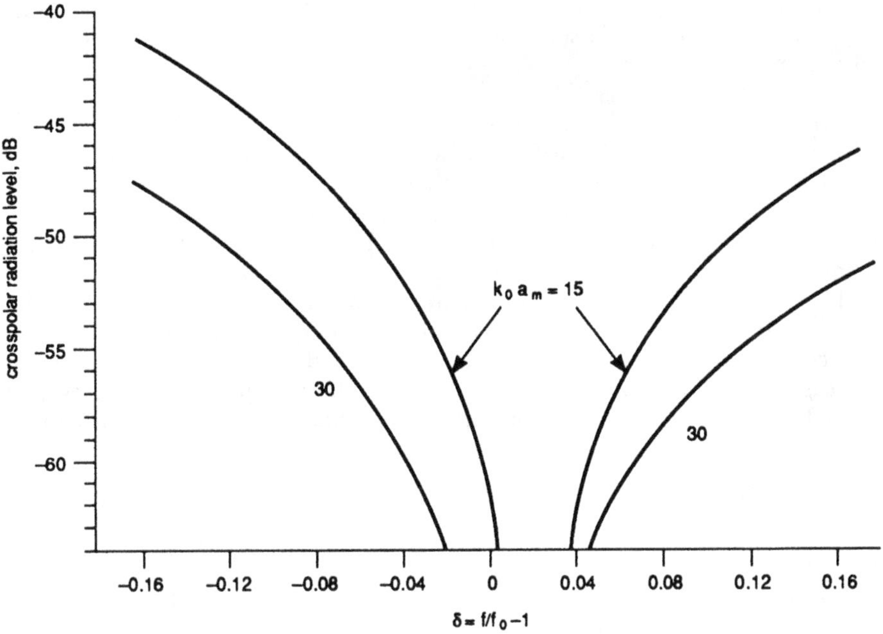

Fig. 5.17 Peak crosspolar level of the HE_{11} mode of a corrugated cone versus normalised frequency

since the normalised wall admittance B is not constant along the horn for a fixed corrugation depth. A detailed study of higher order mode generation on conical horns and the net crosspolar radiation levels has been carried out by Mahmoud and Clarricoats (1982).

Fig. 5.18 shows the various components of relative peak crosspolar radiation from a 30° cone fed by a smooth waveguide versus frequency. The components displayed are those due to (*a*) the HE_{11} mode, (*b*) the EH_{12} mode generated at the throat, and (*c*) the EH_{12} mode generated by conversion of the HE_{11} mode along the horn. Experimental data are also plotted as vertical bars and reasonable agreement with the overall theoretical level is observed. In this case it is seen that the crosspolar level due to mode conversion along the horn is insignificant over all the operating frequency band. It is dominated by the EH_{12} mode generated at the throat at the higher frequencies and by the intrinsic crosspolarisation of the HE_{11} mode at the lower frequencies. As the flare angle increases above 30°, the crosspolar fields of the EH_{12} mode generated at the throat gradually dominate the other two components over all the band. In this case, it may be required to design a matching section between the cylindrical feed and the conical section. A matching section will often be a cone with a small flare and varying depth corrugations, starting at $\lambda/2$ and ending at $\lambda/4$. A more detailed study of a matching section is given by Dragone (1977*b*).

Alternatively, a profiled corrugated horn may be designed so as to achieve better matching at the input and also reduce mode conversion along the horn's length without being excessively long. Such a compact profiled corrugated horn was first suggested by Watson *et al.* (1980). In the following, we study profiled corrugated conical horns.

Profiled corrugated conical horn

The conical horn can be regarded as a device that transforms a narrow aperture at the throat to a wide aperture at the mouth. A compact horn of linear profile will obviously require a large semiflare angle θ_1. However, as the flare angle increases, two undesirable effects occur. Firstly, generation of the higher order modes increases at the throat junction and results in higher levels of crosspolar and side lobe radiation. The increase in crosspolar radiation is caused by the EH_{12} mode and in sidelobe level by the HE_{12} mode. The second effect is a reduction in the boresight gain of the copolar radiation as a consequence of increased quadrature phase changes over the radiating aperture. To appreciate this effect, the gain of a linear conical horn of a fixed aperture radius is plotted

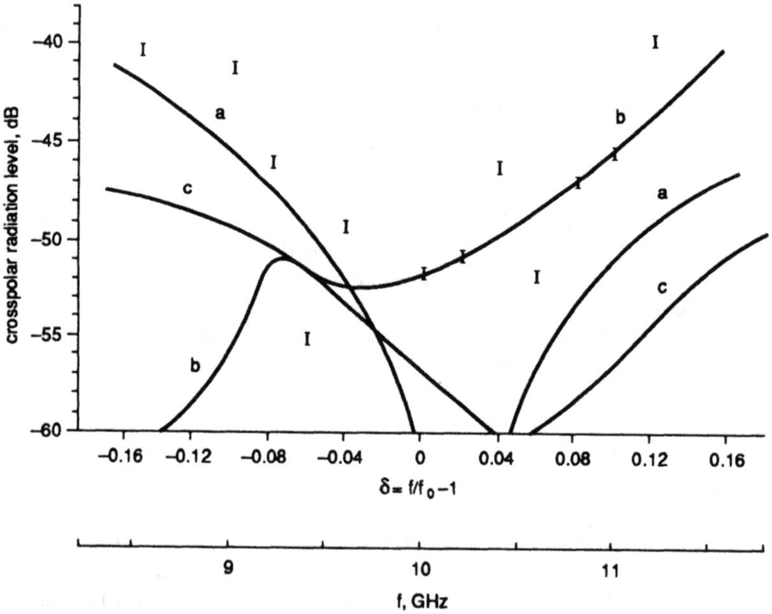

Fig. 5.18 Components of peak crosspolar radiation from a 30° conical corrugated horn fed by a smooth guide (after Mahmoud and Clarricoats, 1982)

a the HE_{11} mode
b EH_{12} mode generated at the throat
c EH_{12} mode converted along the horn
Corresponding experimental data are shown by bars
$k_0 a_{th} = 3$, $k_0 a_m = 15$

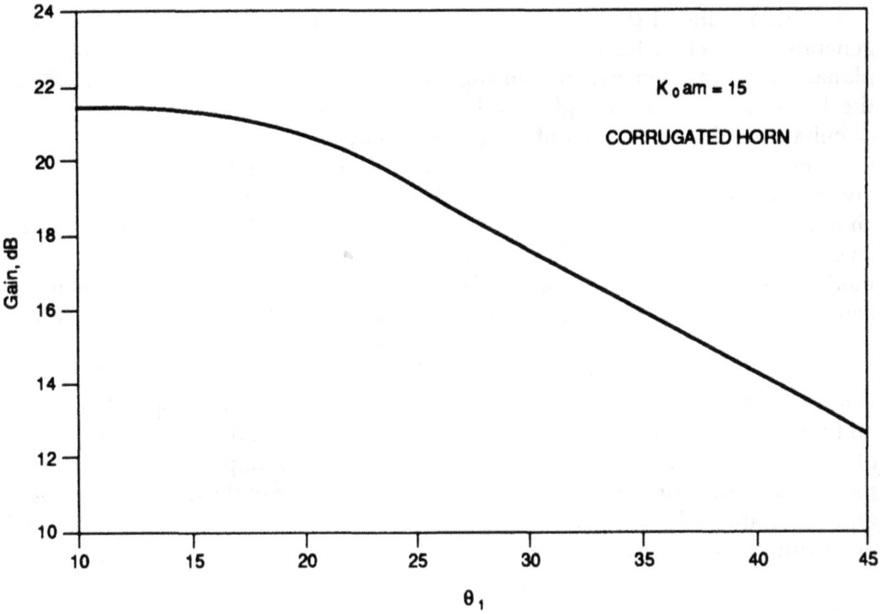

Fig. 5.19 Gain of a corrugated conical horn versus θ_1

Aperture radius is fixed at $k_0 a_m = 15$.

versus θ_1 in Fig. 5.19. The gain is reduced by about 3 dB, from the cylindrical case, at $\theta_1 = 30°$. More drastic reductions occur at higher angles; e.g. 15 dB at 60°. This indicates that the linear profile is not the optimum to achieve a compact horn with high gain and low levels of higher order modes. Now, we introduce a compact profiled conical corrugated horn. Consider the horn of Fig. 5.20. The profile is governed by:

$$a(z) - a_{th} = 2(a_m - a_{th}) \cos^2[\tfrac{1}{2}\pi(1 - z/2L_1)]/(1+\gamma) \quad 0 \leq z \leq L_1$$

$$= \frac{2(a_m - a_{th})}{1+\gamma} \left\{ \gamma \cos^2\left[\tfrac{1}{2}\pi \left(1 - \frac{z + L_2 - L_1}{2L_2} \right) \right] + \frac{1-\gamma}{2} \right\},$$

$$L_1 \leq 2 \leq L_2 \quad (5.49)$$

with $\gamma = L_2/L_1$, and a_{th} and a_m are the cross sectional radii at the horn throat and mouth. The flare angle $\theta_1(z)$ is zero at $z = 0$ and at $z = L = L_1 + L_2$, that is at both the throat and the mouth, and attains a maximum value $\hat{\theta}_1$ at $z = L_1$. Thus the parameter γ determines the location of the maximum flare angle which is closer to the mouth if $\gamma < 1$ and closer to the throat if $\gamma > 1$. However, the maximum value of $\hat{\theta}_1$ is independent of γ and given by:

$$\tan \hat{\theta}_1 = \pi(a_m - a_{th})/2L \quad (5.50)$$

One can easily verify by differentiating (5.49) once and twice that $a(z)$, $\theta_1(z)$ and $d\theta_1(z)/dz$ are all continuous at $z = L_1$.

The zero value of θ_1 at the throat ensures low return loss and low levels of generated higher order modes. The zero value of θ_1 at the mouth means a planar wavefront; hence no loss in copolar gain. However the change of θ_1 along the horn is a source of higher order modes generated by conversion of the dominant mode. The design of the profiled horn starts by the choice of a_m so as to meet a specified radiation beamwidth or boresight gain, and of a_{th} to fit a given waveguide feeder. The horn length L and the profile parameter γ are then chosen so as to achieve a certain crosspolar level and/or a certain side lobe level. These two requirements depend upon the levels of the EH_{12} and HE_{12} modes generated along the horn. In the following a brief analysis of mode conversion on profiled conical horns as introduced by Mahmoud (1983) is given.

Let $dC_n(z)$ be the incremental increase of the n th higher order mode due to conversion from a unit amplitude of the dominant mode in the region between z and $z + dz$. This increase is attributed to a change in θ_1 of $d\theta_1$ and a change in B of dB. Neglecting multiple scattering due to the small amplitudes of converted modes along the horn and assuming that the amplitude of the dominant mode is almost unaffected by conversion, one can write the following for the total amplitude $C_n(L)$ of the n th converted mode at $z = L$:

$$C_n(L) = \exp\left(-i\int_0^L \beta_n(z)\, ds\right)\int_0^L dC_n(z)\, \exp\left\{-i\int_0^z (\beta_1(z') - \beta_n(z'))\, ds\right\} \tag{5.51}$$

where $ds = dz'\ (1 + (da(z')/dz')^2)^{1/2}$, β_1 is the radial phase constant of the dominant HE_{11} mode and β_n is the radial phase constant (generally complex) of the n th higher order mode. The differential coupling terms $dC_n(z)$ are derived explicitly by Mahmoud (1983). It is shown there that $dC_n(z)/d\theta_1$ is a slowly varying function of θ_1; hence the overall coupling level $C_n(L)$ is critically dependent on the second phase integral term in (5.51). This, in turn, is a function of the horn profile.

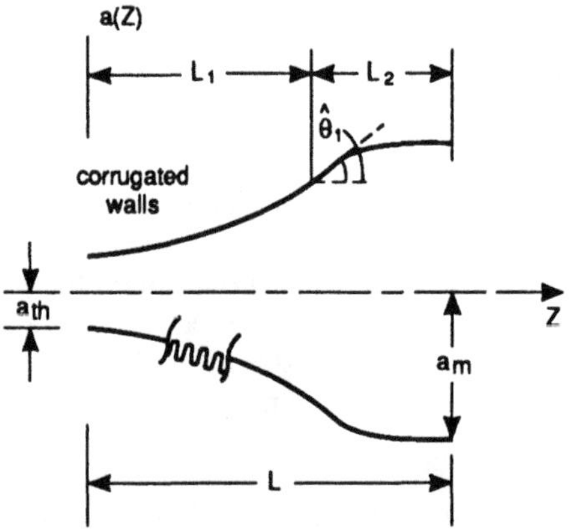

Fig. 20 Compact profiled horn geometry (after Mahmoud, 1983)

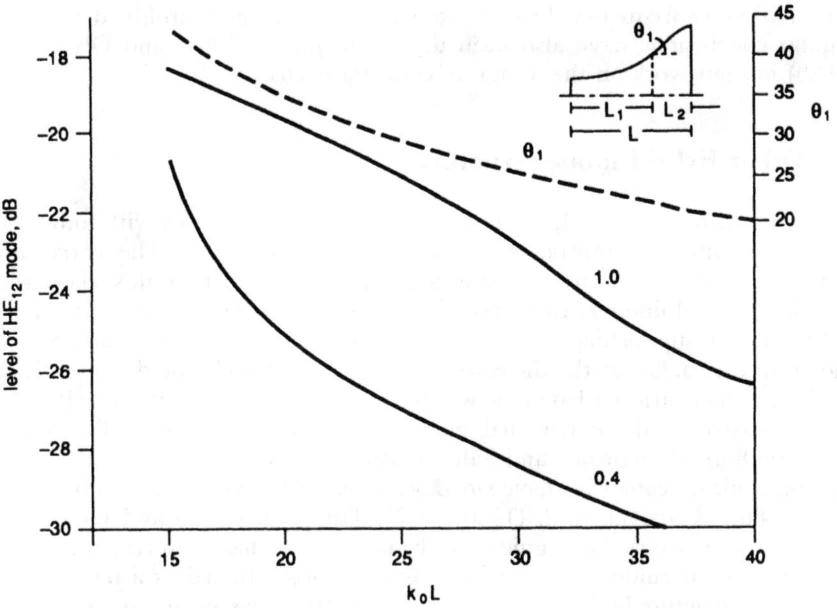

Fig. 5.21 Level of converted HE_{12} mode versus horn's length k_0L (after Mahmoud, 1983)

The parameter $\gamma = 1$ and 0·4

Numerical results for the levels of the converted HE_{12} and EH_{12} modes are shown in Figs. 5.21 and 5.22. The horn dimensions are chosen as follows: $k_0a_{th} = 3{\cdot}0$, $k_0a_m = 12{\cdot}0$, while k_0L is varied between 15·0 and 40·0, and γ takes on the values 1·0 and 0·4. In order to concentrate study on the profile alone, the slot depth is assumed constant so that $Y = 0$ at the mouth. As would be expected, the levels of converted modes decrease monotonically with increased L. The more striking result though is the considerable reduction in levels of converted modes for $\gamma = 0{\cdot}4$ relative to those for $\gamma = 1{\cdot}0$. The first value corresponds to a maximum flare angle $\hat{\theta}_1$ closer to the horn's mouth than to its throat. This can be explained by referring to (5.51) as follows. Near to $\hat{\theta}_1$, $d\theta_1$ changes from positive to negative as the plane $z = L_1$ is crossed. This means that dC_n changes sign and, if the phase integral term is slowly varying, a good deal of cancellation occurs, reducing the overall level of conversion. Now, since $(\beta_1 - \beta_n)$ decreases as the mouth is approached, the condition of cancellation is better satisfied when $\hat{\theta}_1$ is closer to the mouth than the throat, or equivalently, if γ is less than unity. It is found that the optimum value of γ is about 0·4. For smaller values of γ, the flare angle will have to change from $\hat{\theta}_1$ to zero in a fairly short distance, hence causing high conversion levels. A general conclusion may be drawn and stated as follows: a profile which has a maximum flare angle in a region with sufficiently high k_0a is likely to have low mode conversion owing to the occurrence of phase cancellation in that region. It is important to note that this cancellation is frequency broadband since it is mainly profile dependent. Similar conclusions have also been drawn by James (1984) and Olver *et al.* (1988) in their work on the design of compact horns.

5.4 Other hybrid mode structures

As discussed in section 3.4, the class of cylindrical waveguides with finite wall impedance and/or admittance can support hybrid modes. The corrugated waveguide studied in the previous section is a member of this class with vanishing Z and finite Y. In this section, we extend our study to more general hybrid mode supporting structures. Two examples of these structures are shown in Fig. 5.23: (*a*) the dielectric loaded guide and (*b*) the dielectric lined guide. The dielectric loaded guide was introduced by Clarricoats *et al.* (1983) as an alternative to the corrugated guide and as an extension to the conical dielectric horn (Clarricoats and Salema, 1973*a*, *b*). Modal analysis of cylindrical and conical geometries have since been made by several authors including Lier (1986), Raghaven *et al.* (1986, 1987), Tun *et al.* (1987) and Olver *et al.* (1988). The dielectric lined guide, on the other hand, has been recognised as a low attenuation guide for a long time and has been studied as a possible low crosspolar structure by Dragone (1981). In the following we present a common modal analysis for both guides. Note that the difference between the two guides can be stated as: in (*a*) $\varepsilon_1 > \varepsilon_2 = \varepsilon_0$ and in (*b*) $\varepsilon_2 > \varepsilon_1 = \varepsilon_0$. Apart from a common factor $\exp(i\omega t - i\beta z)$, the modal longitudinal fields in the core region $0 \leqslant \rho \leqslant a$ have the forms:

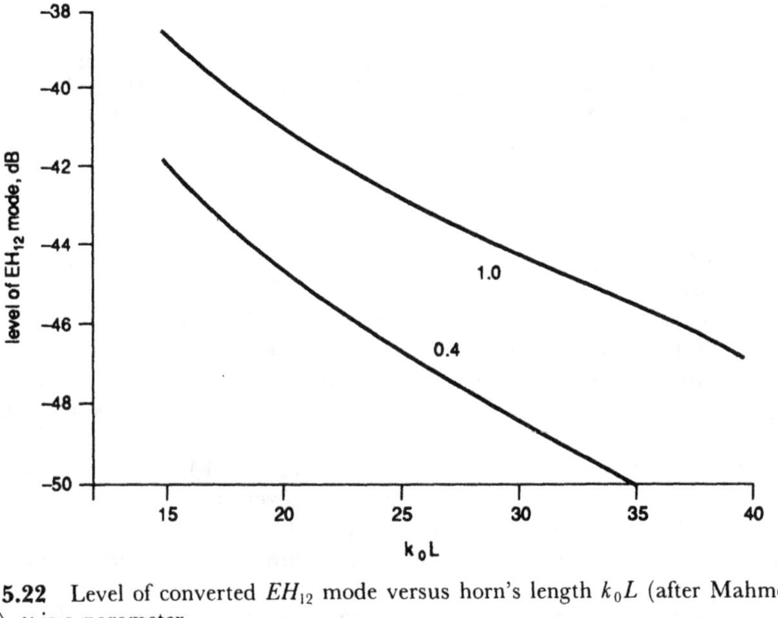

Fig. 5.22 Level of converted EH_{12} mode versus horn's length k_0L (after Mahmoud, 1983). γ is a parameter.

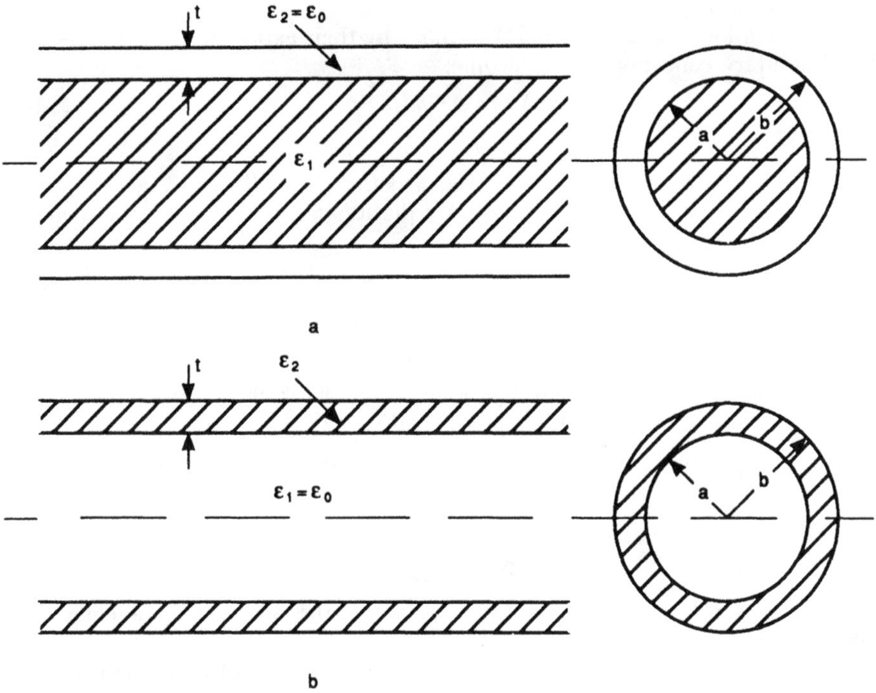

Fig. 5.23 Hybrid mode supporting structures

a Dielectric loaded guide
b Dielectric lined guide

$$e_z = iJ_m(u\rho/a)\cos m\phi \tag{5.52}$$

$$\eta_1 h_z = i\Lambda J_m(u\rho/a)\sin m\phi \tag{5.53}$$

where $\eta_1 = (\mu/\varepsilon_1)^{1/2}$ and Λ is the modal hybrid factor. The transverse fields are obtainable from (1.27)—(1.28) and they resemble (3.124)—(3.127) with appropriate changes; namely η_0 and k_0 are replaced by η_1 and $k_1 = \omega(\mu\varepsilon_1)^{1/2}$. The boundary conditions at the interface $\rho = a$ are expressed rigorously by the forms:

$$e_\varphi = Zh_z + ip(\partial e_z/\partial\phi) \tag{5.54}$$

$$h_\varphi = -Ye_z + ip(\partial h_z/\partial\phi) \tag{5.55}$$

where

$$Z = (i\omega\mu a^2/w^2)\ [(\partial h_z/\partial\rho)/h_z]_{\rho=a+} \tag{5.56}$$

$$Y = (i\omega\varepsilon_2 a^2/w^2)\ [(\partial e_z/\partial\rho)/e_z]_{\rho=a+} \tag{5.57}$$

$$p = \beta a/w^2 \tag{5.58}$$

$$w = (\beta^2 - k_2^2)^{1/2} a \tag{5.59}$$

and $k_2=\omega(\mu\varepsilon_2)^{1/2}$.
Substituting for e_φ and h_φ in (5.54)—(5.55) by their expressions at $\rho=a-$, the two boundary conditions now become:

$$F_m(u)-Xu^2/v=-m(\hat{\beta}+pu^2/v)/\Lambda \tag{5.60}$$

$$F_m(u)-Bu^2/v=-m(\hat{\beta}+pu^2/v)\Lambda \tag{5.61}$$

where $X=-iZ/\eta_1$, $B=-iY\eta_1$, $v=k_1a$, $u^2+(\beta a)^2=v^2$, $\hat{\beta}=\beta/k_1$, and $F_m(u)$ is defined by (3.130). Now, the modal equation for u (or β) is obtained by multiplying (5.60) by (5.61):

$$[F_m(u)-Xu^2/v]\ [F_m(u)-Bu^2/v]=(m\hat{\beta})^2(1+u^2/w^2)^2 \tag{5.62}$$

Subtracting (5.61) from (5.60), an equation for Λ results:

$$\Lambda-\Lambda^{-1}=(u^2/v)\,(B-X)/m\hat{\beta}(1+u^2/v^2) \tag{5.63}$$

which is an algebraic equation of second degree with the solution

$$\Lambda=\pm(1+\delta^2/4)^{1/2}+\delta/2 \tag{5.64}$$

with

$$\delta=(u^2/v)\,(B-X)m\hat{\beta}(1+u^2/w^2)$$

Obviously, the balanced hybrid mode condition ($\Lambda=\pm1$) occurs when

$$B=X \tag{5.65}$$

This is a generalisation of the condition $B=0$ of the corrugated waveguide for which X identically vanishes.

Now, the copolar and crosspolar aperture fields and radiation fields can be derived in the manner used for the corrugated guide; namely, equations (5.10)—(5.17) apply as they are for the present structures. We merely note that the maximum value of $N_2(u)/N_0$, which is a measure of the relative crosspolar radiation, depends on the modal eigenvalue u which varies from one structure to another for the dominant mode. A plot of this maximum value versus u is shown in Fig. 5.24. Next, we find the balanced hybrid mode condition for each of the dielectric loaded and the dielectric lined structures.

The balanced hybrid mode condition

The condition for $\Lambda=1$, or almost zero crosspolar field, is given by $B=X$, as stated in (5.65). In order to derive B and X, or Y and Z, we use (5.56) and (5.57). First, we write down the longitudinal fields in the region $a\leqslant\rho\geqslant b$. Appropriate expressions are:

$$e_z=A[I_1(w\rho/a)K_1(wb/a)-I_1(wb/a)K_1(w\rho/a)]\cos\phi \tag{5.66}$$

$$h_z=C[I_1(w\rho/a)K_1'(wb/a)-I_1'(wba)K_1(w\rho/a)]\sin\phi \tag{5.67}$$

Substituting in (5.56)—(5.57), we immediately get X and B as:

$$X=\frac{-v}{w}\,\frac{I_1'(w)K_1'(wb/a)-I_1'(wb/a)K_1'(w)}{I_1(w)K_1'(wb/a)-I_1'(wb/a)K_1(w)} \tag{5.68}$$

$$B=\frac{-v\varepsilon_2}{w\varepsilon_1}\,\frac{I_1'(w)K_1(wb/a)-I_1(wb/a)K_1'(w)}{I_1(w)K_1(wb/a)-I_1(wb/a)K_1(w)} \tag{5.69}$$

where the prime denotes differentiation with respect to the argument. At this point, it is convenient to treat each of the two structures in Fig. 5.23 separately.

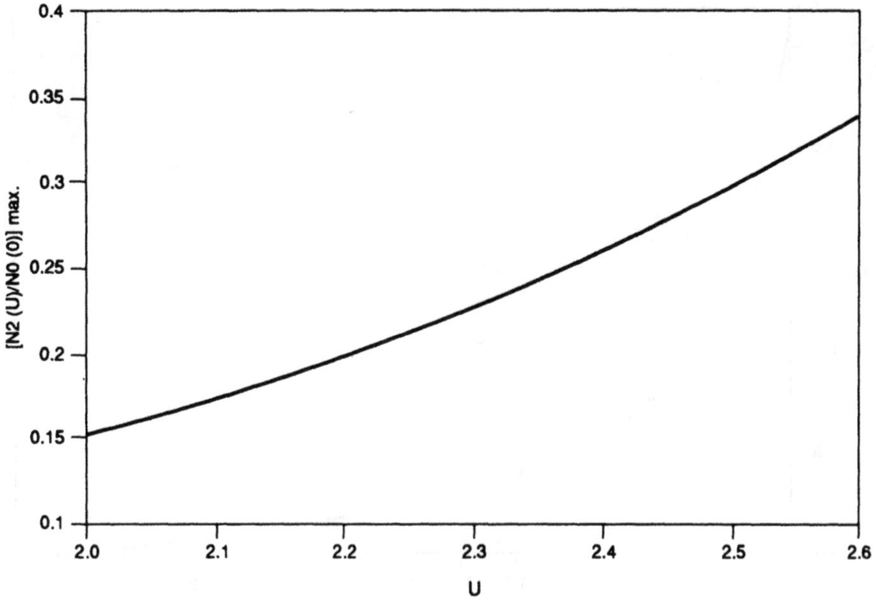

Fig. 5.24 Maximum of $[N_2(u, v \sin\theta)/N_0(u, O)]$ versus u

5.4.1 The dielectric loaded waveguide

For the dielectric loaded waveguide, w is real (see 5.29) and $\varepsilon_1/\varepsilon_2 \equiv \varepsilon_r > 1$. By invoking the condition of large guide radius, $v \gg 1$, which is satisfied near the radiating aperture of a horn, we have $\beta \simeq k_1$, and $w \simeq (\varepsilon_r - 1)^{1/2} v \gg 1$. Therefore, one can use the large argument approximation for the modified Bessel functions in the expressions for X and B to obtain:

$$X \simeq \frac{(v/w)\ \tanh(wt/a)}{1 - \tanh(wt/a)/2w} \tag{5.70}$$

$$B \simeq (v/w\varepsilon_r)[\coth(wt/a) + 1/2w] \tag{5.71}$$

where $t = b - a$ is the air gap thickness. Now, condition (5.56) for hybrid mode balance requires that:

$$\tanh(wt/a) = \varepsilon_r^{-1/2} \tag{5.72}$$

Invoking the approximation $w \simeq (\varepsilon_r - 1)^{1/2} v$, the thickness t is determined from:

$$k_1 t \simeq \tanh^{-1}(\varepsilon_r^{-1/2})/(\varepsilon_r - 1)^{1/2} \tag{5.73}$$

A plot of $k_1 t$ versus ε_r is given in Fig. 5.25 and shows that the optimum air gap thickness decreases monotonically with ε_r. The eigenvalue u and hybrid factor Λ of the HE_{11} mode are plotted versus $k_1 a$ for two values of ε_r in Fig. 5.26 and 5.27. The air gap thickness is chosen to be optimum at $k_1 a = 15$. It is observed that the bandwidth over which Λ is close to unity is greater for the higher dielectric constant. The resulting peak crosspolar radiation levels are computed from (5.17) and shown in Fig. 5.28. The low crosspolar bandwidth is higher for the

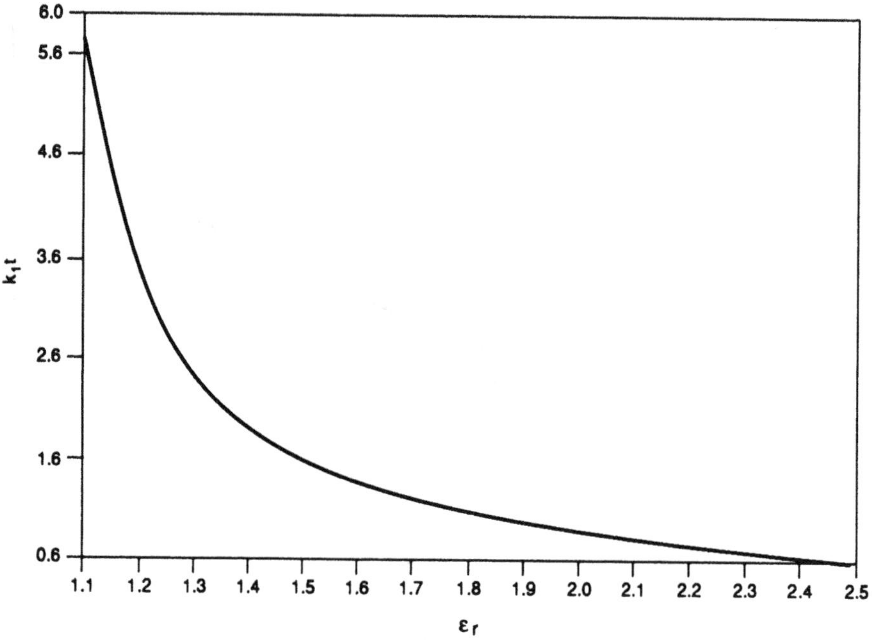

Fig. 5.25 Optimum normalised air gap thickness k_1t versus ε_r for the dielectric loaded guide

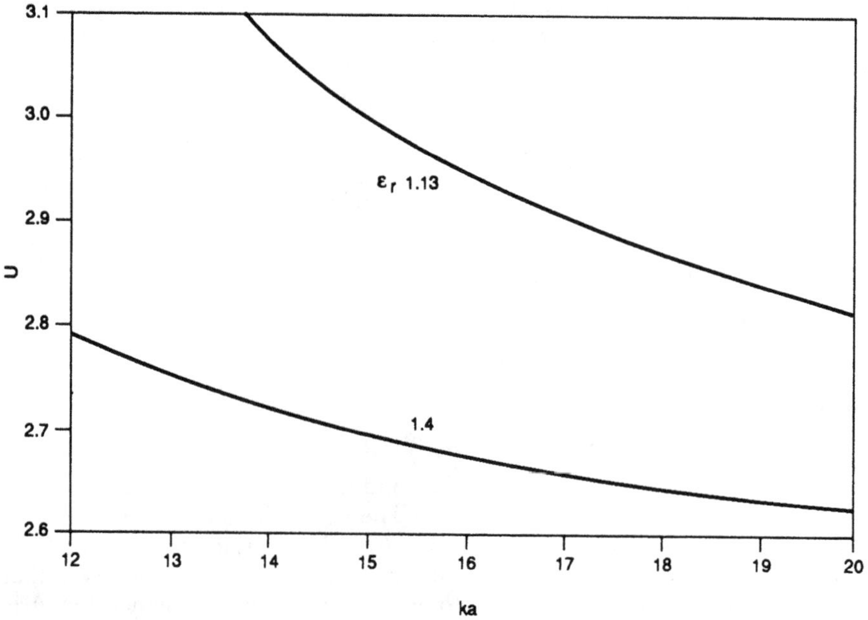

Fig. 5.26 Eigenvalue u of the HE_{11} mode for the dielectric loaded guide with $\varepsilon_r = 1{\cdot}13$ and $1{\cdot}4$

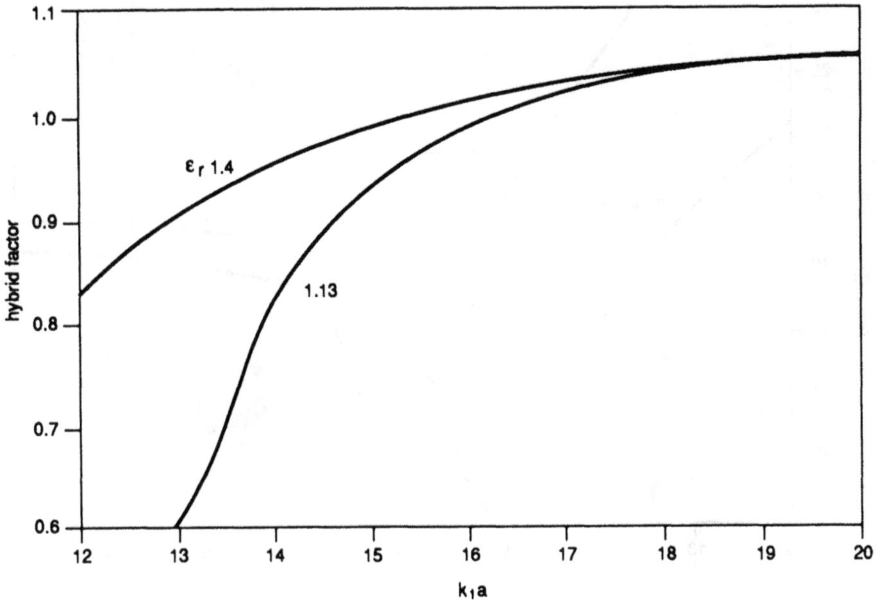

Fig. 5.27 Hybrid factor Λ of the HE_{11} mode versus k_1a for a dielectric loaded guide with $\varepsilon_r = 1{\cdot}13$ and $1{\cdot}4$

higher ε_r. However increasing ε_r will eventually cause a mismatch at the radiating aperture which can result in an increase of crosspolar radiation. It should be emphasised that the results in Fig. 5.28 are predicted for a single HE_{11} mode. In an actual conic horn, higher order modes will have some influence on the overall crosspolar performance as indicated by Olver *et al.* (1988) and Tun *et al.* (1987).

Because of its relative ease of manufacture, the dielectric loaded guide offers an economic alternative to the corrugated waveguide, particularly in the millimetric wave band where the accurate machining of a corrugated surface becomes very expensive. The dielectric material used in the dielectric loaded guide can be expanded polystyrene for low ε_r (1.13—1.4), or the solid polystyrene for higher ε_r (2·5).

5.4.2 The dielectric lined waveguide

For the dielectric lined waveguide, w in (5.66)—(5.67) is purely imaginary and $\varepsilon_2/\varepsilon_1 \equiv \varepsilon_r > 1$. Equations (5.70)—(5.71) still apply, but now we may define $w = iq$, where $q \simeq (\varepsilon_r - 1)^{1/2} v$ is real, and these equations reduce to:

$$X \simeq \frac{\tan(qt/a)}{(\varepsilon_r - 1)^{1/2}} \frac{1}{1 - \tan(qt/a)/2q} \tag{5.74}$$

$$B \simeq \frac{-\varepsilon_r \cot(qt/a)}{(\varepsilon_r - 1)^{1/2}} \left[1 + \tan(qt/a)/2q\right] \tag{5.75}$$

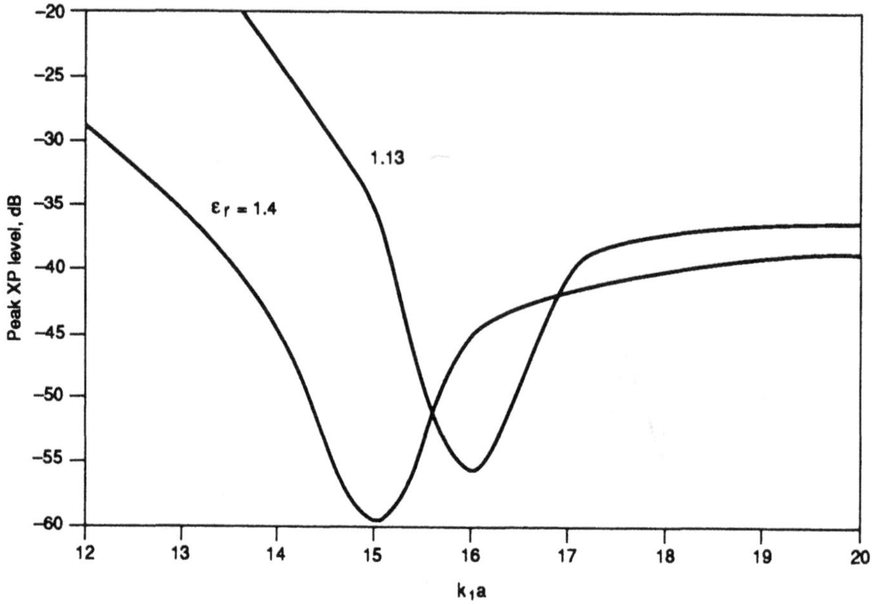

Fig. 5.28 Peak crosspolar radiation level versus k_1a for the dielectric loaded guide

A close look at these expressions reveals that the condition $X=B$ cannot be satisfied at any real frequency. Actually the difference $|B-X|$ is given approximately by:

$$|B-X| \simeq (\varepsilon_r - 1)^{-1/2} \, [\varepsilon_r \cot(qt/a) + \tan(qt/a)] \tag{5.76}$$

which does not have a zero, but has only a minimum value when:

$$\tan^2(qt/a) = \varepsilon_r \tag{5.77}$$

From this, the optimum dielectric thickness is given by:

$$k_0 t = \tan^{-1}(\varepsilon_r^{1/2}) / (\varepsilon_r - 1)^{1/2} \tag{5.78}$$

and the minimum value of $|B-X|$ is $2(\varepsilon_r/(\varepsilon_r-1))^{1/2}$. Using this value in (5.64) and (5.17), the minimum peak crosspolar radiation ratio is given approximately by:

$$XP_{\min} \simeq (u^2/2k_0 a)\,(\varepsilon_r/(\varepsilon_r - 1))^{1/2} (N_2/N_0)_{\max} \tag{5.79}$$

A plot of the peak crosspolar level in dB versus $k_0 a$ for $\varepsilon_r = 2{\cdot}5$ is shown in Fig. 5.31 by the dotted curve. The dielectric thickness is optimised for $k_0 a = 15$. It is seen that the crosspolar level has a broad minimum versus frequency, but the minimum level is rather high. We could reduce this level by increasing $k_0 a$, but this is not a very effective way. For example a reduction of 10 dB would require an increase of a by about three times, which means a very bulky horn.

To alleviate this basic limitation of the dielectric lined guide, a new version has been suggested by Mahmoud and Aly (1987). In this version, the lining is not continuous but is periodically interrupted by air gaps (Fig. 5.29). The resulting guide has rings of dielectric lining (ε_r) of thickness d separated by empty sections of thickness $l-d$. With the period l much less than a wavelength, the dielectric lining can be assumed continuous but effectively anisotropic. This is so since the longitudinal permittivity ε_z will differ from the transverse permittivity ε_t. For the longitudinal electric field the dielectric and air regions appear in series while for the transverse e, they appear in parallel. Hence, one can easily deduce that:

$$\varepsilon_z/\varepsilon_0 = \varepsilon_r/(\varepsilon_r - (\varepsilon_r - 1)\ d/l) \tag{5.80}$$

$$\varepsilon_t/\varepsilon_0 = 1 + (\varepsilon_r - 1)\ d/l \tag{5.81}$$

The fields in the region $a \leqslant \rho \leqslant b$ will now be characterised by two different values for q, say q_1 for TE waves and q_2 for TM waves. For TE waves only ε_t will appear, so $q_1 = (k_0^2\varepsilon_t/\varepsilon_0 - \beta^2)^{1/2}a$, while for TM waves both ε_t and ε_z play roles. It can be shown that $q_2 = q_1 . (\varepsilon_z/\varepsilon_t)^{1/2}$. Since the reactance X is affected by the TE part and the susceptance B by the TM part of the mode, (5.74)—(5.75) are modified to:

$$X \simeq (v/q_1)\tan(q_1 t/a) \tag{5.82}$$

$$B \simeq (\varepsilon_t/\varepsilon_0)(v/q_2)\cot(q_2 t/a) \tag{5.83}$$

Now since the arguments of the tan and cot functions are different, there is a possibility that the condition $B=X$ be satisfied at a real frequency. This, however, depends on the amount of anisotropy or the deviation of the ratio $\varepsilon_t/\varepsilon_z$ from unity. Using (5.80)—(5.81) it is easy to verify that for a given ε_r the maximum of $\varepsilon_t/\varepsilon_z$ occurs when $d=l/2$. Thus, we fix d/l at 1/2 and minimise $|B-X|$ numerically at given ε_r by adjusting $k_0 t$. It has been demonstrated that $|B-X|$ can be nullified only if ε_r exceeds a value of about 7 (Mahmoud and Aly, Aly, 1987). The optimum value of $k_0 t$ is minimum when $\varepsilon_r \simeq 8$ and is then equal to $3\pi/8$. To support these statements, the quantity $B-X$ is plotted versus t/a with $\varepsilon_r = 7$, 8 and 10 in Fig. 5.30. It is clear that $B-X$ has no zero for $\varepsilon_r \leqslant 7$. The resulting peak crosspolar radiation level for the HE_{11} mode is plotted in dB

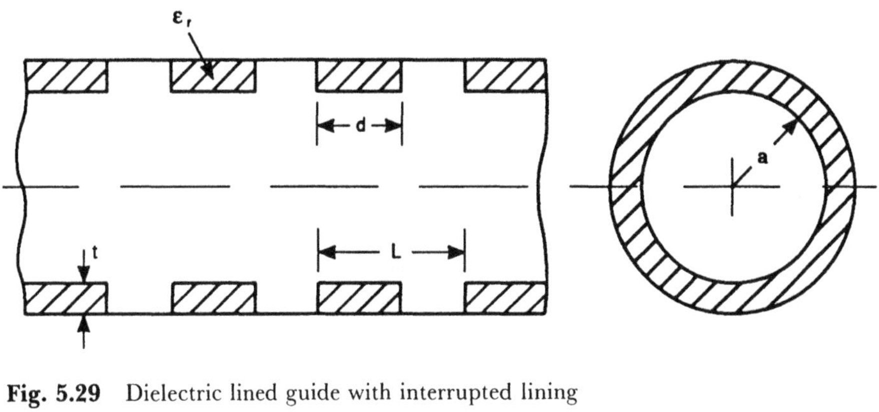

Fig. 5.29 Dielectric lined guide with interrupted lining

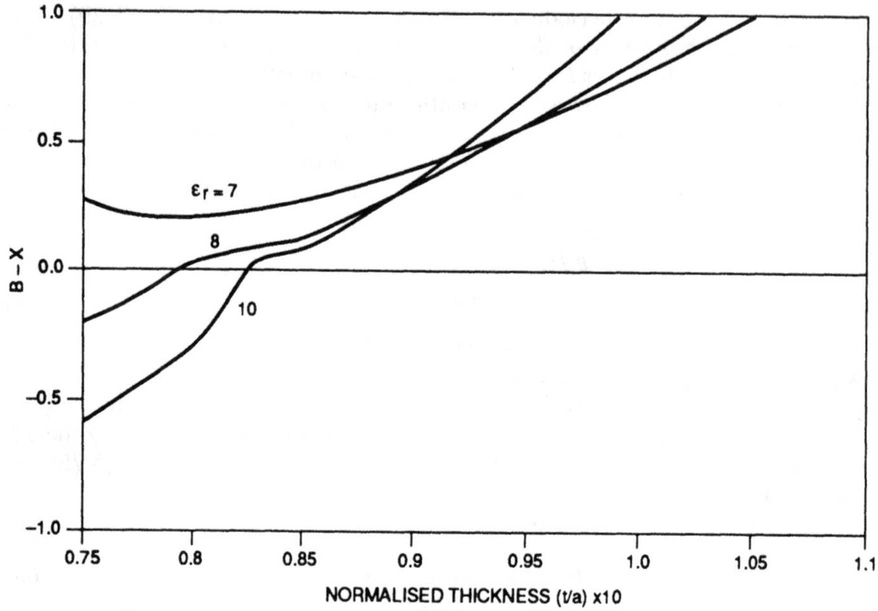

Fig. 5.30 The value of $B-X$ versus k_0t for the guide with interrupted lining

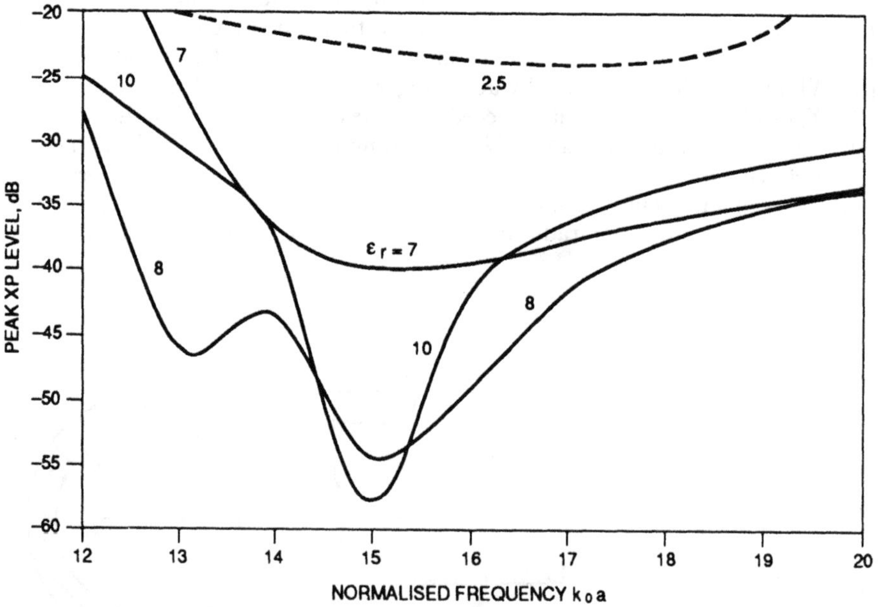

Fig. 5.31 Peak crosspolar level versus k_0a for the guide with interrupted lining

The dotted curve is for a guide with continuous lining having $\varepsilon_r = 2{\cdot}5$. Optimum lining thickness is assumed in each case versus k_0a with ε_r as a parameter in Fig. 5.31. For each value of ε_r the optimum value of t (at $k_0a = 15$) is chosen. It is clear that $\varepsilon_r = 8$ is optimum in the sense that it provides the least crosspolar levels over the widest frequency band. For comparison, the peak crosspolar levels computed for the continuously lined guide are shown by the broken curve. The improvement attained by the interrupted lining is clear.

5.5 Longitudinally slotted waveguides

Longitudinally slotted waveguides form by themselves a class of low crosspolar guides. They differ from transversely corrugated guides in that they allow pure *TE* or *TM* modes to propagate. They are manufactured by first coating the wall of a smooth waveguide by a dielectric layer, then fitting in longitudinal fins which are uniformly spaced around the circumference (see Fig. 5.32). The number of fins is made large enough to reduce the space harmonics to negligible levels. The surface $\rho = a$ can now be treated as having an average impedance $Z = e_\varphi / h_z$ and an infinite admittance $Y = h_\varphi / e_z$. The reason that makes $Y = \infty$ is that e_z must vanish at the fins' edges, and because of the large number of fins, it will also vanish in the slots. Mode description in longitudinally slotted guides has been introduced by Scharten *et al.* (1981) for air filled slots and by Aly and Mahmoud (1985) and Lier *et al.* (1988) for dielectric loaded slots. Considering the *TM* modes, one finds that the vanishing of e_z at $\rho = a$ makes the guide behave exactly as a smooth wall guide with radius a. Therefore the *TM* (or *E*) modes are not

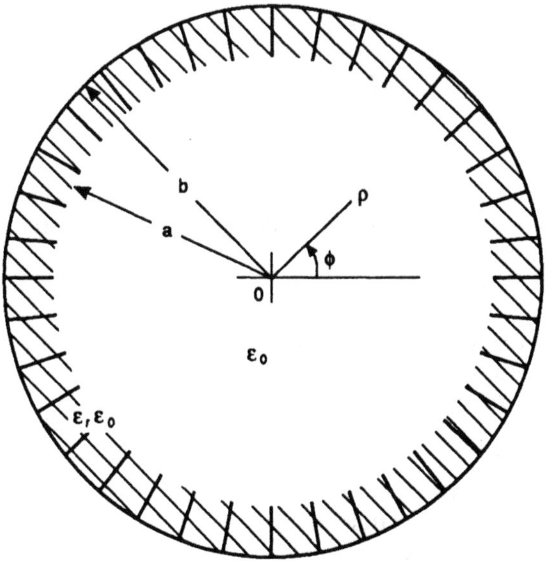

Fig. 5.32 A longitudinally slotted guide

modified by the slotted region. For the *TE* (or *H*) modes, however, the situation is quite different since these modes are drastically affected by the average wall impedance *Z*. To find the properties of the dominant *H* mode, let us express the longitudinal magnetic field as:

$$h_z = J_1(u\rho/a)\ \sin\phi\ \exp(i\omega t - i\beta z) \tag{5.84}$$

Therefore,

$$e_\phi = (i\omega\mu a^2/u^2)\,(\partial h_z/\partial\rho) \tag{5.85}$$

and the boundary condition at $\rho = a$ becomes:

$$X = -ie_\phi/\eta_0 h_z = (k_0 a/u)\,[J_1'(u)/J_1(u)] \tag{5.86}$$

where $X = -iZ/\eta_0$ and $\eta_0/\varepsilon_0)^{1/2}$. The wall impedance Z is given exactly by (3.56) and approximately by (3.58) when $k_0 a \gg 1$. Depending on the slot depth and the dielectric constant of the material filling the slots, X can assume any real value between zero, 0 and $\pm\infty$. Of course, if $X = 0$, the first root u_{11} of (5.86) is equal to 1·841. One can verify that u_{11} is less than 1·841 if X is positive, and greater than 1·841 if X is negative. In the limit $X = \infty$, $u_{11} = 0$ meaning that the dominant *H* mode reduces in this case to a *TEM* mode with purely x polarised field. In fact, as u_{11} decreases the transverse e field lines gradually straighten out, and as a result the crosspolar field is reduced. This can also be seen from equation (5.5) showing that the ratio of crosspolar to copolar fields in the plane $\phi = \pi/4$ of the aperture is:

$$|e_{zp}/e_{a0} = J_2(u_{11}\rho/a)/J_0(u_{11}\rho/a) \tag{5.87}$$

For $u_{11} \leq 1{\cdot}841$, this ratio is monotonically decreasing with decrease of u_{11}. This explains the principle of operation of the longitudinally slotted horn as a low crosspolar structure.

A plot of u_{11} versus the relative slot depth d/a for $k_0 a = 10$ and with ε_r as a parameter is given in Fig. 5.33. The case of air filled slots ($\varepsilon_r = 1$) shows the slowest rate of decrease of u_{11} with d/a, which means that a large slot depth will be required to reduce u_{11} to a given value. As ε_r increases u_{11} is reduced at a higher rate; hence the required slot depth and consequently the guide size and weight are reduced. For example a relative slot depth $d/a = 0{\cdot}2$ is required to get $u_{11} \simeq 1.4$ in air filled slots, while $d/a \simeq 0.1$ if the slots are filled with a dielectric material having $\varepsilon_r = 2{\cdot}5$. This means a relative saving in guide size of 19%. Much more saving in size is attained if u_{11} is required to be further reduced, say below 0·5, in order to reduce crosspolar radiation.

It may appear from (5.87) that a zero crosspolar field is obtained if u_{11} is reduced to zero, i.e. if a *TEM* mode exists. This is actually not true since this

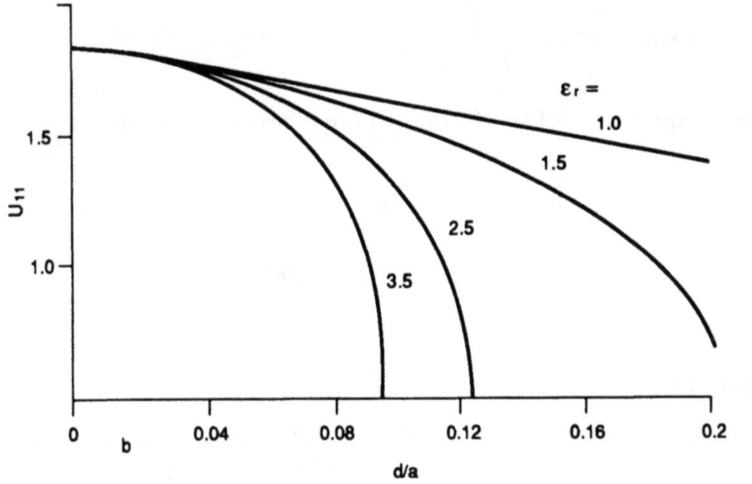

Fig. 5.33 Eigenvalue u_{11} versus the relative slot depth d/a with ε_r a parameter and $k_0a=10$ (after Aly and Mahmoud, 1985)

equation does not account for the ϕ directed electric field component in the slots which will then be high and will contribute to the crosspolar field. One should then expect that the minimum crosspolar radiation will occur at a finite value of u_{11} rather than zero. As indicated by Aly and Mahmoud (1985), the frequency at which $u_{11}=0$ should be set as the upper limit of the operating frequency band. This is so since, at a higher frequency, u_{11} takes imaginary values and the modal fields start to cling to the slots, resulting in increased crosspolar field.

5.6 Problems

5.1 Prove equation (5.5). Find the ratio between power flow associated with e_x and e_y.

5.2 Prove equation (5.12). Hint: express any cartesian component of e as a plane wave spectrum; e.g.

$$e_x(x,y,z)=(1/2\pi)\int\int_{-\infty}^{\infty} A(k_x, k_y)\exp(-ik_xx-ik_yy-ik_zz)\,dk_x\,dk_y.$$

Concentrating on the aperture field at $z=0$, deduce that $A(k_x, k_y)$ is double FT of the aperture field. Now, consider the far zone, $k_0r \gg 1$: use the stationary phase approach to evaluate the integrals, and apply the identity (Clemmow, 1966):

$$\exp(-ik_0 r)/r = \int\int_{-\infty}^{\infty} \exp(-ik_x x - ik_y y - ik_z z)\, dk_x\, dk_y / ik_z$$

5.3 Prove equations (5.14)—(5.16). You may use the identity:

$$\int_0^a J_m(kr) J_m(lr) r\, dr = \frac{a}{k^2 - l^2} \cdot [l J_{m-1}(la) J_m(ka) - k J_{m-1}(ka) J_m(la)]$$

5.4 Prove (5.19)

5.5 Prove (5.23).

5.6 Rederive the modal equations (5.29)—(5.31) for a corrugated rectangular guide by applying the general equations (3.171)—(3.174).

5.7 Find the radiation pattern from the dominant y polarised mode in a corrugated rectangular guide.

5.8 Prove (5.40).

5.9 Show that the modal equation (5.34) for the conical corrugated horn implies that $f_{v+}(\theta_1) = 0$ if $B = 0$ (at hybrid balance) and $\beta \simeq 1$.

5.10 Show that $f_{v-}\ (\theta = 0) = 0$.

5.11 Prove (5.48).

5.12 Use (5.49) to show that $a(z)$, $\theta_1(z)$ and $d\theta_1(z)/dz$ are all continuous at $z = L_1$. Suggest other profiles that can have low mode conversion.

5.13 Prove (5.51) on the basis of low conversion levels.

5.14 Prove (5.60)—(5.62). Compare the two modal equations (5.62) and (3.134) and comment on their difference.

5.15 Prove (5.70)—(5.73).

5.16 Prove (5.76)—(5.79).

5.17 Prove (5.80) and (5.81).

5.18 In a longitudinally slotted circular guide, show that, for $|X| \ll 1$, the first modal eigenvalue u_{11} is given approximately by:

$$u_{11} \simeq u^0 - (X/k_0 a)(u^0)^3/[(u^0)^2 - 1]$$

where $u^0 = 1{\cdot}841$ is the first zero of $J_1'(u)$.

5.7 References

ALY, M.S., and MAHMOUD, S.F. (1985): 'Propagation and radiation behaviour of a longitudinally slotted horn with dielectric filled slots', *IEE Proc.*, **132,** Pt. H, pp. 477–479

BALDWIN, R., and MCINNES, P.A. [1975]: 'A rectangular corrugated feed horn', *IEEE Trans.*, **AP–23**, 814–817

CLARRICOATS, P.J.B., and SAHA, P.K. (1969): 'Radiation from wide flare angle scalar horns', *Elecron. Lett.*, **5,** p. 376

CLARRICOATS, P.J.B., and SAHA, P.K. (1971*a*): 'Propagation and radiation behaviour of corrugated feeds, Pt. 1: Corrugated waveguide feeds', *Proc. IEE*, **118,** pp. 1167–1176

CLARRICOATS, P.J.B., and SAHA, P.K. (1971*b*): 'Propagation and radiation behaviour of corrugated feeds. Pt.2: Corrugated conical horn feeds', *Proc. IEE*, **118,** pp. 1177–1186

CLARRICOATS, P.J.B., OLVER, A.D., and SAHA, P.K. (1971): 'Near field radiation characteristics of corrugated horns', *Electron Lett.*, **7,** pp. 446–448

CLARRICOATES, P.J.B., and SALEMA, C.E.R.C. (1973*a*): Antennas employing conical dielectric horn. Pt.1: Propagation and radiation characteristics of dielectric cones', *IEE Proc.*, **120,** pp. 741–749

CLARRICOATS, P.J.B., and SALEMA, C.E.R.C. (1973*b*): 'Antennas employing conical dielectric horn. Pt.2: The cassegrain antenna', *IEE Proc.*, **120**, pp. 760–766

CLARRICOATS, P.J.B. (1979); 'Hybrid mode feeds of microwave reflector antennas', *in* 'Modern topics in electromagnetics and antennas' (Peter Peregrinus)

CLARRICOATS, P.J.B., MAHMOUD, S.F., and OLVER, A.D. (1981*a*): 'Cross-polar behaviour of wide angle corrugated horns'. *IEEE Symp. on Ant. & Prop.*, Los Angeles, June, pp. 65–68

CLARRICOATS, P.J.B., MAHMOUD, S.F., and OLVER, A.D. (1981*b*): 'Low cross-polar radiation from wide flare angle conical corrugated horns'. *Proc. 11th European Microwave Conf.*, Amsterdam, Sept., pp. 735–739

CLARRICOATS, P.J.B. (1982): 'Feeds for earth-station and spacecraft antennas', in MITTRA, IMBERIALE and MAANDERS (Eds.): 'Satellite communication technology'. *Proc. Summer School. Univ. of Tech.*, Eindhoven, The Netherlands

CLARRICOATS, P.J.B., OLVER, A.D., and RISK, M.S. (1983): A dielectric loaded conical feed with low crosspolar radiation'. *Proc. URSI Symp. on EM Theory*, Spain, pp. 351–354

CLARRICOATS, P.J.B., and OLVER, A.D. (1984): 'Corrugated horns for microwave antennas'. IEEE Electromagnetic wave series 18 (Peter Peregrinus Ltd.)

CLEMMOW, P.C. (1966): 'The plane wave spectrum representation of electromagnetic fields', (Pergamon Press.)

DRAGONE, C. (1977*a*): 'Reflection, transmission and mode conversion in a corrugated feed', *Bell Syst. Tech. J.*, **56,** pp. 835–867

DRAGONE, C. (1977*b*): 'Characterristics of a broadband microwave corrugated feed: a comparison between theory and experiment', *Bell Syst. Tech. J.*, **56,** pp. 869–888

DRAGONE, C. (1980): 'Attenuation and radiation characteristics of the HE_{11} mode', *IEEE Trans.* **MTT–28,** pp. 704–710

DRAGONE, C. (1981): 'High frequency behaviour of waveguides with finite surface impedance', *Bell. Syst. Tech. J.*, **60,** pp. 89–115

DRAGONE, C. (1985): 'A rectangular horn of four corrugated plates', *IEEE Trans.*, **AP–33,** pp. 160–164

GHOSH, S., ADATIA, N., and WATSON, B.K. (1982): 'Hybrid mode feed for multiband applications having a dual depth corrugation boundary', *Electron. Lett.*, **18,** pp. 860–862

GUY, R.F.E., and ASHTON, R.W. (1979): 'Crosspolar performance of an elliptical corrugated born antenna', *Electron. Lett.*, **15,** p. 400

JAMES, G.L. (1982): 'TE_{11} to HE_{11} mode converters for small angle corrugated horns', *IEEE Trans.* **30,** pp. 1057–1062

JAMES, G.L. (1984): 'Design of wide band compact corrugated horns', *IEEE Trans.* **AP–32,** pp. 1134–1138

JANSEN, J.M.K., JEUKEN, M.E.J., and LAMBRECHTSE, C.W. (1972): 'The scalar feed', *Archiv fur Elektronik & Uber.* **26,** pp. 22–30

KAY, A.F. (1962): 'A wide flare angle horn. A novel feed for low noise broadband and high aperture efficiency antennas'. US Air Force Cambridge Research Labs Report 62–757

KNOP, C.M., CHENG, Y.B., and OSTERTAG, E.L. (1986): 'On the fields in a conical horn having an arbitrary wall impedance'. *IEEE Trans.* **AP–34,** pp. 1092–1098

LIER, E., and KILDAL, P.S. (1988): 'Soft and hard horn antenas', *IEEE Trans.*, **AP–36,** pp. 1152–1157

LOVE, A.W. (1976): 'Electromgnetic horn antennas', IEEE (Press, NY.)

LOVE, A.W., RUDGE, A.W., and OLVER, A.D. (1982): 'Primary feed antennas', *in* RUDGE, MILNE, OLVER and KNIGHT (Eds): 'The handbook of antenna design' (Peter Peregrinus) chap. 4

LUDWIG, A.C. (1973): 'The definition of cross-polarisation', *IEEE Trans.*, **AP–21,** pp. 116–119

MAHMOUD, S.F., and CLARRICOATS, P.J.B. (1982): 'Radiation from wide flare-angle corrugated conical horns', *IEE Proc.*, Pt. H, **129,** pp. 221–228

MAHMOUD, S.F. (1983): 'Mode conversion on profiled corrugated conical horns', *IEE Proc.*, Pt. H, **130**, pp. 415—419

MAHMOUD, S.F., and ALY, M.S. (1987): 'A new version of dielectric lined waveguide with low crosspolar radiation', *IEEE Trans.*, **AP–35**, pp. 210–212

MINNETT, H.C., and THOMAS, MacA.B. (1966): 'A method of synthesising radiation patterns with axial symmetry', *IEEE Trans.*, **AP–14,** pp. 654–656

NARASIMHAN, M.S., and RAO, V.V. (1970): 'Hybrid modes in corrugated conical horns', *Electron. Lett.*, **6**, pp. 32–34

NARASIMHAN, M.S., (1971): 'Radiation from conical horns with large flare angles', *IEEE Trans.*, **AP–19**, pp. 678–681

NARASIMHAN, M.S., (1974): 'Radiation from wide flare angle corrugated E plane sectoral horns', *IEEE Trans.*, **AP–22,** pp. 603–608

OLVER, A.D., CLARRICOATS, P.J.B., and RAGHAVEN, K. (1988): 'Dielectric cone loaded born antennas', *IEE Proc.*, **135,** Pt. H, pp. 158–162

OLVER, A.D., and XIANG, J. (1988): 'Design of profiled corrugated horns', *IEEE Trans.*, **AP–36,** pp. 936–940

PARINI, C.G., and OLVER, A.D. (1980): 'Accurate prediction of the crosspolar performance of narrow flare angle corrugated horns'. Int. URSI Symp. on EM Theory, Munich, August

POTTER, P.D. (1963): 'A new horn antenna with suppressed sidelobes and equal beamwidth', *Microwave J.*, **6,** p. 71

RAGHAVAN, K., OLVER, A.D., and CLARRICOATS, P.J.B. (1987): 'Computer aided design of dielectric loaded horns'. Fifth Int. Conf. on Antenn. Propagat., ICAP–87, pp. 434–437

RUDGE, A.W., and ADATIA, N.A. (1975): 'A new class of primary feed antennas for use with offset parabolic reflector antennas', *Electron. Lett.*, **11,** pp. 579–599

RUDGE, A.W., and ADATIA, N.A. (1976): 'Matched feeds for offset parabolic reflector antennas', Proc. European Microwave Conf., pp. 1–5

SATOH, T. (1974): 'Dielectric loaded horn antenna' *Inst. Electronic and Comm. Engs. Japan.* **57B,** pp. 81–86

SCHARTEN, T., NELLEN, J., and BOGAART, V.D. (1981): 'Longitudinally slotted conical horn antenna with small flare angle'. *Proc. IEE*, Pt. H, **128,** pp. 117–123

SILVER, S. (1965): 'Microwave antenna theory and design' (Dover Publications Inc., New York). A more recent edition is published by Peter Peregrinus, 1984)

TAKEICHI, Y., HASHIMOTO, T., and TAKEDA, F. (1971). 'The ring loaded corrugated waveguide', *IEEE Trans.*, **MTT–19,** pp. 947–950

TAKEDA, F., AND HASHIMOTO, T. (1976): 'Broadening of corrugated conical horns by means of the ring loaded corrugated waveguide structure', *IEEE Trans.*, **AP–24,** pp. 786–792

THOMAS, MacA.B. (1969): 'Bandwidth properties of corrugated conical horns', *Electron. Lett.*, **5,** pp. 561–563

THOMAS, MacA.B. (1978): 'Design of corrugated conical horns', *IEEE Trans.*, **AP–26,** pp. 367–372

TUN, S.M., PHILIPPOU, G.Y., and ADATIA, N. (1987): 'Analysis and design of rotationally symmetric dielectric loaded waveguides and horns'. Fifth Int. Conf. on Antenn. Propagat., ICAP–87, p. 438–442

WATSON, B.K., DANG, R., RUDGE, A.W., and OLVER, A.D. (1980): 'Compact low crosspolar corrugated feed for ECS'. IEEE A P Conf. Digest, Vol. 1, Quebec, PQ, Canada, p. 209–212

Chapter 6

Guided waves in subsurface tunnels

6.1 Introduction

The study of electromagnetic wave propagation through the earth has been of growing interest for many years. Applications are diverse and include communication between underground work areas such as in mines, and remote sensing of subsurface resources such as water, oil and minerals. Direct propagation through the earth is basically limited by high signal losses particularly in the high frequency bands. For example, with a conductivity of 10^{-2} mho/m, the attenuation within a homogeneous earth is about 2 nepers/100 m at a frequency f of 10 kHz and increases as $\sqrt{f}$ up to frequencies in the HF range. Therefore communication through the earth is preferred at frequencies lower than few tens of kHz although these frequencies suffer from the narrow bandwidth available for information and from the high dispersion caused by strong dependence of conductivity on frequency. Short range communication can be maintained in the VHF band in the presence of resistive layers with conductivity less than about 10^{-3} mho/m (e.g. Wait, 1971, and Gabillard *et al.*, 1971).

Considerably better conditions for electromagnetic wave propagation through the earth occur wherever straight tunnels exist within the earth. A tunnel acts as a waveguide with imperfectly reflecting walls, but most of the power travels in the lossless interior of the tunnel if this is sufficiently wide. Some power is lost by conduction and/or radiation in the surrounding rocks. Tunnels within the earth are either manmade, such as in the case of road tunnels, or they are natural as in the case of mine tunnels. In the latter case, they are usually found interspersed with highly conductive regions and covering large areas of perhaps several square kilometers. As a waveguide, a tunnel has a cutoff frequency below which no effective propagation occurs. Usually the tunnel cross section varies from circular to rectangular with linear dimensions of a few meters. The cutoff frequency will therefore lie somewhere in the VHF band and free propagation is limited to the VHF and part of the UHF band, or say from 30 to 1000 MHz. Higher frequencies can also propagate, but with increasing attenuation due to roughness of the tunnel walls.

Often, longitudinal conductors, such as pipes and power cables, are located inside the tunnel running parallel to the walls. Such conductors alter drastically the nature of electromagnetic wave propagation in the tunnel by extending the useful range of frequencies to much lower values. This is so because of the appearance of a coaxial like mode in which currents flow along the axial conductor and return in the tunnel walls. This mode, which is known as the monofilar mode, has a zero cutoff frequency. So it can propagate at low frequencies when all the free tunnel modes are cutoff. The monofilar mode is characterised by fields which are accessible throughout the tunnel so that they can easily couple to fixed or mobile antennas therein. The attenuation of the monofilar mode is mainly dependent on the conductivity of the axial conductor as well as its proximity to the tunnel wall.

If a two wire transmission line is installed in the tunnel, two modes with zero cutoff frequencies are supported. The first is the monofilar mode in which current is carried in the same direction by the two wires and returns in the tunnel walls. The second mode, termed the bifilar mode, has almost balanced currents in the two wires with negligible current carried by the tunnel walls. For a well designed line, the bifilar mode has a much lower attenuation than the monofilar mode. However, the former is much less accessible to antennas within the tunnel since its fields are confined to the very near vicinity of the line. Other leaky cables which are more durable than the simple two wire line have been used in mine tunels for communication. They are mostly coaxial lines with imperfect shield such as braided shield, perforated shield or longitudinally slotted shield (Delogne, 1982).

A continuous access communication scheme can now be visualised as one which strikes a proper balance between the monofilar and bifilar modes in such a way as to ensure signal accessibility within the tunnel and, meanwhile, low net attenuation of the signal. A scheme proposed by Prof. P. Delogne and used in Belgian coal mine tunnels achieves this goal by having mode convertors installed on the cable at specified intervals (e.g. Delogne *et al.*, 1973, Delogne, 1974). This scheme takes advantage of the low attenuation of the bifilar mode and still makes the signal accessable in the tunnel by converting some of its power to the monofilar mode at each mode convertor. The Belgian scheme, known as the INIEX/Delogne system, uses a conventional coaxial cable with mode convertors in the form of an annular slot which is a complete interruption of the outer conductor. The annular slot is designed to radiate about 10% of the power from the bifilar mode into the monofilar mode which provides signal accessability within the tunnel. Another type of mode convertor consists of a short length of a leaky cable inserted in a well shielded main cable. Typically mode convertors are installed every few hundred meters. A comprehensive coverage of leaky feeder systems is found in Delogne's book (1982).

In this chapter we study the electromagnetic wave propagation in free tunnels and in tunnels with axial conductors. Thus the attenuation of free modes in tunnels of circular and rectangular shapes is derived and plotted versus frequency in section 6.2. The effect of curvature of the tunnel walls is then investigated and it is shown that curvature can cause a considerable increase of the mode attenuation. In section 6.3 the theory of modal propagation in a tunnel containing an axial conductor is presented and in section 6.4 the character of the monofilar and bifilar modes is investigated. Mode conversion in tunnels is discussed in section 6.5.

6.2 Mode propagation in free tunnels

We begin this section by studying mode characteristics in straight tunnels of a circular shape. We then consider tunnels of a rectangular shape. We end the section by a study of the effect of curvature of tunnel walls by using a simplified

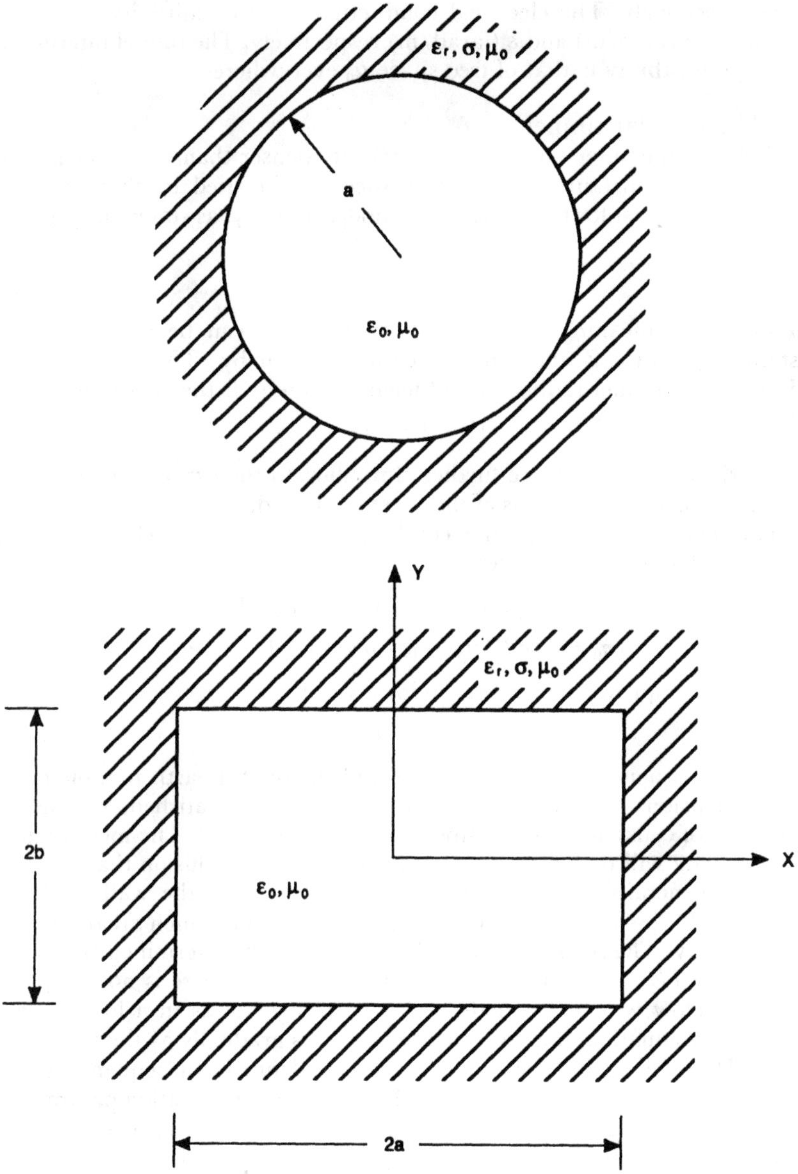

Fig 6.1 *a* A circular tunnel
b A rectangular tunnel

model. The geometry of a circular and a rectangular cross section tunnel is given in Fig. 6.1*a*, *b*. The tunnel walls are assumed smooth, the radius is a for the circular tunnel and the width and height of the rectangular tunnel are $2a$

and $2b$ respectively. The electrical conductivity and permittivity of the outer medium are σ (mho/m) and ε (farad/m) respectively. The tunnel interior is air and the permeability is that of free space μ_0 everywhere.

6.2.1 The circular tunnel

Provided that the outer medium is electrically denser than the inner medium, the concept of constant surface impedance can be used to characterise the tunnel walls, at least for the lower order modes; namely, as stated in section 3.2, if

$$|k^2 - k_0^2| \gg |k_0^2 - \beta^2| \tag{6.1}$$

with the usual meaning of the symbols, the wall can be characterised by constant impedance Z and admittance Y as given by (3.32) and (3.33). In addition, if the tunnel radius a is sufficiently large that the condition:

$$|k^2 - k_0^2|^{1/2} a \gg 1 \tag{6.2}$$

is satisfied, then Z and Y are further simplified to the expressions (3.35) and (3.36). More convenient forms of these expressions display explicitly the tunnel wall parameters, which are the conductivity σ and the relative dielectric constant ε_r. These are as follows:

$$Z/\eta_0 = iX = F^{-1}[1 + 1/2ik_0aF] \tag{6.3}$$

$$Y\eta_0 = iB = F^{-1}(\varepsilon_r - i\sigma/\omega\varepsilon_0)\,[1 + 1/2ik_0aF] \tag{6.4}$$

where F is a frequency dependent factor given by

$$F = [\varepsilon_r - 1 - i\sigma/\omega\varepsilon_0]^{1/2} \tag{6.5}$$

Within the frequency ranges of interest and for typical earth parameters and tunnel dimensions, conditions (6.1) and (6.2) are easily satisfied, and therefore the concept of constant surface impedance is well justified. The modal analysis of section 3.4.2 will therefore apply and leads to the conclusion that the tunnel modes are generally of hybrid nature. However, since the wall parameters $|X| \ll 1$ and $|B^{-1}| \ll 1$, the mode hybrid factors are either much greater or much less than unity. The modes can then be appropriately determined by modified or perturbed TE_{mn} and TM_{mn} modes. For these modes there is no sharp cutoff frequency below which the mode ceases to propagate as in a lossless guide. Instead the atenuation of a given mode increases gradually as the frequency is reduced. However a cutoff frequency can still be defined as a frequency at which the attenuation reaches a preset value. The complex propagation parameter β_{mn} of the TE_{mn} modes has been derived in section 3.4.2 and given by (3.144) along with (3.143). The attenuation is merely given by the negative of the imaginary part of β_{mn}. We have computed the attenuation factor multiplied by the tunnel radius and plotted it in dB versus k_0a for both the TE_{11} and TE_{01} modes in Figs. 6.2 and 6.3 respectively. The relative permittivity of the tunnel wall is taken as 10. The parameter x equals $\sigma/\omega_c\varepsilon_0$ where ω_c is the angular frequency at which the TE_{11} mode would be cutoff in a tunnel with perfect walls; i.e. $\omega_c a/c = 1{\cdot}841$. The gradual cutoff of modes is clear in these two Figures. The curves for $x \leqslant 1$ in Fig. 6.2 correspond to low conducting walls. As a result, the attenuation is mainly due to radiation in the outer medium of the tunnel, and it shows a continuous decrease with increasing frequency. For $x = 100$, the conduction currents start to dominate over the displacement currents and the characteristic increase of attenuation with frequency above a certain value due to ohmic losses can then be noticed. To give a feeling for the order of magnitude of attenuation in tunnels, the outer scales in Figs. 6.2 and 6.3 give the attenuation in dB/100 m at a given frequency in MHz for a tunnel radius of 2·5 m. The conductivity σ is then related to x by σ (mho/m) $\simeq 2 \times 10^{-3} x$. As an example, for a tunnel radius of 2·5 m and outer medium having $\varepsilon_r = 10$ and $\sigma = 0{\cdot}2$ mho/m, the attenuation of the TE_{11} mode is 25 dB/100 m at 200 MHz. As expected, the attenuation of the TE_{01} mode is lower, but this mode cannot be easily excited.

6.2.2 The rectangular tunnel

Now consider the rectangular tunnel with four lossy walls. As discussed in chapter 3, one cannot satisfy all the boundary conditions by using simple wave functions. However, we shall find modal solutions which are approximate in the sense that one weak boundary condition is ignored. This approximate approach will have sufficient accuracy for the dominant modes of the tunnel. We start by considering vertically polarised modes. So, with reference to Fig. 6.1*b*, let us try

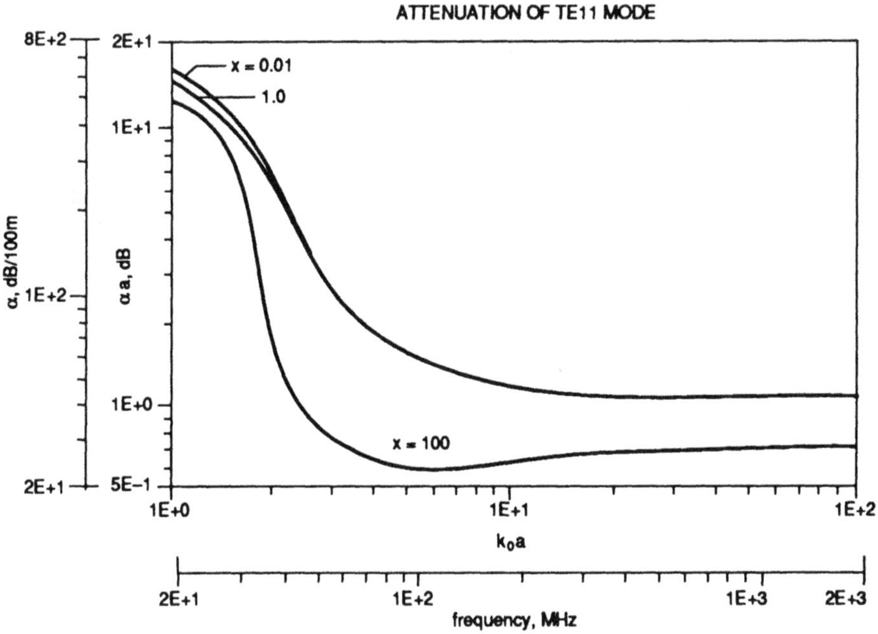

Fig 6.2 Normalised attenuation rate of modified TE_{11} mode in a circular tunnel

The parameter $x = \sigma/\omega_c\varepsilon_0$ and $\omega_c = 1{\cdot}8411 c/a$
Outer scales correspond to a tunnel radius $a = 2{\cdot}5$ m

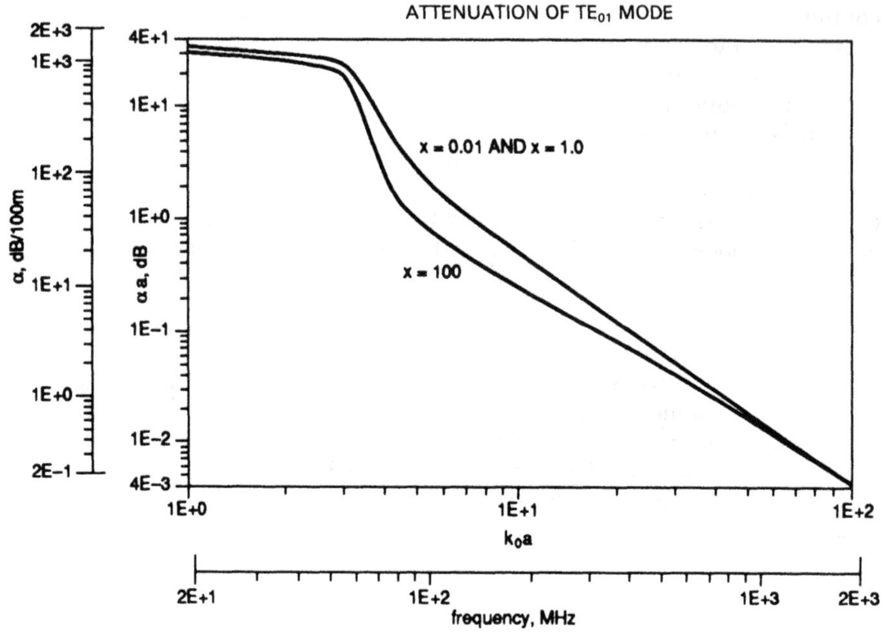

Fig 6.3 Normalised attenuation rate of modified TE_{01} mode in a circular tunnel.

The parameter $x = \sigma/\omega_c\varepsilon_0$ and $\omega_c = 1{\cdot}8411c/a$
Outer scales correspond to a tunnel radius $a = 2{\cdot}5$ m

a TM to y modal solution. All fields are derivable from an e_y field component which, apart from an $\exp(i\omega t - i\beta z)$ common factor, can take the form

$$\begin{aligned} e_y &= \cos k_y y\, f(k_x x) && |y| \leq b \\ &= A \exp(-\gamma(|y| - b)) f(k_x x) && |y| \geq b \end{aligned} \tag{6.6}$$

where only even modes in y are considered. The function $f(.)$ is a sum of sin and cos terms of the argument. At the moment, we are assuming that the walls at $x = \pm a$ recede to $\pm\infty$. From the wave equation, one has:

$$\beta^2 + k_x^2 = k_0^2 - k_y^2 = k^2 + \gamma^2 \tag{6.7}$$

From relations (5.25)—(5.28), one can verify that the continuity of h_x, h_z, e_x and e_z at $y = b$ merely require that $(\varepsilon - i\sigma/\omega)\ e_y$ and $\partial e_y/\partial y$ be continuous. This leads to the modal equation

$$k_y \tan(k_y b) = \gamma/(\varepsilon_r - i\sigma/\omega\varepsilon_0) \tag{6.8}$$

This along with the relation

$$\gamma = k_0[1 - \varepsilon_r + i\sigma/\omega\varepsilon_0 - (k_y/k_0)^2]^{1/2},\ \text{Real}(\gamma) > 0 \tag{6.9}$$

determine a set of eigenvalues for k_y, say k_{y0}, k_{y1}, . .

Now, let us turn attention to the side walls $x = \pm a$. The boundary conditions at these walls are:

$$e_y/h_z = \pm Z_s \equiv \pm i\eta_0 X_s \tag{6.10}$$

$$e_z/h_y = \mp Y_s^{-1} \equiv \pm i\eta_0 B_s^{-1} \tag{6.11}$$

where Z_s and Y_s are the effective constant impedance and admittance of the side walls and may be given by (3.14)—(3.15). Now, since $h_y = 0$ for the modes considered, it turns out that condition (6.11) cannot be satisfied for finite values of Y_s. However, since e_z is proportional to $\partial e_y/\partial y$ it is expected to be small for the important low order modes and hence the boundary condition (6.11) may be ignored without incurring much error. Let us now proceed to satisfy condition (6.10). On using (5.28) and (6.6), this becomes

$$[(k_0^2 - k_y^2)/k_0 k_x][f(k_x a)/f'(k_x a)] = -X_s \tag{6.12}$$

which is a modal equation for k_x. For modes of every symmetry in x, $f(.)$ is a cosine function and then (6.12) reduces to

$$[(k_0^2 - k_y^2)/k_0 k_x] \cot(k_x a) = X_s \tag{6.13}$$

For a given k_y this determines a set of k_x values, k_{x1}, k_{x2}, ... On determination of both sets of eigenvalues of k_x and k_y all modal characteristics become known. Essentially, the approach outlined above was suggested by Wait (1980) who noted that the solution becomes exact in the special case of perfectly conducting side walls. This can be seen if we let $X_s = B_s^{-1} = 0$, whence (6.12) gives $f(k_x a) = 0$, satisfying both (6.10) and (6.11) simultaneously.

Now, for a more realistic case in which $|k| \gg k_0$, a perturbation analysis can be used to get a first order solution to (6.8) and (6.13) for the dominant TM_{y10} mode. First, the RHS of (6.8) is recognised as equal to $k_0 B^{-1}$, where B is the normalised surface susceptance of the lower and upper walls (see equation 3.15). With $|B^{-1}|/k_0 b \ll 1$, a first order solution of (6.8) is

$$(k_y/k_0) \simeq (Bk_0 b)^{-1/2} \tag{6.14}$$

To solve (6.13) for $X_s \ll 1$, we set $k_x a = \pi/2 + \delta$ where δ is a small quantity. Neglecting $|k_y^2|$ relative to k_0^2, we get

$$k_x/k_0 \simeq (\pi/2k_0 a)(1 - \pi X_s/k_0 a) \tag{6.15}$$

Using (6.14) and (6.15) to get β, we arrive at

$$\beta \simeq k_0(1 - \pi^2/4k_0^2 a^2)^{1/2} - B^{-1}/2b + X_s \pi^2/4k_0^2 a^3 \tag{6.16}$$

The modal attenuation constant α is extracted as the negative of the imaginary part of β; therefore

$$\alpha \simeq \mathrm{Imag}(B^{-1})/2b - \mathrm{Imag}(X_s)\pi^2/4k_0^2 a^3 \tag{6.17}$$

We immediately notice that the attenuation rate of the dominant vertically polarised (*VP*) mode has two parts; the first part, caused by upper and bottom walls, is inversely proportional to the tunnel height b and the second caused by the side walls, is inversely proportional to the cube of the tunnel width a and the frequency squared. This result has been indicated by Emslie *et al.* (1975) by using a different approach. Since $a > b$ and $k_0 a \gg 1$ in the frequency ranges of interest, we conclude that the bottom and upper walls contribute much more to the attenuation of the *VP* mode than do the two side walls. Obviously, the reverse is true for the horizontally polarised (*HP*) mode which has a lower attenuation rate relative to the *VP* mode according to the ratio b/a. To support these results, the normalised attenuation αa is plotted for each of the *VP* and *HP* modes over a wide frequency range in Fig. 6.4. The relative permittivity of the outer medium is taken equal to 10 and the conductivity σ is normalised by $x=\sigma/\omega_c\varepsilon_0$, where ω_c is the cutoff frequency of the TE_{10} mode in a tunnel on the assumption of perfectly conducting walls. The outer scales in Fig. 6.4 correspond to a tunnel of dimensions $2a=4b=4{\cdot}25$ m and therefore $\sigma \simeq 1{\cdot}96\times10^{-3}x$.

6.2.3 The curved tunnel

It has been observed by several workers that curvature in mine tunnels causes a pronounced increase of wave propagation loss (e.g., Emslie *et al.*, 1973). In this section, we will investigate quantitatively the effect of curvature by using a fairly simple model of a rectangular tunnel (Mahmoud and Wait, 1974*b*). With reference to Fig. 6.5, a rectangular tunnel of height $2h$ and width $2b$ is assumed to have a mean radius of curvature equal to a. To concentrate on the effect of curvature, the upper and lower flat surfaces are assumed to be either electric or magnetic walls; hence they contribute nothing to the mode losses. The two

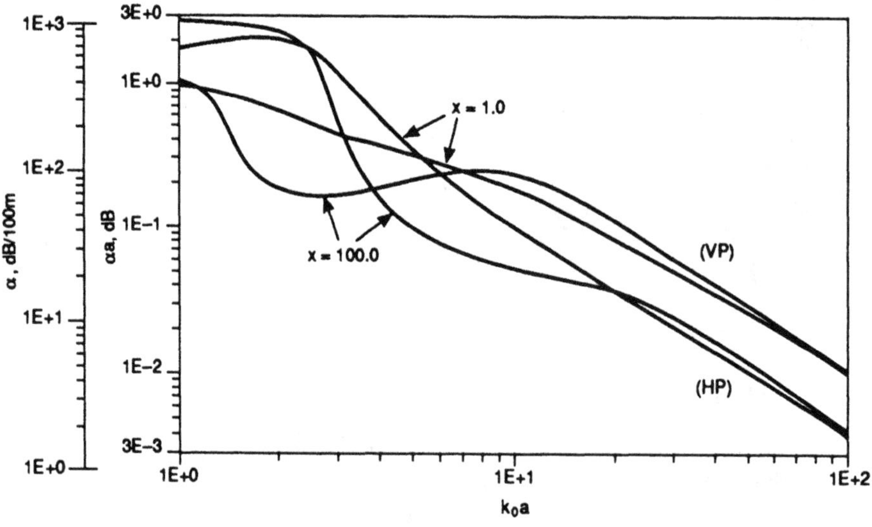

Fig 6.4 Normalised attenuation rate of vertically polarised (VP) and horizontally polarised (HP) modes in a rectangular tunnel with $b/a=1/2$

The parameter $x=\sigma/\omega_c\varepsilon_0$ and $\omega_c=\pi c/2a$
The outer scales correspond to a tunnel having $2a=4b=4{\cdot}25$ m.

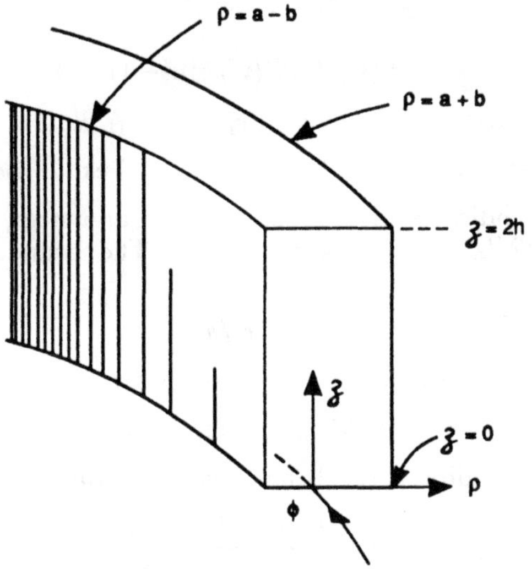

Fig 6.5 A curved tunnel model (adapted from Mahmoud and Wait, 1974*b*).

Mean radius of curvature is *a*. The upper and lower walls are assumed either perfect electric or perfect magnetic conductors. The curved walls are characterised by a surface impedance *Z* and a surface admittance *Y*.

curved walls are smooth walls separating the outside homogeneous earth's medium, with parameters ε_r and σ, from the inside. These two walls coincide with the surfaces $\rho = a \pm b$. For the dominant modes and with condition (6.1) satisfied, the wall can be treated as constant impedance surfaces with Z and Y given by (3.35) and (3.36). One can easily show that the modes are generally hybrid, i.e. having both e_z and h_z. However, when their fields do not vary with z, the modes reduce to either *TE* or *TM* to z modes. Some authors also use the terms *LSE* or *LSM* modes which stand for Longitudinal Section Electric, $e_z = 0$ or Longitudinal Section Magnetic, $h_z = 0$, modes respectively. In the following, we find the properties of these modes under the assumption of a large radius of curvature, i.e. $a \gg b$ and $k_0 a \gg \pi$. Starting with the *TE* or *LSE* modes, we find that the nonzero fields are h_z, e_ϕ and e_ρ. The magnetic field component h_z has the form:

$$h_z = f(k_0\rho)\,\exp(-i\nu\phi) \tag{6.18}$$

where the flat walls have been assumed to act as perfect magnetic conductors to allow the fields to be independent of z. One should expect ν to be close to $k_0 a$ for the dominant modes. So both ν and $k_0\rho$ within the tunnel are large numbers with a small difference between them. This allows the use of the Airy functions to express the radial dependence of the fields (see section 2.8), and hence (6.18) can be rewritten as:

$$h_z = [Aw_1(t) + Bw_2(t)]\,\exp(-i\nu\phi) \tag{6.19}$$

where

$$t=(k_0\rho/2)^{2/3}(\nu^2/k_0^2\rho^2-1) \tag{6.20}$$

The associated electric field component $e_\phi=(i/\omega\varepsilon_0)\ \partial h_z/\partial\rho$; hence

$$e_\phi=-i\eta_0(2/k_0\rho)^{1/3}[Aw_1'(t)+Bw_2'(t)]\ \exp(-iv\phi) \tag{6.21}$$

where the prime indicates differentiation w.r.t. argument, and the relation $\partial/\partial k_0\rho\approx(\partial/\partial t)\ (2/k_0\rho)^{1/3}$ has been invoked. The boundary conditions at $\rho=a\pm b$ are expressed by

$$Z=\pm e_\phi/h_z \tag{6.22}$$

Suitably normalised impendances may be defined by:

$$q_\pm=(iZ/\eta_0)\,(k_0(a\pm b)/2)^{1/3} \tag{6.23}$$

and the boundary conditions (6.22) now take the form

$$q_\pm=\pm\frac{Aw_1'(t_\pm)+Bw_2'(t_\pm)}{Aw_1(t_\pm)+Bw_2(t_\pm)} \tag{6.24}$$

where $t_\pm$ are given by (6.20) with $\rho=a\pm b$ respectively. By eliminating the constants A and B in these two equations, one gets the modal equation for ν as:

$$\begin{aligned}&[w_2'(t_+)-q_+w_2(t_+)]/[w_1'(t_+)-q_+w_1(t_+)]\\&\quad=[w_2'(t_-)+q_-w_2(t_-)]/[w_1'(t_-)+q_-w_1(t)]\end{aligned} \tag{6.25}$$

The TM_{0n} or LSM_{0n} modes can be treated in the same fashion to get an equation similar to (6.25), but with the parameter q redefined as:

$$q_\pm=(iY/\eta_0)\,(k_0(a\pm b)/2)^{1/3}$$

Solution of (6.25) yields the complex eigenvalues ν_n for the modes of the structure. An azimuthal phase constant β radians/m can be defined by $\beta=\text{Real }(\nu)/a$ and an attenuation rate by $\alpha=-\text{Imag}(\nu)/a$. These quantities have been computed by Mahmoud and Wait (1974*b*) for the first few low order modes and are plotted here versus the curvature parameter b/a in Figs. 6.6 and 6.7. The most important observation about these Figures is the substantial increase of attenuation rate with curvature. This increase is attributed to the tendency of the fields to cling towards the outer wall as the tunnel curvature increases. As a result, the modes change from fast to slow modes with a curvature increase as indicated by Fig. 6.6.

6.3 Wave propagation in tunnels containing an axial wire

Most subsurface tunnels contain longitudinal conductors such as pipes, tracks or electric power lines. In some cases, special types of transmission lines are strung along the tunnel in order to improve communication, as explained briefly in section 6.1. The presence of a longitudinal conductor inside the tunnels creates a coaxial like mode which has a zero cutoff frequency, and hence allows the propagation of low frequencies with low attenuation rates (e.g. Wait and Hill, 1974*a*, *b*; Mahmoud and Wait, 1974*a*; and Mahmoud, 1974). In the following, we outline the theory of mode propagation in a tunnel containing an axial wire. First a tunnel of a circular cross section is treated and is followed by one of a rectangular cross section.

The geometry is shown in Fig. 6.8*a* for a circular tunnel of radius a that contains an axial wire of small radius c_w and conductivity σ_w located at (ρ_0, ϕ_0) in the cylindrical coordinate frame shown. We look for the permitted modes in the tunnel. Starting with an axial current $I \exp(i\omega t - i\beta z)$ flowing in the wire, we can get the modal fields everywhere in the tunnel. Of course, these fields must satisfy the boundary conditions at the tunnel walls and at the wire surface.

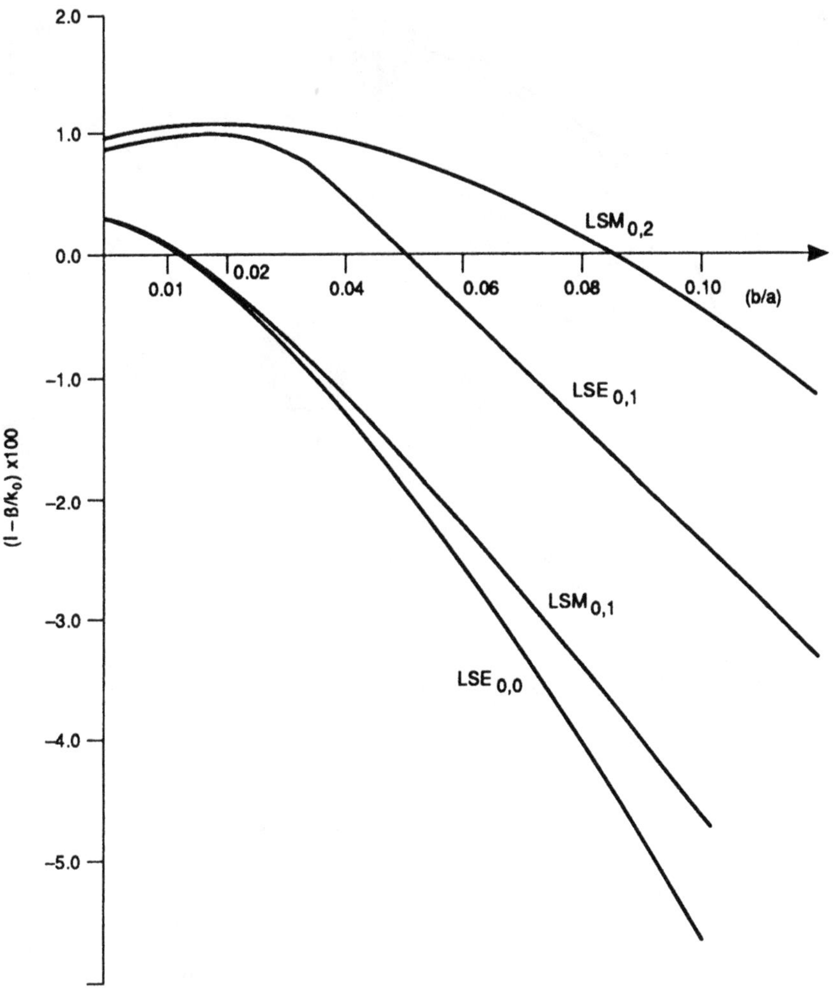

Fig 6.6 Normalised phase constants versus b/a in a curved tunnel model for $k_0 b = 22{\cdot}4$, $\varepsilon_r = 10$ and $\sigma/\omega\varepsilon_0\varepsilon_r = 0{\cdot}018$ (after Mahmoud and Wait, 1974*b*)

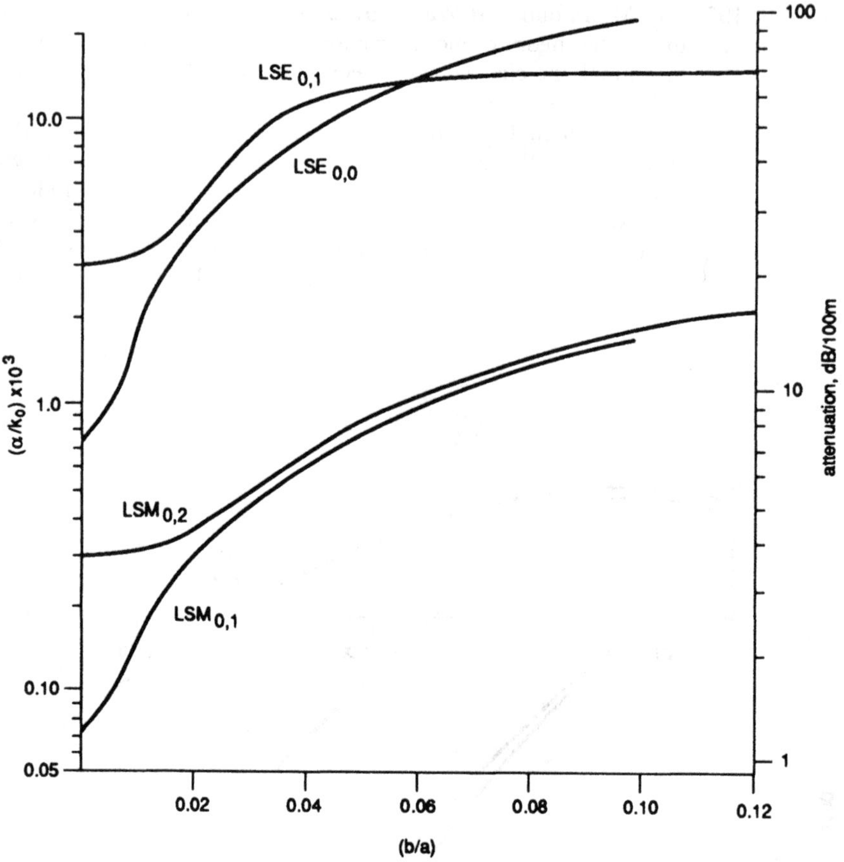

Fig 6.7 Normalised attenuation rate versus b/a in a curved tunnel model. Same parameters as in Fig. 6.6. The scale on the right hand side gives the attenuation in db/100 m at 1000 MHz when 2b = 2·133 m (after Mahmoud and Wait, 1974*b*)

Ignoring the azimuthal variation of fields on the wire by virtue of the smallness of its radius, the boundary condition on the wire becomes:

$$e_z(\rho_0, \phi_0)/I = Z_s \tag{6.26}$$

where Z is the series impedance of the wire, per unit length, which is given by Wait (1959): see also probl. 6.2:

$$Z_s \simeq (i\omega\mu_w/2\pi)\,[I_0(\gamma_w c_w)/\gamma_w c_w I_1(\gamma_w c_w)] \tag{6.27}$$

where $\gamma_w^2 = i\omega\mu_w(\sigma_w + i\mu\varepsilon_w)$. For a highly conducting wire $|\gamma_w c_w| \gg 1$ and Z_s tends to the more familiar form:

$$Z_s \simeq (\omega\mu_w/2\sigma_w)^{1/2}(1+i)/2\pi c_w \tag{6.28}$$

Application of the boundary condition (6.26) results in the modal equation for the mode propagation parameter β. Now, we concentrate attention on the derivation of the modal fields in terms of the axial travelling wave current I. The primary field e_{zp} produced by I, i.e. in the absence of the tunnel walls, is given (see section 1.5.3):

$$e_{zp} = -I(u^2/2\pi i\omega\varepsilon_0)K_0(u\rho_d)\exp(-i\beta z) \quad (6.29)$$

where $u = (\beta^2 - k_0^2)^{1/2}$, and ρ_d is the distance between the observation point (ρ, ϕ) and the wire axis at (ρ_0, ϕ_0) where the current is assumed to be concentrated. Using the addition theorem for the modified Bessel function (e.g. Harrington, 1961), (6.29) takes the form

$$e_{zp} = -I(u^2/2\pi i\omega\varepsilon_0)\sum_{m=-\infty}^{\infty} I_m(u\rho_<)K_m(u\rho_>)\exp(-im(\phi-\phi_0))$$
$$\exp(-i\beta z) \quad (6.30)$$

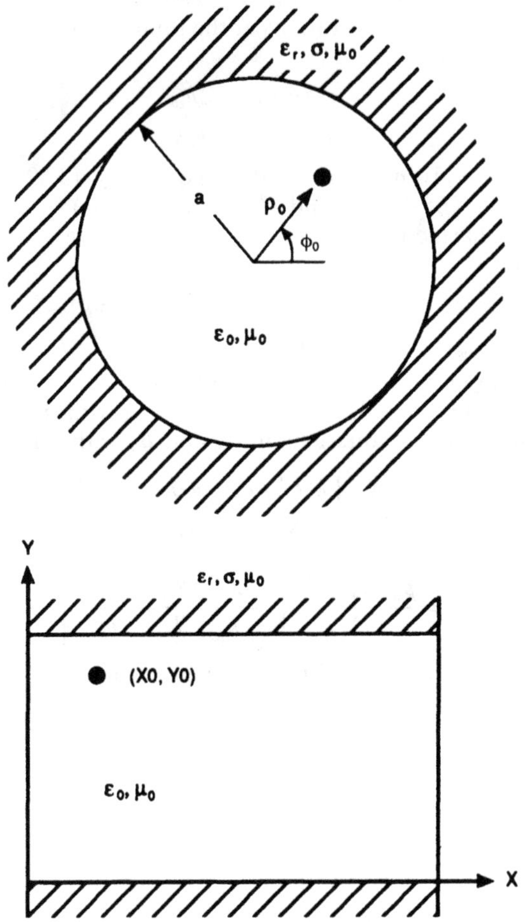

Fig 6.8 *a* A circular tunnel model containing an axial wire
b A rectangular tunnel model containing an axial wire. The side walls are assumed either perfect electric or perfect magnetic conductors

where $\rho_>$ and $\rho_<$ are the greater and the smaller of ρ and ρ_0. Now, the tunnel walls will produce a secondary field inside the tunnel. Because this field is source free, it has the appropriate form:

$$e_{zs} = -I(u^2/2i\pi\omega\epsilon_0)\sum_m A_m I_m(u\rho_0)I_m(u\rho)\exp(-im(\phi-\phi_0))\exp(-i\beta z) \quad (6.31)$$

Besides the reflected *TM* wave, a source free *TE* wave is also needed to satisfy the boundary conditions imposed by the tunnel walls. Thus, an h_{zs} exists and takes the form:

$$h_{zs} = I(u^2/2\pi\omega\varepsilon_0)\sum_m B_m I_m(u\rho_0)I_m(u\rho)\exp(-im(\phi-\phi_0))\exp(-i\beta z) \quad (6.32)$$

In the above two equations A_m and B_m are coefficients that need to be determined by the boundary conditions at the tunnel walls. These state that for every azimuthal harmonic m;

$$e_{\phi m} = p_m e_{zm} + Z_m h_{zm} \quad (6.33)$$

$$h_{\phi m} = -Y_m e_{zm} + p_m h_{zm} \quad (6.34)$$

where Z_m and Y_m are given by the RHS of equations. (3.32) and (3.33), and $p_m = m\beta/v^2a$, with $v=(\beta^2-k^2)^{1/2}$ and $k^2 = -i\omega\mu(\sigma+i\omega\varepsilon)$.

The azimuthal harmonics $e_{\phi m}$ and $h_{\phi m}$ at $\rho = a$ are also obtained from (1.27)—(1.28) by:

$$e_{\phi m} = (m\beta/u^2a)\, e_{zm} - (i\omega\mu/u^2)(\partial h_{zm}/\partial\rho)_{\rho=a} \quad (6.35)$$

$$h_{\phi m} = (i\omega\varepsilon_0/u^2)(\partial e_{zm}/\partial\rho)_{\rho=a} + (m\beta/u^2a)h_{zm} \quad (6.36)$$

Now we use the mth terms of the sums in (6.30)–(6.32) to express $e_{\phi m}$ and $h_{\phi m}$ of (6.35) and (6.36) in terms of A_m and B_m. Comparing with (6.33) and (6.34), we obtain two inhomogeneous equations in A_m and B_m with solution:

$$A_m = \frac{K_m}{\Delta_m I_m}\left[\frac{k_0^2 K'_m}{u^2 K_m}\left(\frac{I'_m}{I_m} + \frac{uZ_m}{ik_0\eta_0}\right) - q_m^2\right] \quad (6.37)$$

$$B_m = \frac{p_m(Ik_0/u)K_m}{\eta_0\Delta_m I_m}\left[-\frac{K'_m}{k_m} + \frac{I'_m}{I_m} + \frac{uY_m\eta_0}{ik_0}\right] \quad (6.38)$$

where the argument of I_m, K_m and their derivatives is ua,

$$\Delta_m = q_m^2 - (k_0^2/u^2)\left[\frac{I'_m}{I_m} + \frac{uZ_m}{ik_0\eta_0}\right]\left[\frac{I'_m}{I_m} + \frac{uY_m\eta_0}{ik_0}\right] \quad (6.39)$$

and $q_m = (m\beta/a)(u^{-2} - v^{-2})$

A β dependent external impedance for the wire can now be defined as follows:

$$Z_{ex} = [e_{zp}(\rho_d = c_w) + e_{zs}(\rho = \rho_0,\ \phi = \phi_0)]/I \quad (6.40)$$

On using (6.29) and (6.31), Z_{ex} becomes:

$$Z_{ex}(\beta) = -(u^2/2\pi i\omega\varepsilon_0)\left[K_0(uc_w) + \sum_{m=-\infty}^{\infty} A_m I_m^2(ua)\right] \quad (6.41)$$

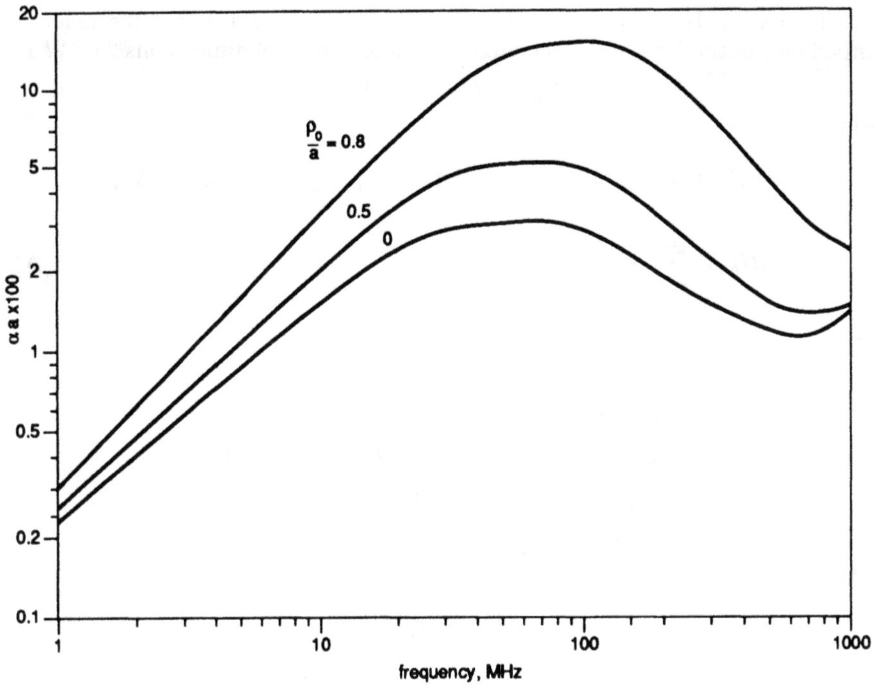

Fig 6.9 Normalised attenuation rate of the coaxial like mode in a circular tunnel model containing an axial wire (after Wait and Hill, 1974*a*)

and the boundary condition (6.26) at the wire surface now takes on the compact form:

$$Z_{ex}(\beta) = Z_s \tag{6.42}$$

This is the modal equation for the wire loaded tunnel. It has been derived and solved numerically for the coaxial like mode by Wait and Hill (1974*a*). In the quasistatic limit, where the free space wavelength is not small relative to the tunnel dimensions, the modal equation is greatly simplified and leads to an explicit formula for β as demonstrated by Wait (1977).

The normalised mode attenaution obtained by solving (6.42) is plotted versus frequency in the range 1 MHz—1 GHz in Fig 6.9. The following parameters are assumed: tunnel radius = 2 m, wire radius = 1 cm, wire conductivity = 10^5 mho/m, $\varepsilon_r = 10$ and σ of earth = 10^{-2} mho/m. It is noticed that the attenuation rate increases appreciably according to the proximity of the wire from the tunnel walls.

The rectangular geometry was treated by Mahmoud and Wait (1974*a*) and Mahmoud (1974). To make the analysis tractable, the side walls were assumed to behave as perfect electric or perfect magnetic conductors (Fig 6.8b). In spite of this simplification, the results have been indicative of the main modal characteristics. It is shown by Mahmoud and Wait (1974*a*) that the external impedance of the wire in the rectangular tunnel model of dimensions $2a \times 2b$ is:

$$Z_{ex} = -(2i\omega\mu/a)D(\beta) \tag{6.43}$$

where

$$D(\beta) = (1-\beta^2/k_0^2)\sum_m \sin^2(m\pi x_0/2a)\,[f_m(y_0)f_m(d-c_w)/u_m\Delta_m] +$$

$$(\beta/k_0)^2\sum_m \sin^2(m\pi x_0/2a)R_m^*[f_m(y_0)g_m(y_0+c_w)+f_m(d)g_m(d-c_w)]/\Delta_m\Delta_m^*$$

and

$$d = 2b - y_0$$
$$f_m(y) = (1/2)\,[\exp(u_m y) + R_{mh}\exp(-u_m y)]$$
$$g_m(y) = (1/2)\,[\exp(u_m y) - R_{me}\exp(-u_m y)]$$
$$\Delta_m = \exp(2u_m b)\,[1 - R_{mh}^2\exp(-4u_m b)]$$
$$\Delta_m^* = \exp(2u_m b)\,[1 - R_{me}^2\exp(-4u_m b)]$$
$$R_{mh} = (u_m - v_m)/(u_m + v_m)$$
$$R_{me} = [(k/k_0)^2 u_m - v_m]/[(k/k_0)^2 u_m + v_m]$$
$$R_m^* = 2i\beta[(u_m+v_m)^{-1} - ((k/k_0)^2 u_m + v_m)^{-1}]$$
$$u_m = (\beta^2 + (m\pi/a)^2 - k_0^2)^{1/2},\ v_m = (\beta^2 + (m\pi/a)^2 - k^2)^{1/2},\ Real(u_m, v_m) > 0.$$

R_{mh} and R_{me} are recognized as the specular reflection coefficients at the upper and lower walls for horizontally and vertically polarised waves, while R_m^* is the reflected *TM* to *y* wave due to a unit incident *TM* to *z* wave on any of the two walls.

Numerical data on the attenuation rate of the coaxial mode is plotted versus frequency in Fig. 6.10. The tunnel dimensions are given by $2a = 4b = 4{\cdot}266$ m. The wire is located at $x_0 = a/2$ while y_0 is allowed to vary. Other parameters are σ of earth $= 10^{-2}$ mho/m and $\varepsilon_r = 10$, and the wire conductivity $\sigma_w = 10^5$ mho/m. The results generally confirm those obtained by Wait and Hill (1974*a*) and shown in Fig. 6.9; namely, the attenuation increases monotonically with frequency below about 100 MHz, and increases with wire proximity to the tunnel wall.

A dipole antenna located inside the tunnel will generally excite the coaxial mode and all the natural tunnel modes which are perturbed by the wire. Levels of excitation of the coaxial mode and the first vertically polarised and horizontally polarised modes due to a $\lambda/2$ dipole have been calculated by Mahmoud (1974*a*) for a rectangular tunnel and by Hill and Wait (1974) for a circular one. It is generally concluded that the coaxial mode is dominant at frequencies below 100 MHz in tunnels of typical dimensions. Conversely the perturbed natural tunnel modes dominate over the coaxial mode above 1000 MHz.

Tunnels with elliptical cross sections containing an axial wire have been considered by Seidel and Wait (1979*a*) who conclude that the exact shape of the cross section has little influence on the dominant mode propagation constants if the cross sectional area is kept the same. This is also asserted by Keuster and Seidel (1979) who treat a tunnel with arbitrary cross sectional shape in the quasistatic regime.

6.4 Tunnels with leaky transmission lines

In this section we consider tunnels in which a leaky transmission line or cable is installed. By leaky cable, we mean one that has its modal field accessible to the outside medium. Actually, an axially uniform leaky cable does not radiate; it supports guided modes but with fields extending in the vicinity of the cable and hence accessible to outside antennas. A good discussion of this point is found in Beal *et al* (1973), and Gale and Beal (1980) in their discussion on continuous access guided communication systems using coaxial cables. Leaky lines studied in the literature include the two wire open line and several types of coaxial leaky cables. The fundamental mode on the two wire open line (the bifilar mode) has almost equal and opposite currents on the two lines; hence the field decay is fast

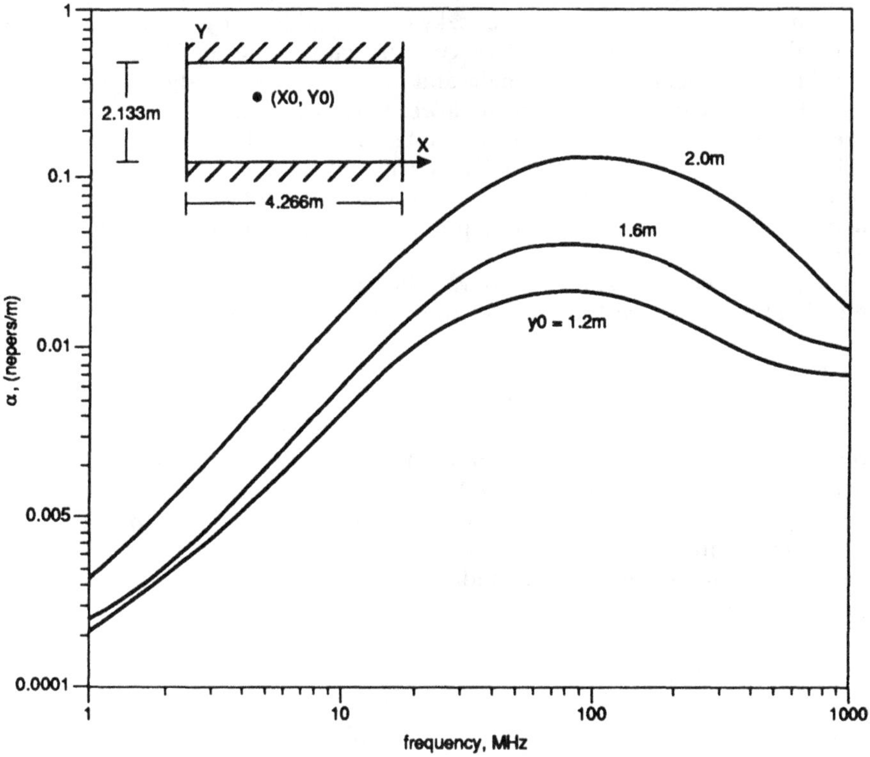

Fig 6.10 Normalised attenuation rate of the coaxial like mode in rectangular tunnel model containing an axial wire (after Mahmoud, 1974*a*)

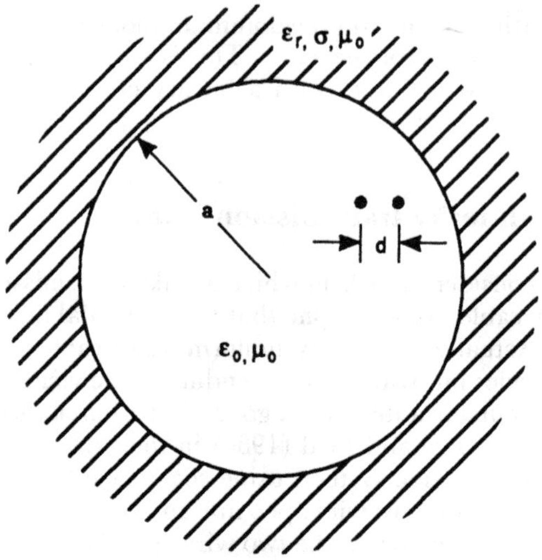

Fig 6.11 A two wire open line inside a tunnel

in the outer medium. The field accessibility outside this line is low if only the bifilar mode is present. Considering leaky coaxial cables, we note that, owing to their inherent unbalance, the difference in the currents in the inner conductor and the outer shield is not too small, and hence will cause rather high outside field. This is particularly true at the lower frequency end, say at $f \leqslant 30$ MHz, when a bifilar mode of a coaxial leaky cable can provide field access in a tunnel of 3 or 4 m width (Delogne, 1982). Thus, while the two wire line requires mode convertors to provide continuous access communication via the monofilar mode, the leaky coaxial cables can provide this by virtue of the bifilar mode alone.

In the following we study the mode characteristic of a two wire line in a tunnel. This is followed by a similar study for a leaky coaxial cable.

6.4.1 The two wire open line

We consider a tunnel which contains a two wire open line as shown in Fig. 6.11. Let the spacing d between the two wires be small relative to their separation from the tunnel wall. The currents in the two wires are $I_{1,2} \exp(i\omega t - i\beta z)$ for a mode with propagation parameter β. Without specialising our discussion to any particular tunnel shape, one can define an external impedance $Z_{ex}(\beta)$ for a wire as the net longitudinal electric field on its surface due to unit current flow in the wire (see equation 6.40). The boundary conditions at the surfaces of the two wires take the forms:

$$Z_{ex1}(\beta)I_1 + Z_m(\beta)I_2 = Z_{s1}I_1 \tag{6.44}$$

$$Z_{ex2}(\beta)I_2 + Z_m(\beta)I_1 = Z_{s2}I_2 \tag{6.45}$$

where $Z_m(\beta)$ is the mutual impedance between the two wires; that is the longitudinal electric field on the surface of one line due to an axial unit current in the other. Of course, we are assuming that the wire diameters are sufficiently small that aximuthal variation on any of the wires is neglected. The quantities $Z_{s1,2}$ are the series impedances of the wires per unit length. Now, if the two wires are identical and their separation d is small relative to their distances from the wall, we expect that $Z_{ex1} \simeq Z_{ex2} = Z_{ex}$ say and, of course, $Z_{s1} = Z_{s2} = Z_s$. Therefore (6.44) and (6.45) yield the two solutions:

$$Z_{ex} - Z_s \pm Z_m = 0 \tag{6.46}$$

with currents related by:

$$I_1/I_2 = \pm 1 \tag{6.47}$$

These two solutions describe the monofilar and bifilar modes respectively. Obviously if the difference between Z_{ex1} and Z_{ex2} cannot be ignored, the ratio of currents will deviate from unity (see probl. 6.3). Numerical results for the attenuation of the two modes in a rectangular tunnel model are shown in Fig. 6.12. The following physical parameters are assumed: $2a = 4$m, $2b = 3$ m, $x_0 = a/2$, $d = 0{\cdot}01a$, $c_w = 2$ mm, $\sigma_w = 10^6$ mho/m, $\varepsilon_r = 10$ and $\sigma_{\text{earth}} = 10^{-2}$ mho/m. In order to demonstrate the effect of proximity to the tunnel walls, the parameter b_0/b is varied between 0·6 and 0·9 and is seen to affect the monofilar mode, but it hardly affects the bifilar mode since the fields of the latter are confined to the near vicinity of the line. The frequency range from 20 to about 50 MHz is useful for continuous access communication. In this range, the bifilar mode provides low signal attentuation and the monofilar mode provides access of the field for antennas in the tunnel. The only disadvantage with the open two wire line is its exposure to the tunnel environment causing a possible increase of attenuation, particularly for the bifilar mode. Leaky coaxial cables are superior in this respect.

6.4.2 Leaky coaxial cables

A leaky coaxial cable is depicted in Fig. 6.13*a*. The inner conductor has a radius a and conductivity σ_w. The filling dielectric with permittivity ε has an outer radius b and is covered by an imperfect shield. For definiteness, we consider a thin metal braided shield of radius b. This is surrounded by a coating layer or a jacket of outer radius c and permittivity ε_c. This, in turn, can be covered by a thin lossy film of mine dust or conducting fluid. Now it is required to obtain the net surface impedance Z_s of the cable at $\rho = c$. To this end, we note that the surface impedance at $\rho = a$, say Z_a, is simply that of the bare inner conductor and is given by the form (6.27) or (6.28). The effect of the insulation of radius b has been shown by Wait and Hill (1975) to be equivalent to a series impedance $Z'(\beta)$ ohm/m given by (see probl. 6.4):

$$Z'(\beta) = -(k^2 - \beta^2)\,(\ln(b/a)/2\pi i\omega\varepsilon \tag{6.48}$$

under the condition of a thin cable; that is $kb \ll 1$, $k = \omega(\mu\varepsilon)^{1/2}$. Similarly, the insulating jacket acts as a series impedance $Z_c(\beta)$, where:

$$Z_c(\beta) = -(k_c^2 - \beta^2)\ln(c/b)/2\pi i\omega\varepsilon_c \tag{6.49}$$

A thin uniform braided shield can be modelled by a transfer impedance Z_T defined by:

$$E_z(b) = 2\pi b Z_T(H_\phi(b+) - H_\phi(b-)] \tag{6.50}$$

Z_T has been estimated by a few authors (Fontaine *et al.*, 1973 and Delogne, 1982) and is expected to behave as an inductance with value depending on the braid holes, that is, $Z_T = i\omega L$, where L ranges from 1 to 100 nH/m. The larger L, the more leaky is the cable. Of course if $L = 0$, the shield acts as a short circuit and becomes a perfect shield. Finally the lossy film of dust or conducting fluid on the jacket behaves as a shunt impedance equal to:

$$Z_f = (2\pi c\sigma d)^{-1} \tag{6.51}$$

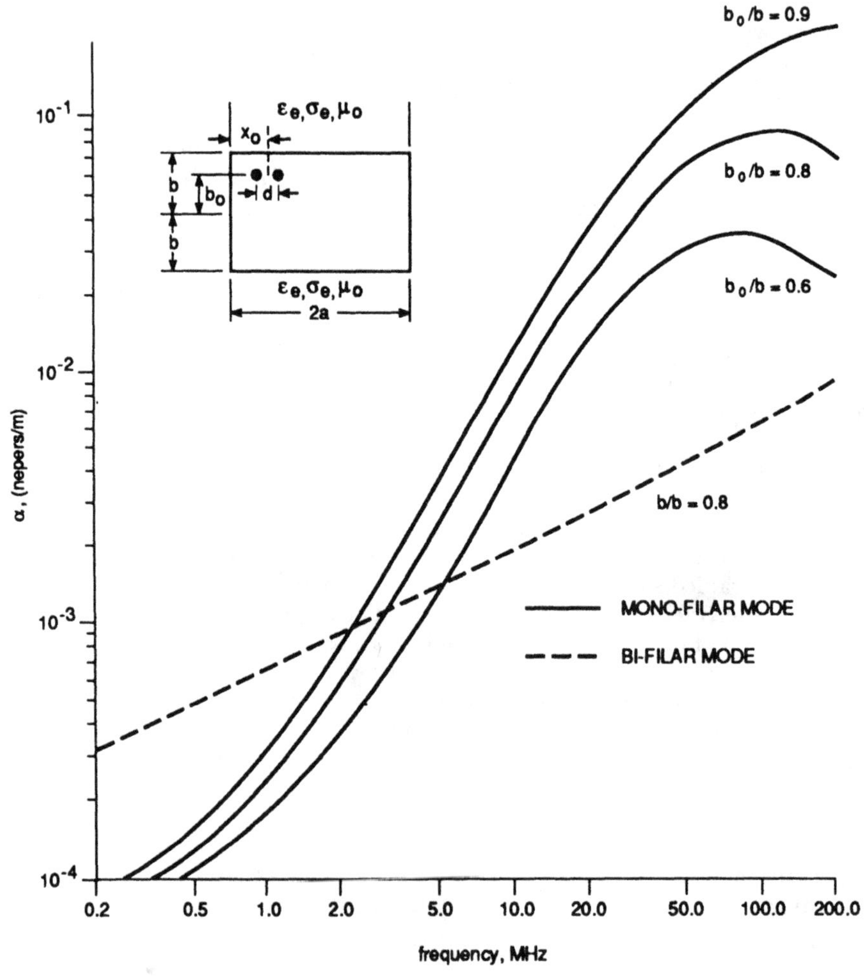

Fig 6.12 Attenuation rates of the monofilar (——) and bifilar (– – –) modes in a rectangular tunnel model containing a two wire open line (after Mahmoud, 1974*b*)

b_0/b is a parameter

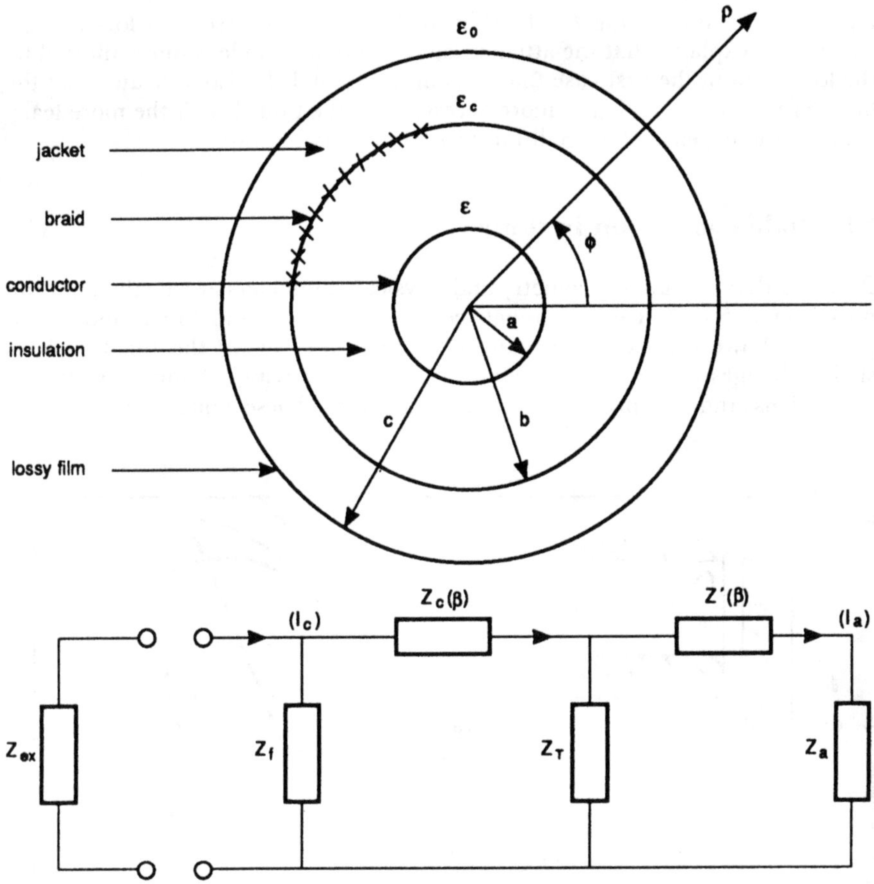

Fig 6.13 *a* A braided coaxial cable with an insulating jacket and an outer lossy film. *b* An equivalent circuit for the braided coaxial cable

where σd is the conductivity–thickness product of the film. The electrically thin cable can thus be represented by an equivalent circuit as shown in Fig. 6.13*b*. The effective series impedance at the outer surface of the cable can be calculated and inserted in (6.46) to determine the propagation parameters of the modes of propagation. Besides the monofilar and bifilar modes, Seidel and Wait (1978*a*) have found a third mode which they called the jacket mode. In this mode, the current flows primarily in the braided shield and returns in the lossy film. The jacket mode has an attenuation rate that is several orders of magnitude higher than the monofilar or bifilar modes; hence it has no effect on communication in the tunnel except for increasing the total losses.

Representative results for the attenuation of the monofilar and bifilar modes in a rectangular tunnel model for two values of the braid transfer inductance are shown in Fig. 6.14*a*, *b*. The dotted curves acccount for a lossy film with the indicated values of σd. The value $L = 40$ nH/m in Fig. 6.14*a* represents a rather highly leaky cable, while $L = 4$ nH/m in Fig. 6.14*b* represents a low leakage cable. This explains that the attenuation of the bifilar mode is more affected by the lossy film in the first case than it is in the second. It also indicates that the fields of the bifilar mode are more accessible in the tunnel with the more leaky cable at the expense of a small increase in attenuation rate.

6.5 Mode conversion in tunnels

Provided that the tunnel geometry and physical parameters are axially uniform, no power exchange can occur between the monofilar and bifilar modes. In practice, however, there are many nonuniformities along the tunel such as slight changes in rock conductivity and permittivity, changes in tunnel dimensions and the presence of wall roughness. These nonuniformities will

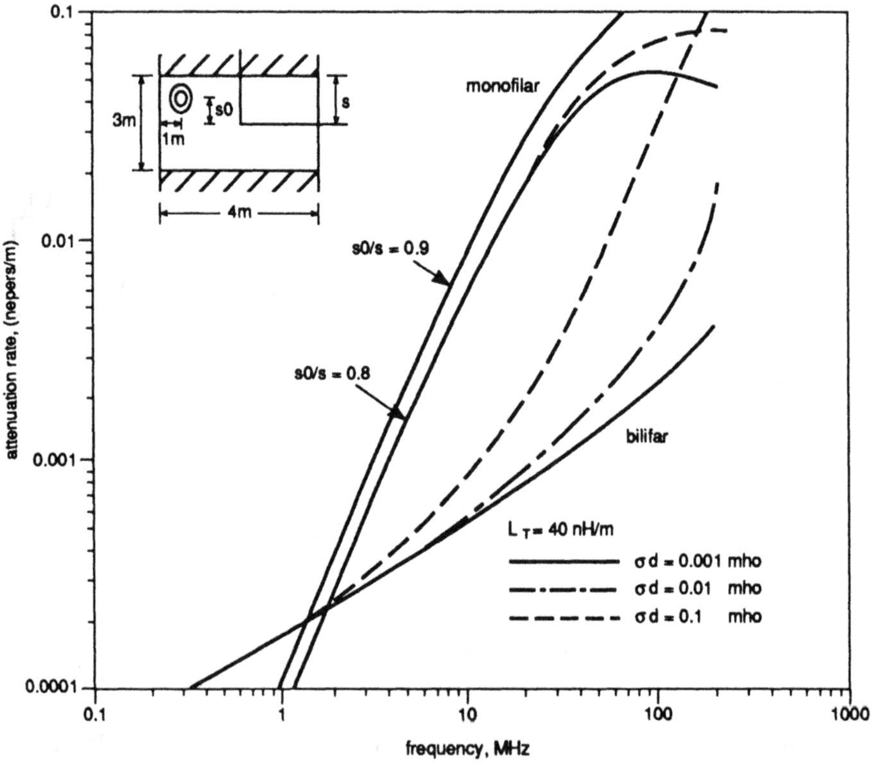

Fig 6.14 Attenuation rates of monofilar and bifilar modes in a rectangular tunnel model containing a braided coaxial cable (after Mahmoud and Wait, 1976)

The cable parameters are: $a_w = 1{\cdot}5$ mm, $b = 10$ mm, $c = 11{\cdot}5$ mm, $\sigma_w = 5{\cdot}7 \times 10^7$ mho/m, $\varepsilon/\varepsilon_0 = 2{\cdot}5$ and $\varepsilon_c/\varepsilon_0 = 3{\cdot}0$ and s_0/s is a parameter. The braide transfer inductance $L_T = 40$ nHm. An outer lossy film of indicated σd is assumed.

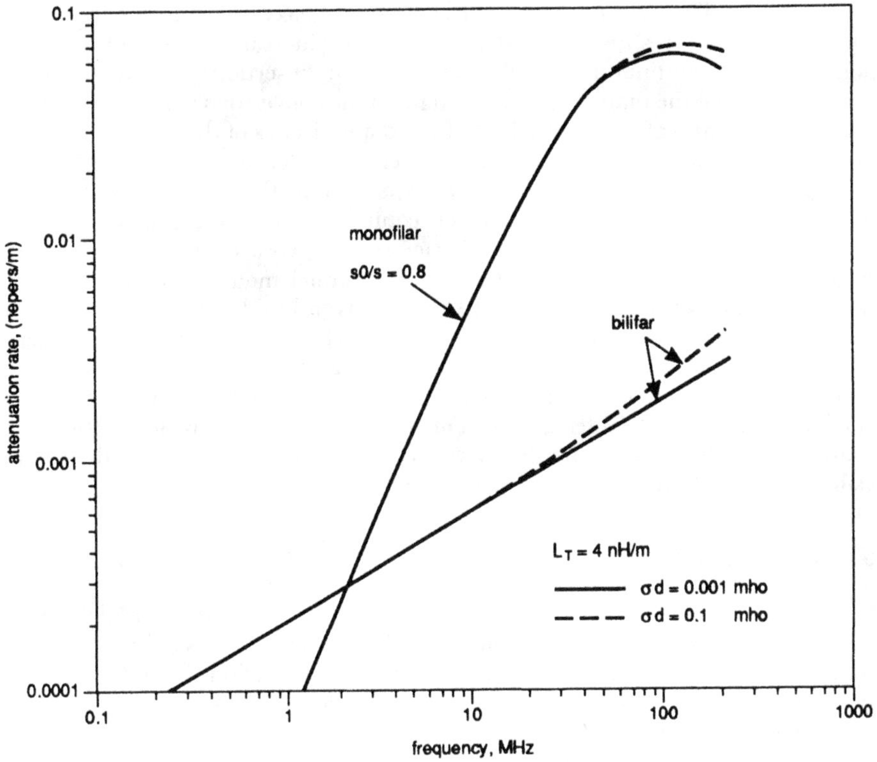

Fig 6.14b As Fig. 6.14*a* except that $L_T = 4$ nH/m (after Mahmoud and Wait, 1976)

cause continuous mode conversion among the monofilar, bifilar, as well as the free tunel modes. The useful communication range depends to a great extent on such a mode conversion mechanism. A leaky cable with reasonable leakage can often provide a higher range of continuous access communication than would be expected from the same cable in a perfectly uniform tunnel. One can also depend on controlled mode conversion by installing specially designed mode convertors at discrete positions along a well shielded cable. In the successful Delogne–INIEX scheme in the Belgian mines, a conventional coaxial line is used and complete annular slots in the perfect shield act as mode convertors. An analysis of the performance of this system is found in Delogne and Laloux (1980) and Hill and Wait (1975). In another system a mode convertor is made of a section of leaky cable inserted along the main shielded cable. In a two wire line system, a mode convertor is a circuit which breaks the balance of currents in the two conductors (Deryck, 1975). In all these systems, matching of mode convertors to the main line is an important consideration (Delogne, 1982).

Analytical and numerical study of controlled mode conversion for a leaky cable in a tunnel model has been presented by Seidel and Wait (1978*b*). The main cable is taken as a braided cable with a low transfer inductance of

$L_T = 1\,\text{nH/m}$. This is interrupted by a more leaky section having $L_T = 100\,\text{nH/m}$. Their results demonstrate clearly the increase in the communication range of continuous signal access by the insertion of several mode convertors along the main cable. The length of the convertor section is chosen to maximise the ratio of transmitted to reflected amplitudes of the monofilar mode due to an incident bifilar mode. The criterion governing the choice of interspacing between the mode convertors is to limit the interference between the reflected monofilar mode from one connector and the same transmitted mode from the previous convertor. Other related work by the same authors study the inadvertent mode conversion in a tunnel model caused by lateral nonuniformities (Seidel and Wait, 1979*b*). Thus, a braided coaxial cable with $L_T = 40\,\text{nH/m}$ is considered to exist in a tunel which is axially uniform except for a small section with different ε_e or σ_e. It is demonstrated that the resulting mode conversion can be the dominant mechanism for communication in systems with loosely braided cables. On the other hand, if it is desired to have controlled mode conversion, this inadvertent conversion can be reduced by using cables with high dielectric permittivity.

6.6 Problems

6.1 Compare the attenuation rates of the dominant mode in a circular and a rectangular tunnel having the following dimensions: radius of the circular tunnel = 2·5 m and the rectangular tunnel cross section is 5 × 3 m. The surrounding earth has $\varepsilon_r = 6{\cdot}0$ and $\sigma = 10^{-2}$ mho/m. Consider the range of frequencies 50—200 MHz.

6.2 Consider a bare coaxial wire of radius c_w and conductivity $\sigma_w \gg \omega\varepsilon_0$. Show that the surface series impedance Z_s, defined as e_z/I_z, is given by 6.27, and for $|\gamma_w c_w| \gg 1$, by (6.28).

6.3 Consider a two wire open line with identical wires which is located close to the walls of a tunnel. As a result, the bifilar mode is slightly unbalanced. So, let $Z_{ex1,2} = Z_{ex} \pm \delta$, with $\delta \ll Z_{ex}$. Using (6.44)—(6.45) show that the ratio of currents is given by:

$$I_2/I_1 \simeq -(1 + \delta/(Z_{ex} - Z_s))$$

where Z_s is the series impedance of either wire.

6.4 Consider a dielectric coated wire of inner and outer radii a and b. Show that, in the quasistatic limit, the surface series impedance at $\rho = b$ is:

$$Z_s(b) = Z_s(a) + Z'(\beta)$$

where $Z'(\beta)$ is given by (6.48).

6.5 With reference to the equivalent circuit (6.13*b*), show that for a low leakage cable, i.e., Z_T is sufficiently less than both Z_{ex} and $Z(\beta)$, the modal equations for the monofilar and bifilar modes can be approximated by the forms:

$$Z_{ex}(\beta) = Z_f(Z_c(\beta) + Z_T)/[Z_f + Z_c(\beta) + Z_T]$$

$$Z_a + Z'(\beta) + Z_T = 0$$

respectively. For the bifilar mode the latter equation results in the following explicit expression for β:

$$\beta^2 = k^2 - 2\pi i\omega\varepsilon(Z_a + Z_T)/ln(b/a)$$

6.6 Find an expression for the ratio I_c/I_a in the circuit of Fig. 6.13*b*. Show that under the approximate modal equation derived in probl. 6.5 for the bifilar mode, $I_c/I_a \simeq -Z_T/Z_f$. Note that I_c represents the unbalance of currents for this mode.

6.7 References

BEAL, J.C., JOSIAK, J., MAHMOUD, S.F. and RAWAT, V. (1973): Continuous-access guided communication (CAGC) for ground transportation systems', *Proc. IEEE* (Special issue on ground transportation) **61**, pp. 562–568

DELOGNE, P. (1972): 'Systeme INIEX ???gne de telecommunications et de telecommande par radio', *Ann. Mines de Belg.*, (ii), pp. 1061–1081

DELOGNE, P. (1974): 'The INIEX mine communications systems', Intern. Conference Radio, Roads, Tunnels and Mines, Leige, Belgium 1–5 April

DELOGNE, P. (1976): 'Basic mechanisms of tunnel propagation', *Radio Science,* **11**, pp. 295–303

DELOGNE, P., and LALOUX, A. (1981): 'Theory of the slotted cable', *IEEE Trans.* **MTT–28**, pp. 1102–1107

DELOGNE, P. (1982): 'Leaky fears and subsurface radio communications'. IEE Elecromagnetic Waves Series 14 (Peter Peregrinus Ltd.)

DERYCK, L. (1975): 'Control of mode conversions on bifilar line in tunnels', *Radio and Electron Eng.*, **45**, pp. 241–247

EMSLIE, A.G., LAGACE, R.L., and STRONG, P.F. (1975): Theory of propagation of UHF radio waves in coal mine tunnels', *IEEE Trans.* **AP–23**, pp. 192–205

FONTINE, J., DeMOULIN, B., DeGAUQUE, P., and GABILLARD, R. (1973): 'Feasibility of radio communication in mine galleries by means of a coaxial cable having a high coupling impedance'. Proc. Through the earth Electromagnetics workshop, Colorado School of Mines, US Bureau of Mines Contract G-133023. Final report, pp. 130–139

GABILLARD, R., DeGAUQUE, P., and WAIT, J.R. (1971): 'Subsurface electromagnetic telecommunications: A review', *IEEE Trans.* **COM–19**, pp. 1217–1228

GALE, D.J., and BEAL, J.C. (1980): 'Comparative testing of leaky coaxial cables for communications and guided radar', *IEEE Trans.*, **MIT–28**, pp. 1006–1013

HILL, D.A., and WAIT, J.R. (1974): 'Examination of monofilar and bifilar modes on a transmission line in a circular tunnel', *J. Applied Physics*, **45**, pp. 3402–3406

HILL, D.A., and WAIT, J.R. (1975): 'Electromagnetic fields of a coaxial cable with an interrupted shield located in a circular tunnel', *J. Appl. Phys.* **46**, pp. 4352–4356

KEUSTER, E.F., and SEIDEL, D.B. (1979): 'Low fequency behaviour of the propagation constant along a wire in an arbitrarily shaped tunnel', *IEEE Trans.*, **MTT–27**, pp. 736–741

MAHMOUD, S.F. (1947*a*): 'Characteristics of electromagnetic guided waves for communication in coal mine tunnels', *IEEE Trans.* **COM–22**, pp. 1547–1554

MAHMOUD, S.F. (1974*b*): 'On the attenuation of monofilar and bifilar modes in mine tunnels', *IEEE Trans.*, **MTT–22**, pp. 845–847

MAHMOUD, S.F., and WAIT, J.R. (19??): 'Theory of wave propagation along a thin wire inside a rectangular waveguide', *Radio Science*, pp. 417–420

MAHMOUD, S.F., and WAIT, J.R. (1977*b*): 'Guided electromagnetic waves in a curved rectangular mine tunnel', *Radio Science*, pp. 567–572

MAHMOUD, S.F., and WAIT, J.R. (1976): 'Calculated channel characteristics of a braided coaxial cable in a mine tunnel', *IEEE Trans.*, **COM–24**, pp. 82–87

SEIDEL, D.B., and WAIT, J.R. (1978*a*): 'Transmission modes in a braided coaxial cable and coupling to a tunnel environment', *IEEE Trans.*, **MTT–26**, pp. 494–499

SEIDEL, D.B., and WAIT, J.R. (1978*b*): 'Price of controlled mode conversion in leaky feeder mine communication systems', *IEEE Trans.* **AP–26**, pp. 690–694

SEIDEL, D.B., and WAIT, J.R. (1971*a*): 'Radio transmission in an elliptical tunnel with a contained axial conductor', *J. Appl. Phys.*, **50**, pp. 602–605

SEIDEL, D.B., and WAIT, J.R. (1979*a*): 'Mode conversion by tunnel nonuniformities in leaky feeder communication systems', *IEEE Trans.*, **AP–27**, pp. 560–563

WAIT, J.R. (1959): 'Electromagnetic radiation from cylindrical structures' (Pergamon Press, New York). Corrected reprint edition (Peter Peregrinus Ltd., 1988)

WAIT, J.R. (1971): 'On radio propagation through earth' *IEEE Trans.*, **AP–19**, pp. 796–798

WAIT, J.R. (1977): 'Quasistatic limit for the propagation along a thin wire in a circular tunnel', *IEEE Trans.*, **AP-25**, pp. 248–253

WAIT, J.R., and HILL, D.A. (1974*a*): 'Guided electromagnetic waves along an axial conductor in a circular tunnel', *IEEE Trans.*, **AP–22**, pp. 627, 630

WAIT, J.R., and HILL, D.A. (1974*b*): 'Coaxial and bifilar modes on a transmission line in a circular tunnel', *Appl. Phys.*, **4**, pp. 307–312

WAIT, J.R., and HILL, D.A. (1975): Propagation along a braided coaxial cable in a circular tunnel', *IEEE Trans.*, **MTT–23**, pp. 401–405

Appendix 1

Generalised orthogonal coordinate systems

A1.1 Metric coefficients

Consider an orthogonal coordinate system (u, v, w). Unit vectors along the u, v and w axes are denoted by $\hat{u}$, $\hat{v}$ and $\hat{w}$ and they form a mutually orthogonal set of basis vectors. Taken in pairs, the cross products of these basis vectors are given by:

$$\hat{u} \times \hat{v} = \hat{w}, \; \hat{v} \times \hat{w} = u \text{ and } \hat{w} \times \hat{u} = \hat{v}$$

Differential elements of length along the basis vectors are given by h_1 du, h_2 dv, and h_3 dw where h_1, i = 1, 2, 3 are the metric coefficients. They are given by:

$$(h_1 \, du)^2 = (\partial x)^2 + (\partial y)^2 + (\partial z)^2 \big|_{v, \, w = \text{constant}}$$

i.e.

$$h_1 = \left[\left(\frac{\partial x}{\partial u} \right)^2 + \left(\frac{\partial y}{\partial u} \right)^2 + \left(\frac{\partial z}{\partial u} \right)^2 \right]^{1/2} \tag{A1.1}$$

with similar expressions for h_2 and h_3. In the above x, y and z are the cartesian rectangular coordinates and each is generally a function of u, v and w. The metric coefficients are also functions of u, v and w in the general case. In the special cases of cylindrical coordinate systems, $w = z$ is the axial coordinate with $\hat{z}$ being a constant unit vector. Therefore $h_3 = 1$ and both h_1 and h_2 are independent of z.

A1.2 Differential operators

Let A be a vector function of u, v and w. Its components along the three basis vectors are A_u, A_v and A_w. The divergence of A at a point is defined as the flux of A over a closed surface divided by its enclosed volume as the latter collapses to the point of interest. The divergence operator is denoted by $\nabla.$ and is given by

$$\nabla.\mathrm{A}=\frac{1}{h_1h_2h_3}\left(\frac{\partial}{\partial u}(h_2h_3A_u)+\frac{\partial}{\partial v}(h_3h_1A_v)\right.$$
$$\left.+\frac{\partial}{\partial w}(h_1h_2A_w)\right) \tag{A1.2}$$

Obviously $\nabla.\mathrm{A}$ is a scalar quantity.

The curl of a vector A is another vector whose u component is equal to the counterclockwise integral around a closed loop orthogonal to $\hat{u}$, divided by the area of the loop as this goes to zero. Similar definitions apply to the v and w components. Denoting the curl operator by $\nabla\times$

$$\nabla\times\mathrm{A}=\frac{1}{h_1h_2h_3}\begin{vmatrix} h_1\hat{u} & h_2\hat{v} & h_3\hat{w} \\ \partial/\partial u & \partial/\partial v & \partial/\partial w \\ h_1A_u & h_2A_v & h_3A_w \end{vmatrix} \tag{A1.3}$$

Let ψ be a scalar function of u, v and w. The gradient of ψ is a vector whose direction is that of maximum rate of increase of ψ, and whose magnitude is equal to that maximum rate of increase. Denoting gradient of ψ by $\nabla\psi$, then

$$\nabla\psi=(\partial\psi/h_1\partial u)\hat{u}+(\partial\psi/h_2\partial v)\hat{v}+(\partial\psi/h_3\partial w)\hat{w} \tag{A1.4}$$

Finally the Laplacian of a scaler ψ is defined by

$$\nabla^2\psi=\nabla.\nabla\psi=\frac{1}{h_1h_2h_3}\left(\frac{\partial}{\partial u}\left(\frac{h_2h_3}{h_1}\frac{\partial\psi}{\partial u}\right)+\frac{\partial}{\partial v}\left(\frac{h_3h_1}{h_2}\frac{\partial\psi}{\partial v}\right)\right.$$
$$\left.+\frac{\partial}{\partial w}\left(\frac{h_1h_2}{h_3}\frac{\partial\psi}{\partial w}\right)\right) \tag{A1.5}$$

For example in circular cylindrical coordinates (ρ, ϕ, z), $h_1=1$, $h_2=\rho$ and $h_3=1$. The Laplacian is given by:

$$\nabla^2\psi=\frac{1}{\rho}\frac{\partial}{\partial\rho}\left(\rho\frac{\partial\psi}{\partial\rho}\right)+\frac{1}{\rho^2}\frac{\partial^2\psi}{\partial\phi^2}+\frac{\partial^2\psi}{\partial z^2} \tag{A1.6}$$

In elliptical cylindrical coordinates (u, v, z), $h_1=h_2=h$ and $h_3=1$, and the Laplacian is:

$$\nabla^2\psi=\frac{1}{h^2}\left(\frac{\partial^2\psi}{\partial u^2}+\frac{\partial^2\psi}{\partial v^2}\right)+\frac{\partial^2\psi}{\partial z^2} \tag{A1.7}$$

where $h=c(\cosh^2 u-\cos^2 v)^{1/2}$; c is equal to half the distance between the two focii.

A1.3 Vector identities

$$A_1.A_2 \times A_3 = A_2.A_3 \times A_1 = A_3.A_1 \times A_2 \tag{A1.8}$$

$$A_1 \times (A_2 \times A_3) = (A_1.A_3)A_2 - (A_1.A_2)A_3 \tag{A1.9}$$

$$\nabla(\psi_1\psi_2) = \psi_1\nabla\psi_2 + \psi_2\nabla\psi_1 \tag{A1.10}$$

$$\nabla.(\psi A) = \psi\nabla.A + A.\nabla\psi \tag{A1.11}$$

$$\nabla.(A_1 \times A_2) = A_2.\nabla \times A_1 - A_1.\nabla \times A_2 \tag{A1.12}$$

$$\nabla \times \psi A = \psi\nabla \times A - A \times \nabla\psi \tag{A1.13}$$

$$\nabla \times \nabla\psi = 0 \tag{A1.14}$$

$$\nabla.(\nabla \times A) = 0 \tag{A1.15}$$

$$\nabla \times \nabla \times A = \nabla(\nabla.A) - \nabla^2 A \tag{A1.16}$$

where

$$\nabla^2 A = \nabla^2 A_x \hat{x} + \nabla^2 A_y \hat{y} + \nabla^2 A_z \hat{z} \tag{A1.17}$$

A1.4 Divergence and Stoke's theorems

The divergence theorem relates the integration of the divergence of a vector field A over a volume to the flux of the field out of a closed surface enclosing that volume. Symbolically it states that

$$\int_{\text{vol}} \nabla.A \, dV = \int_s A.\hat{n} dS \tag{A1.18}$$

where n̂ is the unit outward normal to the closed surface S.

Stoke's theorem relates the integration of the curl of a vector field over an open surface to the line integration of the vector around a closed line prescribing the edge of the surface. Symbolically Stoke's theorem states.

$$\int_s \nabla \times A.\hat{n} \, dS = \oint_C A.\hat{t} dC \tag{A1.19}$$

where n̂ is a unit outward normal to the surface S and $\hat{t}$ is a unit tangential vector along the closed line C taken in the anticlockwise direction.

Appendix 2

Cylindrical wave functions

In this appendix, we review some of the basic properties of Bessel functions, Mathieu functions and Airy functions. These functions occur as wave functions in circular and elliptical cylindrical coordinates. For more comprehensive coverage, the reader is referred to mathematical books such as Abramowitz and Stegun (1965), Erdelyi *et al.* (1953) and Mclachlan (1955).

A.2.1 Bessel functions

The Bessel differential equation of order ν is:

$$z\frac{d}{dz}(z\,dR/dz)+(z^2-\nu^2)R=0 \tag{A2.1}$$

Solutions of this equation are $J_\nu(z)$ and $J_{-\nu}(z)$, where $J_\nu(z)$ is given by the series expansion

$$J_\nu(z)=\sum_{n=0}^{\infty}(-1)^n(z/2)^{\nu+2m}/\{m!(\nu+m)!\} \tag{A2.2}$$

In case $\nu=$ an integer n, $J_{-n}(.)=(-)^nJ_n(.)$ and a second solution must be defined; this is $N_n(z)$ defined by

$$N_n(z)=\lim_{\nu\to n}(J_\nu(x)\cos\nu\pi-J_{-\nu}(x))/\sin\nu\pi \tag{A2.3}$$

Small and large argument approximations for $J_\nu(z)$ and $N_n(z)$ are given below:
For $z\ll 1$

$$J_0(z)\simeq 1-(z/2)^2 \tag{A2.4}$$

$$N_0(z)\simeq(2/\pi)\ln(0{\cdot}891z)J_0(z) \tag{A2.5}$$

$$J_\nu(z)\simeq\frac{(z/2)^\nu}{\Gamma(\nu+1)}\left(1-\frac{z^2}{4(\nu+1)}\right) \tag{A2.6}$$

$$N_n(z)\simeq-(1/\pi)(2/z)^n,\ n=1,2,3\ldots \tag{A2.7}$$

For $z\gg\nu$, and with $x=z-(\nu+1/2)\pi/2$ and $\mu=4\nu^2$:

$$J_\nu(z) \simeq (2/\pi z)^{1/2}\left[\cos x - \sin x \frac{\mu-1}{8z}\right] \tag{A2.8}$$

$$N_n(z) \simeq (2/\pi z)^{1/2}\left[\sin x + \cos x \frac{\mu-1}{8z}\right] \tag{A2.9}$$

The derivatives with respect to z are given by:

$$J_\nu'(z) \simeq (2/\pi z)^{1/2}\left(-\sin x - \cos x \frac{\mu+3}{8z}\right) \tag{A2.10}$$

$$N_n'(z) \simeq (2/\pi z)^{1/2}\left(\cos x - \sin x \frac{\mu+3}{8z}\right) \tag{A2.11}$$

The Wronskian of J_ν and N_n is given by:

$$J_\nu(z)N_\nu'(z) - J_\nu'(z)N_\nu(z) = 2/\pi z \tag{A2.12}$$

The Hankel functions $H_\nu^{(1)}(z)$ and $H_\nu^{(2)}(z)$ are linear combinations of J_ν and N_ν and defined by:

$$H_\nu^{(1),(2)}(z) = J_\nu(z) \pm iN_\nu(z) \tag{A2.13}$$

The Hankel functions represent travelling waves for real z. Large argument approximations of the Hankel functions follow from (A2.8)—(A2.13); Namely:

$$H_\nu^{(1),(2)}(z) \simeq (2/\pi z)^{1/2} e^{\pm ix}\left[1 \pm i\frac{\mu-1}{8z} - \frac{(\mu-1)(\mu-9)}{2!(8z)^2}\right] \tag{A2.14}$$

$$H_\nu'^{(1),(2)}(z) \simeq \pm i(2/\pi z)^{1/2} e^{\pm ix}\left[1 \pm i\frac{\mu+3}{8z} - \frac{(\mu-1)(\mu+15)}{2!(8z)^2}\right] \tag{A2.15}$$

In many cases the logarithmic derivative function defined below is of interest:

$$\begin{aligned} h_\nu(z) &= i(\mathrm{d}/\mathrm{d}z)\ \ln H_\nu^{(2)}(z) = iH_\nu'^{(2)}(z)/H_\nu^{(2)}(z) \\ &\simeq 1 - i/(2z) - (\mu-1)/(8z^2) \end{aligned} \tag{A2.16}$$

For $\nu = 1/2$, $\mu - 1 = 0$; hence

$$H_{1/2}^{(1),(2)}(z) = (2/\pi z)^{1/2} e^{\pm i(z-\pi/2)} \tag{A2.17}$$

which is *exact* !. Exact formulas for $\nu = n + 1/2$, $n = 1, 2 \ldots$ can be obtained by applying the following recurrence relations

Recurrence relations

Recurrence relations for all functions satisfying (A2.1) are given below. The $B_\nu(z)$ stands for either J_ν, N_ν, $N_\nu^{(1)}$ or $H_\nu^{(2)}$:

$$\begin{aligned} B_\nu'(z) &= (\nu/z)B_\nu(z) - B_{\nu+1}(z) \\ &= [B_{\nu-1}(z) - B_{\nu+1}(z)]/2 \end{aligned} \tag{A2.18}$$

$$(2\nu/z)B_\nu(z) = B_{\nu-1}(z) + B_{\nu+1}(z) \tag{A2.19}$$

Assymptotic relations for large orders (Debye's expansions)

When ν is large, tending to ∞ through positive values, and $\alpha > 0$,

$$J_\nu(\nu \operatorname{sech} \alpha) \simeq e^{-\nu(\alpha - \tanh \alpha)} / (2\pi\nu \tanh \alpha)^{1/2} \tag{A2.20}$$

$$N_\nu(\nu \operatorname{sech} \alpha) \simeq -e^{\nu(\alpha - \tanh \alpha)} / (\tfrac{1}{2}\pi\nu \tanh \alpha)^{1/2} \tag{A2.21}$$

The logarithmic derivative function is defined by:

$$\begin{aligned} h_\nu(z) &= i(\mathrm{d}/\mathrm{d}z) \ln H_\nu^{(2)}(z) = iH_\nu'^{(2)}(z)/H_\nu^{(2)}(z) \\ &\simeq -i \sinh(\alpha)\,[1 + i \exp(-2\nu(\alpha - \tanh \alpha))] \end{aligned} \tag{A2.22}$$

where $\nu \gg 1$ and

$$\sinh(\alpha) = (\nu^2/z^2 - 1)^{1/2}$$

In case $\alpha \ll 1$, and $\nu\alpha^3/3 \gg 1$, h_ν takes the form (Wait, 1967):

$$h_\nu(z) \simeq -i(\nu^2/z^2 - 1)^{1/2}(1 + i \exp(-2\nu\alpha^3/3)) \tag{A2.23}$$

Modified Bessel functions

The modified Bessel functions $I_\nu(z)$ and $K_\nu(z)$ are defined by:

$$I_\nu(z) = i^\nu J_\nu(-iz) \tag{A2.24}$$

$$K_\nu(z) = (\pi/2)\,(-i)^{\nu+1} H_\nu^{(2)}(-iz) \tag{A2.25}$$

Small and large argument approximations for these functions can therefore be deduced from (A2.4)—(A2.16).

Recurrence relations for $I_\nu(z)$ and $K_\nu(z)$ are:

$$\begin{aligned} L_\nu'(z) &= (\nu/z)L_\nu(z) \pm L_{\nu+1}(z) \\ &= -(\nu/z)L_\nu(z) \pm L_{\nu-1}(z) \end{aligned} \tag{A2.26}$$

$$\pm(2\nu/z)L_\nu(z) = L_{\nu-1}(z) - L_{\nu+1}(z) \tag{A2.27}$$

where the $\pm$ signs apply to I_ν and K_ν respectively

A2.2 Mathieu functions

The angular elliptic wave functions satisfies the differential equation

$$\mathrm{d}^2V/\mathrm{d}\nu^2 + (a - 2q \cos 2\nu)V = 0 \tag{A2.28}$$

where a is a separation variable. Try a series solution of the form (1.84) which is repeated here for convenience:

$$V(\nu) = \sum_n A_n(q) \cos n\nu \tag{A2.29}$$

where the summation runs over $n = 0, 2, 4. \ldots$ or $n = 1, 3 \ldots$ In the first case, V is periodic with period $= \pi$ and in the second case, the period is 2π. Substituting from (A2.29) in (A2.28) and equating the coefficients of cos $m\nu$ on both sides, we get

$$(a - m^2)A_m = q(A_{m-2} + A_{m+2}), \; m \geq 2$$

and

$$aA_0 = 2qA_2$$

for functions of period π, and

$$(a - m^2)A_m = q(A_{m-2} + A_{m+2}),\ m \geqslant 3$$

and

$$(a - 1 - q)A_1 = qA_3$$

for functions with period 2π.

Now defining $X_m = (a - m^2)/q$ and $G_m = A_m/A_{m-2}$, we have for functions with period π:

$$G_2 = X_0/2 \tag{A2.30}$$

$$G_m = 1/(X_m - G_{m+2}),\ m \geqslant 2 \tag{A2.31}$$

which are recurrence relations for the coefficients in (A2.29). Furthermore equating G_2 between (A2.30) and (A2.31) we get an equation for the characteristic values a:

$$X_0 = \frac{2}{X_2-}\,\frac{1}{X_4-}\,\frac{1}{X_6-} \tag{A2.32}$$

Similarly with functions with period 2π, we get

$$G_3 = X_1 - 1 \tag{A2.33}$$

$$G_m = 1/(X_m - G_{m+2}),\ m \geqslant 3 \tag{A2.34}$$

and equating both equations for G_3, we get the characteristic equation:

$$X_1 - 1 = \frac{1}{X_3-}\,\frac{1}{X_5-}\,\frac{1}{X_7-} \tag{A2.35}$$

The solution (A2.29) is an even function of V. Trying an odd solution we write

$$V(\nu) = \sum_n A_n(q)\sin n\nu \tag{A2.36}$$

where n runs over the values $n = 2, 4 \ldots$ (periodic function with period π) or over the values 1, 3, . . . (period 2π). A similar treatment gives relations equivalent to (A2.30)—(A2.35) (see Abramowitz and Stegun, 1964, chap. 20).

A2.3 Airy functions

The Airy functions $Ai(t)$ and $Bi(t)$ are solutions of the differential equation

$$(\partial^2/\partial t^2 - t)R(t) = 0 \tag{A2.37}$$

Series solutions are given by:

$$Ai(t) = c_1 f(t) - c_2 g(t) \tag{A2.38}$$

$$Bi(t) = (3)^{1/2}(c_1 f(t) + c_2 g(t)) \tag{A2.39}$$

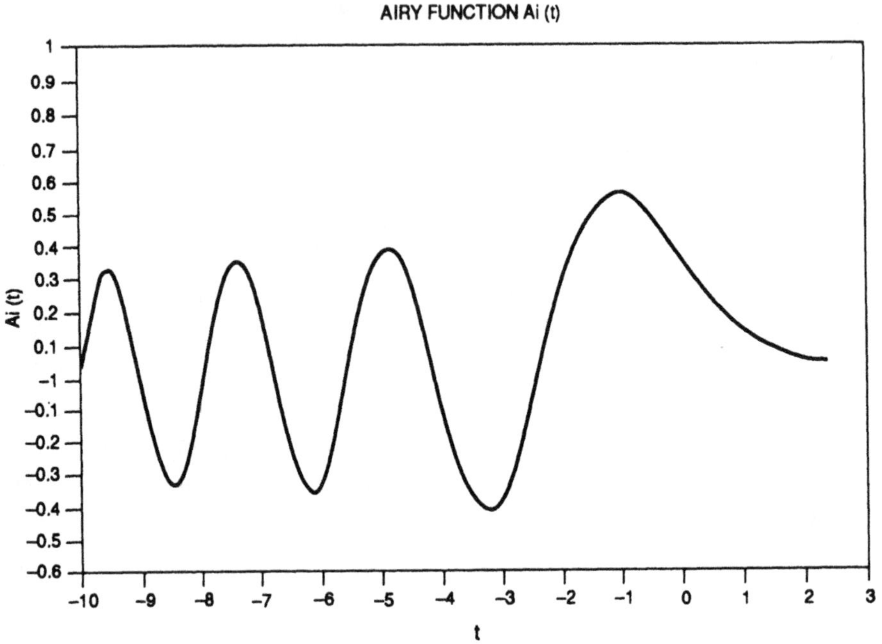

Fig A.2.1 Airy functions on $Ai(t)$

where

$$f(t) = 1 + \sum_{n=1}^{\infty} (z^{3n}/(3n)!)\,(1.4.7.(3n-2))$$

$$g(t) = z + \sum_{n=1}^{\infty} (z^{3n+1}/(3n+1)!)\,(2.5.\,.\,.(3n-1))$$

$$c_1 = 3^{-2/3}/\Gamma(2/3) = 0.355028$$

$$c_2 = 3^{-1/3}/\Gamma(1/3) = 0{\cdot}258819$$

The Wronskian

$$Ai(t)Bi'(t) - Ai'(t)Bi(t) = 1/\pi \qquad \text{(A2.40)}$$

Large argument approximations
For z positive and $\gg 1$, let $\zeta = (2/3)z^{3/2}$, then

$$A_i(z) \simeq \tfrac{1}{2}\pi^{-1/2}z^{-1/4}e^{-\zeta} \qquad \text{(A2.41)}$$

$$A_i(-z) \simeq \pi^{-1/2}z^{-1/4}\sin(\zeta + \pi/4) \qquad \text{(A2.42)}$$

$$B_i(z) \simeq \pi^{-1/2}z^{-1/4}e^{\zeta} \qquad \text{(A2.43)}$$

$$B_i(-z) \simeq \pi^{-1/2}z^{-1/4}\cos(\zeta + \pi/4) \qquad \text{(A2.44)}$$

Plots of $Ai(z)$ and $Bi(z)$ for z between -10 to $+2{\cdot}5$ are given in figs. A2.1, A2.2. It is seen that, for negative z, Ai and Bi represent standing waves with nonconstant wavenumber; i.e. the period of oscillation is a function of z. For positive z, Ai is an evanescent wave and Bi is a growing wave.

Travelling wave solutions are found, in the usual way, as linear combinations of the standing waves; namely, the travelling wave Airy functions are defined by:

$$w_{1_2}(z) = \pi^{-1/2}[Bi(z) \mp iAi(z)] \tag{A2.45}$$

For positive $z \gg 1$, and with $\zeta = (2/3)\, z^{3/2}$, large argument approximation for w_1 and w_2 are given by:

$$w_1(-z) \simeq z^{-1/4} \exp(-i\zeta - i\pi/4) \tag{A2.46}$$

$$w_2(-z) \simeq z^{-1/4} \exp(i\zeta + i\pi/4) \tag{A2.47}$$

These two equations can be explained physically on the basis that the Airy differential equation (A2.37) is a wave equation whose wavenumber squared is a linear function of the distance of travel, namely, for negative z the wavenumber is equal to $(-z)^{1/2}$; hence, the phase delay $= \int_0^z (-z)^{1/2}\, d(-z) = (2/3)(-z)^{3/2}$, hence the term $\exp(\pm i\zeta)$ in (A2.46)—(A2.47). The term $z^{-1/4} \exp(\pm i\pi/4)$ is typical for cylindrical wave spreading.

The logarithmic derivatives of w_1 and w_2 occur often and are related to the phase delay factor. They are given for large argument by (Wait, 1967):

$$w_1'(-z)/w_1(-z) \simeq iz^{1/2}(1 - i/6\zeta) \tag{A2.48}$$

$$w_2'(-z)/w_2(-z) \simeq -iz^{1/2}(1 + i/6\zeta) \tag{A2.49}$$

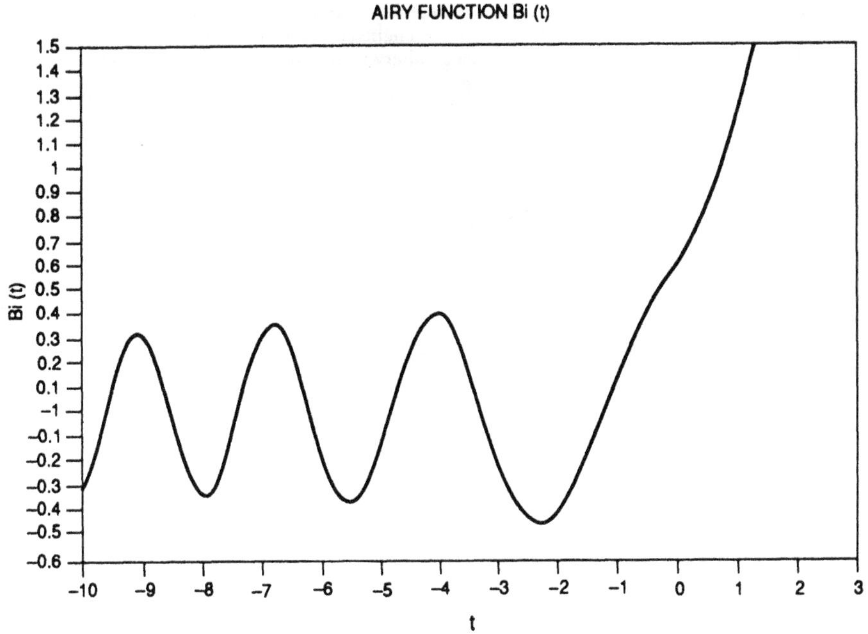

Fig A.2.2. Airy functions on $Bi(t)$

Relation to Bessel functions

The Airy functions are related in an approximate way to the Bessel functions of large order and argument. So, if both ν and z are much larger than unity, but their difference is small so that $|\nu - z| \ll z^{2/3}$, then the following approximate relations are valid:

$$J_\nu(z) \simeq (2/z)^{1/3} Ai(t) \tag{A2.50}$$

$$N_\nu(z) \simeq -(2/z)^{1/3} Bi(t) \tag{A2.51}$$

$$H_\nu^{(2)}(z) \simeq i\pi^{-1/2}(2/z)^{1/3} w_1(t) \tag{A2.52}$$

$$H_\nu^{(1)}(z) \simeq -i\pi^{-1/2}(2/z)^{1/3} w_2(t) \tag{A2.53}$$

where

$$-t = (z/2)^{2/3}(1 - \nu^2/z^2) \tag{A2.54}$$

It follows that the logarithmic derivatives of the Hankel functions are approximated by:

$$H_\nu^{(2)\prime}(z)/H_\nu^{(2)}(z) \simeq -(2/z)^{1/3} w_1'(t)/w_1(t) \tag{A2.55}$$

$$H_\nu^{(1)\prime}(z)/H_\nu^{(1)}(z) \simeq -(2/z)^{1/3} w_2'(t)/w_2(t) \tag{A2.56}$$

A2.4 References

ABRAMOWITZ, M., and STEGUN, I.A. (1965): 'Handbook of mathematical fuctions' (Dover Publications, New York)

ERDELYI, A. (1953): 'Higher transcedental functions' (McGraw Hill, New York)

McLACHLAN, N.W. (1955): 'Bessel functions for engineers' (Oxford Univ. Press)

WAIT, J.R. (1967): 'Electromagnetic whispering gallery modes in a dielectric rod', *Radio Science*, **2**, pp. 1005–1017

Index

www.ingramcontent.com/pod-product-compliance
Lightning Source LLC
Chambersburg PA
CBHW031949180326
41458CB00006B/1673